A Textbook of
Digital Electronics

A Textbook of Digital Electronics

S.S. Bhatti
Director-Principal
Adesh Institute of Engineering and Technology
Faridkot, Punjab

Rahul Malhotra
Professor
Adesh Institute of Engineering and Technology
Faridkot, Punjab

I.K. International Publishing House Pvt. Ltd.

NEW DELHI

Published by

I.K. International Publishing House Pvt. Ltd.
4435-36/7, Ansari Road, Daryaganj
New Delhi – 110 002 (India)
E-mail: info@ikinternational.com
Website: www.ikbooks.com

ISBN 978-93-81141-51-9

Reprint 2019

Published by Krishan Makhijani for I.K. International Publishing House Pvt Ltd., 4435-36/7, Ansari Road, Daryaganj, New Delhi – 110 002 and Printed by Rekha Printers Pvt. Ltd., Okhla Industrial Area, Phase II, New Delhi – 110 020

Preface

The world is going digital. Every aspect of life nowadays has a digital component. The tremendous growth of digital technology can be analyzed from the wide variety of industrial machines, medical equipment, household appliances, e-banking, e-booking, e-governance etc., which are blessed with digital applications. The applications of digital technology have been increasing day-by-day. The important reason behind the tremendous growth of digital technology is the advent of Integrated Circuits and the growth of very large scale integration techniques.

This book is a study of modern digital systems. It teaches fundamental principles of digital systems in a blended way. The text provides a thorough knowledge of digital electronics for students in electronics, electrical, computer science and allied disciplines. The text discusses in depth the working of digital electronic systems, and forms a comprehensive course in digital systems for the proper understanding of subject matter to the young aspirants of undergraduate courses. This book is also helpful for competitive examinations.

Chapter 1 provides a complete and comprehensive study of the basic number systems and conversions.

Chapter 2 introduces the arithmetic of number systems viz. binary, octal and hexadecimal. Simple methods have been incorporated to perform arithmetic operations of number systems.

Chapter 3 gives the introduction of binary codes and explores various applications of these codes in digital systems.

Chapter 4 discusses the basic building blocks for a digital system. It covers basic gates, basic postulates and theorems of Boolean algebra with examples.

Chapter 5 discusses operation and characteristics of various digital logic families and their applications.

Chapter 6 represents the various methods to reduce the complexity of digital circuits using Karnaugh maps.

Chapter 7 explores the various sequential circuits, discusses various flip-flops and their applications in digital circuits.

Chapter 8 introduces the shift registers and describes the various methods to design counters.

Chapter 9 is devoted to the step-by-step procedures to design various types of combinational logic circuits viz. multiplexer, demultiplexers, adders and subtractors, etc.

Chapter 10 gives the introduction about the basic analog and digital conversion methods.

Chapter 11 discusses the different types of semiconductor memories for digital circuits.

Chapter 12 gives a brief introduction of various types of microprocessor systems.

We would appreciate any suggestions for further improvement of the text.

We are thankful to the management of Adesh Foundation, the Chairman Dr. H.S. Gill and the Director Technical Er. Gurfateh Singh Gill for their support during the preparation of the book. We are also thankful the Addl. Director BMS, Er. Janjit Pal Singh for the appreciation of this work.

Words are not sufficient to thank Er. Sukhwinder Singh, who burnt midnight oil in the preparation of this book.

We are also grateful to our family members for their love, support, affection and patience throughout the preparation of this book.

S. S. Bhatti
e-mail: surjitsingh.bhatti@gmail.com
Rahul Malhotra
e-mail: blessurahul@gmail.com

Contents

1

Number Systems and Conversions

INTRODUCTION

In the era of Information & Communication Technology (ICT), the term digital has become a part of our daily life. Digital circuits are widely used in almost all consumer goods like computers, cameras, mobiles and medical equipment and in transportation, telecommunication and several other technologies. The aim of this book is to introduce the basic concepts of these digital systems.

1.1 ANALOG AND DIGITAL REPRESENTATIONS

An analog representation displays information in a *continuous* way. Mercury-in-glass thermometers use analog representation to measure temperature. Similarly, the outdoor thermometers use a pointer which rotates around the dial as a metal coil attached to it expands or contracts with the rise or fall in temperature. The position of the pointer varies continuously with the change in temperature.

In digital representation, the quantities displayed vary in a *discrete* way using numbers or digits. Fingers were the first device used for counting. Digit (Latin word for finger) means both finger and a discrete number (below ten). The difference between analog and digital representations is that the analog representation is continuous whereas the digital representation is discrete.

ANALOG AND DIGITAL SYSTEMS

An analog system deals with the physical quantities that are represented in analog form. Conversely, a digital system is a combination of devices which handle information represented in the digital form.

Advantages of Digital Systems

1. These systems are easier to design compared to analog circuits.
2. Information storage is easier in digital systems than in the analog ones.
3. For the same set of inputs, digital systems provide more exact results.
4. Digital systems are more flexible than the analog ones.

1.2 NUMBER SYSTEMS AND CONVERSIONS

Digital technology uses many number systems. The most common are decimal, binary, octal and hexadecimal systems. We are most familiar with decimal number system, because it is a tool that we use everyday. This chapter is an introduction to the number systems, their representation, and conversion from one form to another. The conversion procedures are illustrated with several examples.

1.3 DECIMAL, BINARY, OCTAL, AND HEXADECIMAL SYSTEMS

The decimal number system has *base* or *radix* 10. It is known as base 10 because it uses ten digits 0, 1, 2, 3, 4, 5, 6, 7, 8 and 9. These digits are known as the coefficients of the decimal system. In the decimal system the coefficients are multiplied by the powers of 10 to form a number. For example, the decimal number 68482.46 is interpreted as:

$$\begin{aligned} 68482.46 &= 60,000 + 8.000 + 400 + 80 + 2 + 0.4 + 0.06 \\ &= 6 \times 10^4 + 8 \times 10^3 + 4 \times 10^2 + 8 \times 10^1 + 2 \times 10^0 + 4 \times 10^{-1} + 6 \times 10^{-2} \end{aligned}$$

In general representation any number may be given by a series of coefficients as given below:

$$A_n A_{n-1} A_{n-2} \,........\, A_2 A_1 A_0 \,........\, A_{-1} A_{-2} \,........\, A_{-n}$$

In the decimal system, the A_k coefficients are the ten coefficients (zero through nine), and the subscript value denotes the power of ten by which the coefficient must be multiplied. Thus, the last expression above can also be written as:

$$A_n \cdot 10^n + A_{n-1} \cdot 10^{n-1} + A_{n-2} \cdot 10^{n-2} + ... + A_2 \cdot 10^2 + A_1 10^1 + A_0 \cdot 10^0 + A_{-1} \cdot 10^{-1} + A_{-n} \cdot 10^{-n}$$

Digital computers use the binary (base 2) system which has only two coefficients, 0 and 1. In the binary system each coefficient A_k is multiplied by 2^k. In general, a number of base or radix r with coefficients A_k is expressed as

$$A_n \cdot r^n + A_{n-1} \cdot r^{n-1} + A_{n-2} \cdot r^{n-2} + ... + A_2 \cdot r^2 + A_1 \cdot r^1 + A_0 \cdot r^0 + A_{-1} \cdot r^{-1} + A_{-n} \cdot r^{-n}$$

The number 101010.11 could be interpreted as binary, or decimal or any other base number since the coefficients 0 and 1 are valid in any number with base 2 or above. Therefore, it is recommended to enclose the number in parentheses and write a subscript representing the base of the number. Thus, if the number 101010.11 is binary, it is represented as

$$(101010.11)_2$$

But a decimal number should be represented as

$$(101010.11)_{10}$$

The other two numbers in digital system are octal (base 8) and hexadecimal (base 16).

The octal number system is composed of eight coefficients 0 through 7. The digit 8 and 9 of decimal number system are not used in octal number system. Hence, the number 676.3 can be either an octal number or a decimal number. If it is an octal number, it must be represented as

$$(676.3)_8$$

If it is a decimal number, it must be represented as

$$(676.3)_{10}$$

The hexadecimal number is composed of sixteen numbers 0, 1, 2, 3, 4, 5, 6, 7, 8 and 9 for the remaining six numbers use the letters A, B, C, D, E and F corresponding to the decimal numbers 10, 11, 12, 13, 14, 15 and 16 respectively. Table 1.1 shows the first 16 numbers of the decimal, binary, octal, and hexadecimal systems for the better understanding of digital number systems.

TABLE 1.1 First sixteen decimal, binary, octal, and hexadecimal numbers

Decimal (Base 10)	*Binary (Base 2)*	*Octal (Base 8)*	*Hexadecimal (Base 16)*
0	0	0	0
1	1	1	1
2	10	2	2
3	11	3	3
4	100	4	4
5	101	5	5
6	110	6	6
7	111	7	7
8	1000	10	8
9	1001	11	9
10	1010	12	A
11	1011	13	B
12	1100	14	C
13	1101	15	D
14	1110	16	E
15	1111	17	F

1.4 BINARY, OCTAL, AND HEXADECIMAL TO DECIMAL CONVERSIONS

Converting a binary, octal, hexadecimal number to a decimal one is a very straightforward procedure. A number with base other than base 10, can be converted to its decimal equivalent using the following steps:

1. Manipulate the given number as given in Ex (1.1).
2. Add the terms following the rules of decimal addition.

Example 1.1 Convert the binary number $(1101101)_2$ to its decimal equivalent.

Solution

$$(1101101)_2 = 1 \times 2^6 + 1 \times 2^5 + 0 \times 2^4 + 1 \times 2^3 + 1 \times 2^2 + 0 \times 2^1 + 1 \times 2^0$$
$$= 64 + 32 + 0 + 8 + 4 + 0 + 1 = (109)_{10}$$

Example 1.2 Convert the octal number $(457.3)_8$ to its decimal equivalent.

Solution

$$(457.3)_2 = 4 \times 8^2 + 5 \times 8^1 + 7 \times 8^0 + 3 \times 8^{-1}$$
$$= 4 \times 64 + 5 \times 8 + 7 \times 1 + 3 \times 8^{-1} = (303.375)_{10}$$

Example 1.3 Convert the hexadecimal number $(BOD \cdot A)_{16}$ to its decimal equivalent.

Solution

$$(BOD \cdot A)_{16} = B \times 16^2 + 0 \times 16^1 + D \times 16^0 + A \times 16^{-1}$$
$$= 11 \times 256 + 0 \times 16 + 13 \times 1 + 10 \times 16^{-1} = (2829.625)_{10}$$

1.5 DECIMAL TO BINARY, OCTAL, AND HEXADECIMAL CONVERSIONS

In the previous section, we have discussed the conversion procedure of binary, octal and hexadecimal numbers into decimal numbers. Now we will see how to convert a decimal number to another (binary, octal and hexadecimal number systems).

The procedure is common, which is as under:

- A decimal number in integer form can be converted to any other number system, say n, by repeatedly dividing the given decimal number by base n of the number system until the quotient becomes zero. The first remainder obtained becomes the least significant digit (lsd), and the last remainder becomes the most significant digit (msd) of the base n number.
- A decimal number in fractional form can be converted to any other number system, say n, by repeatedly multiplying the given decimal number by n until a number with zero fractional part is obtained. This condition may or may not be possible, i.e., the conversion may be endless, but it must be optimized as shown in examples below.
- A hybrid (integer and fractional) decimal number can be converted to any other base number, say n, by first converting the integer part, then converting the fractional part, and finally combining these two parts.

Example 1.4 Convert the decimal number $(34)_{10}$ to its binary equivalent.

Solution

Number	Quotient	Remainder
34/2	17	0 (lsd)
17/2	8	1
8/2	4	0
4/2	2	0
2/2	1	0
1/2	0	1 (msd)

In the last step, the quotient is 0, hence, the conversion is complete and the binary equivalent of $(34)_{10}$ is $(100010)_2$. (Read msd .... lsd)

Example 1.5 Convert the decimal number $(11.2875)_{10}$ to its binary equivalent.

Solution This is a hybrid number composed of integer part and fractional part. It can be converted into any other base number by converting the integer part first and then converting the fractional part. The number will then be combined to represent the conversion.

Integer part	Quotient	Remainder
11/2	5	1 (lsd)
5/2	2	1
2/2	1	0
1/2	0	1 (msd)

Fractional part:

$0.2875 \times 2 = 0$ (msd of binary number) + 0.575
$0.575 \times 2 = 1$ (next binary digit) + 0.15
$0.15 \times 2 = 0$ (next binary digit) + 0.3

and so on.

We find that the conversion is endless and hence, it is to be optimized upto some extent.

Hence, the conversion for the problem will be represented as:

$$(11.2875)_{10} = (1011.010\ldots)_2$$

Example 1.6 Convert the decimal number $(0.79298)_{10}$ to its binary equivalent.

Solution

$0.79298 \times 2 = 1$ (msd of binary number) + 0.58596
$0.58596 \times 2 = 1$ (next binary digit) + 0.17192
$0.17192 \times 2 = 0$ (next binary digit) + 0.34384

and so on.

We find that the conversion is endless and hence, it is to be optimized upto some extent. Hence, the conversion for the problem will be represented as $(0.79298)_{10} = (0.110\ldots)_2$. (Read msd ... lsd)

Conversion from decimal-to-octal is accomplished by repeated division by 8 for the integer part, and by repeated multiplication by 8 for the fractional part.

Example 1.7 Convert the decimal number $(235.165)_{10}$ to its octal equivalent.

Solution First convert the integer part and then convert the fractional part. Combine these to represent the octal conversion of the given number.

Integer part	Quotient	Remainder
235/8	29	3 (lsd)
29/8	3	5
3/8	0	3 (msd)

Fractional part:

$0.165 \times 8 = 1$ (msd of fractional part) + 0.32

$0.32 \times 8 = 2$ (next octal digit) + 0.56

$0.56 \times 8 = 1$ (next octal digit) + 0.12

and so on.

We observe that the fractional part conversion is endless; therefore,

$$(235.165)_{10} = (353.121\ldots)_8$$

Conversion from decimal to hexadecimal is accomplished by repeated division by 16 for the integer part, and by repeated multiplication by 16 for the fractional part.

Example 1.8 Convert the decimal number $(294.125)_{10}$ to its hexadecimal equivalent.

Solution As before, we first convert the integer part, next the fractional part, and then we combine these.

Integer part	Quotient	Remainder
294/16	18	6 (lsd)
18/16	1	2
1/16	0	1

Fractional part:

$0.125 \times 16 = 2$ (msd of fractional part) + 0.00

We find that the conversion of the fractional part of the above decimal number into hexadecimal form is exact and need not be optimized, Hence, the hexadecimal equivalent of $(294.125)_{10} = (126.2)_{16}$.

1.6 INTER-CONVERSIONS (BINARY-OCTAL-HEXADECIMAL)

Since each octal digit has three binary digits and each hexadecimal digit has four binary digits ($2^3 = 8$ and $2^4 = 16$).

Accordingly, to convert binary to octal, we partition the binary number into groups of three digits. For the integer part conversion the partitioning will start from the binary point and proceed to the left. For the conversion of fractional part the partitioning will start from the binary point and proceed towards the right of the binary point.

Example 1.9 Convert the binary number $(11010001101110.10101)_2$ to its octal equivalent.

Solution Since a binary number is to be converted into its octal equivalent and the number is composed of integer and fractional part, partition the integer part into the groups of three digits starting from the binary point towards left and the fractional part towards right as shown below. During partitioning we have observed that two binary numbers are available at the left end of the integer part and right end of the fractional part, which may not represent an octal number. Add zero(s) to complete it as an octal number.

011	010	001	101	110.	101	010
3	2	1	5	6	5	2

Hence, the number will be

$$(11010001101110.10101)_2 = (32156.52)_8$$

To convert octal-to-binary we just reverse the procedure used in the previous example to convert binary-to-octal, i.e. each octal digit is converted to its binary equivalent as it is shown in the following example.

Example 1.10 Convert the octal number $(435.216)_8$ to its binary equivalent.

Here, the conversion is done by converting the given number into its binary equivalent

4	3	5.	2	1	6
100	011	101.	010	001	110

Hence, the number will be

$$(435.216)_8 = (100011101.010001110)_2$$

To convert a binary number to its hexadecimal equivalent or hexadecimal to binary number, the binary number is partitioned into groups of four digits for binary to hexadecimal conversion. For converting hexadecimal number to its binary equivalent the four binary digits represent a hexadecimal digit starting from left for integer part and right for fractional part.

Example 1.11 Convert the binary number $(100110001111001.101001)_2$ to its hexadecimal equivalent.

Solution To convert a binary number to its hexadecimal equivalent partition the binary number into groups of four digits each starting from the binary point and proceeding to the left for the integer part and to the right of the binary point for the fractional part. As the zeros to the left of the integer part of the number and zeros added to the right of the last digit of the fractional part of the number do not alter the value of the number, we partition the number in groups of four digits by inserting a zero to the left of the number, and one zero to the right of the given number, and then assign an equivalent hexadecimal value to each group as given below:

0100	1100	0111	1001.	1010	0100
4	C	7	9.	A	4

Hence,

$$(100110001111001.101001)_2 = (4C79.A4)_{16}$$

Example 1.12 Convert the hexadecimal number $(2A3.C)_{16}$ to its binary equivalent.

Solution

2	A	3.	C
0010	1010	0011.	1010

Hence,

$$(2A3.C)_{16} = (001010100011.1010)_{16}$$

SOLVED EXERCISES

1. Convert the binary number $(1011.011)_2$ to its decimal equivalent.

Solution

$$(1011.011)_2 = 1 \times 2^3 + 0 \times 2^2 + 1 \times 2^1 + 1 \times 2^0 + 0 \times 2^{-1} + 1 \times 2^{-2} + 1 \times 2^{-3}$$
$$= 8 + 0 + 2 + 1 + 0 + 0.25 + 0.125 = (11.375)_{10}$$

2. Convert the octal number $(775.1)_8$ to its decimal equivalent.

Solution $(775.1)_2 = 7 \times 8^2 + 7 \times 8^1 + 5 \times 8^0 + 1 \times 8^{-1}$

$= 7 \times 64 + 7 \times 8 + 5 \times 1 + 1 \times 8^{-1} = (509.125)_{10}$

3. Convert the hexadecimal number $(A2F.9)_{16}$ to its decimal equivalent.

Solution $(A2F.9)_{16} = A \times 16^2 + 2 \times 16^1 + F \times 16^0 + 9 \times 16^{-1}$

$= 10 \times 256 + 2 \times 16 + 15 \times 1 + 9 \times 16^{-1} = (2607.5625)_{10}$

4. Convert the decimal number $(73)_{10}$ to its binary equivalent.

Solution

Number	Quotient	Remainder
73/2	36	1 (lsd)
36/2	18	0
18/2	9	0
9/2	4	1
4/2	2	0
2/2	1	0
1/2	0	1 (msd)

In the last step, the quotient is 0, hence, the conversion is complete and the binary equivalent of $(73)_{10}$ is $(1001001)_2$. (Read msd … lsd)

5. Convert the decimal number $(301.72)_{10}$ to its binary equivalent.

Solution

Integer part	Quotient	Remainder
301/2	150	1 (lsd)
150/2	75	0
75/2	37	1
37/2	18	1
18/2	9	0
9/2	4	1
4/2	2	0
2/2	1	0
1/2	0	1 (msd)

Fractional part:

$0.72 \times 2 = 1$ (msd of binary number) + 0.44

$0.44 \times 2 = 0$ (next binary digit) + 0.88

$0.88 \times 2 = 1$ (next binary digit) + 0.76

$0.76 \times 2 = 1$ (next binary digit) + 0.52

and so on.

We find that the conversion is endless and hence, it is to be optimized upto some extent. Hence, the conversion for the problem will be represented as:

$$(301.72)_{10} = (100101101.1011...)_2$$

6. Convert the decimal number $(0.625)_{10}$ to its binary equivalent.

Solution

$0.625 \times 2 = 1$ (msd of binary number) + 0.25

$0.25 \times 2 = 0$ (next binary digit) + 0.5

$0.5 \times 2 = 1$ (next binary digit) + 0.0

Hence, the conversion of $(0.625)_{10}$ is $(0.101)_2$. (Read msd ... lsd)

7. Convert the decimal number $(422.53125)_{10}$ to its octal equivalent.

Solution First convert the integer part and then the fractional part. Combine these to represent the octal conversion of the given number.

Integer part	Quotient	Remainder
422/8	52	6 (lsd)
52/8	6	4
6/8	0	6 (msd)

Fractional part:

$0.53125 \times 8 = 4$ (msd of fractional part) + 0.25

$0.25 \times 8 = 2$ (next octal digit) + 0.0

Hence, the conversion of $(422.53125)_{10} = (646.42)_8$

8. Convert the decimal number $(757.546875)_{10}$ to its hexadecimal equivalent.

Solution As before, we first convert the integer part, next the fractional part, and then we combine these.

Integer part	Quotient	Remainder
757/16	47	5 (lsd)
47/16	2	15
2/16	0	2

Fractional part:

$0.546875 \times 16 = 8$ (msd of fractional part) + 0.75

$75 \times 16 = 12$ (next hexadecimal digit) + 0.00

Hence, the hexadecimal equivalent of $(757.546875)_{10}$ is $(2F5.8C)_{16}$.

9. Convert the binary number $(100110110011010.1101)_2$ to its octal equivalent.

Solution

100	110	110	011	010.	110	100
4	6	6	3	2	6	4

Hence, the number will be

$$(100110110011010.1101)_2 = (46632.64)_8$$

10. Convert the octal number $(273.15)_8$ to its binary equivalent.

Solution Here, the conversion is done by converting the given number into its binary equivalent.

2	7	3.	1	5
010	111	011.	001	101

Hence, the number will be

$$(273.15)_8 = (010111011.001101)_2$$

11. Convert the binary number $(1101000101001101.101111)_2$ to its hexadecimal equivalent.

Solution

1101	0001	0100	1101.	1011	1100
D	1	4	D.	B	C

Hence, the number will be

$$(1101000101001101.101111)_2 = (D14D.BC)_{16}$$

12. Convert the hexadecimal number $(B295.AF)_{16}$ to its binary equivalent.

Solution

B	2	9	5.	A	F
1011	0010	1001	0101.	1010	1111

Hence, the number will be

$$(B295.AF)_{16} = (1011001010010101.10101111)_2$$

EXERCISES

1. Define the following:
   (a) Analog system (b) Digital system
2. Convert the following binary numbers to their decimal equivalents:
   (a) 1 (b) 100 (c) 110 (d) 1101
   (e) 10101 (f) 10000 (g) 11010011 (h) 11111111
3. Encode the following decimal numbers into binary numbers:
   (a) 10 (b) 5 (c) 3 (d) 16
   (e) 2 (f) 132 (g) 145 (h) 1002
4. Convert the following binary numbers to hexadecimal numbers:
   (a) 11011110 (b) 11101011
   (c) 00010011 (d) 10101010
5. Express the following numbers in decimal equivalents:
   (a) $(11010 \cdot 0111)_2$ (b) $(16 \cdot 9)_{16}$
   (c) $(28 \cdot 20)_8$

6. Convert the hexadecimal number 69AB to binary and then from binary convert it to octal.
7. Convert the following binary numbers to octal numbers.
   (a) 111010 (b) 101010
   (c) 111011 (d) 100010
8. Give the next three numbers in each of the following sequences:
   (a) 4A5, 4A6, 4A7, 4A8, ...,
   (b) B998, B999, ...
9. Show that:
   (a) $(13A7)_{16} = (5031)_{10}$
   (b) $(3F2)_{16} = (1111110010)_2$
10. Find the octal equivalent of $(2F \cdot C4)_{16}$ and the hex equivalent of $(762 \cdot 013)_8$.
11. Find the binary equivalent of $(17E \cdot F8)_{16}$ and hex equivalent of $(1011101101 \cdot 011011101)_2$.
12. Convert $(374.26)_8$ to its binary equivalent.
13. Convert $(1101100 \cdot 0100111)_2$ to its octal equivalent.
14. What is $(256)_7$ in base -10?
15. What is $(100110)_2$ in base -10?
16. What is $(A4E4)_4$ in binary?
17. Convert 0.640625 decimal number to its octal equivalent.
18. List the different number systems.
19. Convert A3B4 and 2F34 into binary and octal respectively.
20. Find the hexadecimal equivalent of the octal number 153.4.

2

Arithmetic of Number Systems

INTRODUCTION

Arithmetic operations in number systems are usually done in binary because designing of logic networks is much easier than decimal.

In this chapter we will discuss arithmetic operations in binary, octal, and hexadecimal number systems. The 1's-complement and 9's-complements in the decimal system and the 2's-complement and 1's-complements in the binary system will also be discussed, which are the key elements for designing a logic circuit.

2.1 BINARY SYSTEM ARITHMETIC

The basic arithmetic in binary number system is binary addition. Binary subtraction is done by using 1's or 2's complements. Multiplication and division are discussed with shift registers in the later section.

The addition of numbers in any numbering system is accomplished as in decimal system, that is, the addition starts in the least significant position (rightmost position), and any carries are added in other positions to the left as it is done in the decimal system.

The binary number system has two bits 0 and 1 only, therefore, the possible binary additions are:

$$0 + 0 = 0 \quad 0 + 1 = 1 \quad 1 + 0 = 1 \quad 1 + 1 = 0 \text{ with a carry of } 1$$

The binary addition can be illustrated by the following example:

Example 2.1 Add $(110011101)_2$ and $(1011011)_2$.

Solution

```
 0011111    Carries
110011101   +
  1011011
---------
111111000
```

When we perform the same addition in decimal system the result will be same as shown below:

$$(110011101)_2 = (413)_{10}$$
$$(1011011)_2 = (91)_{10}$$

Hence, $(111111000)_2 = (413 + 91)_{10} = (504)_{10}$

Example 2.2 Add a string of binary numbers $(110110)_2$. $(101001)_2$, $(111000)_2$, $(10101)_2$, and $(100010)_2$.

Solution

```
11
 11
 111111    Carries
  110110
  101001
  111000
   10101
  100010
--------
11001110
```

When we perform the same addition in decimal system the result will be same as shown below:

$$(110110)_2 = (54)_{10}$$
$$(101001)_2 = (41)_{10}$$
$$(111000)_2 = (56)_{10}$$
$$(10101)_2 = (21)_{10}$$
$$(100010)_2 = (34)_{10}$$

Hence, $(11001110)_2 = (54 + 41 + 56 + 21 + 34)_{10} = (206)_{10}$

2.2 OCTAL SYSTEM ARITHMETIC

The octal number system ranges from 0 through 7, i.e. 0, 1, 2, 3, 4, 5, 6 and 7. The process of addition of octal numbers is analogous to decimal numbers addition with an exception that when the sum of two or more octal numbers is more than seven, a carry occurs just as it occurs in the decimal number system when the sum is more than nine. Table 2.1 summarizes octal addition. This table can also be used for octal subtraction as given by Example 2.4.

TABLE 2.1 Table for addition and subtraction of octal numbers

0	1	2	3	4	5	6	7
0	1	2	3	4	5	6	7
1	2	3	4	5	6	7	10
2	3	4	5	6	7	10	11
3	4	5	6	7	10	11	12
4	5	6	7	10	11	12	13
5	6	7	10	11	12	13	14
6	7	10	11	12	13	14	15
7	10	11	12	13	14	15	16

When the given table is used for addition, locate the least significant digit of the first number (augend) in the upper row of the table first, and then locate the least significant digit of the second number (addend) in the leftmost column of the table. The sum of these two numbers will be located at the intersection point of the augend with the addend. Now apply the same procedure for all other digits from right to left.

Example 2.3 Add $(2427)_8$ and $(3165)_8$.

Solution

```
 011   Carries
 2427 +
 3165
 ----
 5614
```

- Starting with the least significant digit, add 5 with 7.
  - The Table gives us 14 i.e., 4 with a carry of 1.
- Now add 6 and 2, with a carry of 1, or 6 and 3.
  - The Table gives us 11 i.e., 1 with a carry of 1.
- Now add 1 and 4 with a carry of 1, or 1 and 5.
  - The Table gives 6 with no carry.
- Finally, we add 3 and 2
  - The Table gives 5 and no carry.
- The sum can be checked for correctness by converting the numbers to their equivalent decimal numbers.

Octal Number Subtraction

While subtracting of digital number system, subtrahend and minuend are taken up. Subtrahend is referred to as the smaller number and the minuend is referred to as larger number. When the given Table 2.1 is used for subtraction of octal numbers, find the least significant digit of the subtrahend (the smaller number) in the first row of the table. Then, locate the least significant digit of the minuend (the larger number) in the same column. A borrow occurs if the least significant digit of the minuend is less than the least significant digit of the subtrahend, and when visualized from Table 2.1, choose the number from the numbers 10 through 16 whose least significant digit matches the least significant digit of the minuend. Now find the difference by visualizing to the leftmost column. The same procedure will be followed for all others from right to left.

Example 2.4 Subtract $(314)_8$ from $(713)_8$.

Solution

$$\begin{array}{r} 713 \\ 314 \; - \\ \hline 377 \end{array}$$

- Least significant digit of subtrahend is 4, locate it in the first row of Table 2.1. Go downwards in that column where 4 appears, choose 13 as the least significant digit of minuend is 3. Visualizing from the leftmost column the difference will be 7 with a borrow.
- Then, add the borrow to the next digit of the subtrahend, 1 in this case, hence we will subtract 2 from 1. Locate 2 in the first row of the Table and go downwards in the same column, choose 11 because the next digit of the minuend is 1, visualizing from the leftmost column the difference will be 7 with another borrow.
- Now, add that borrow to 3 and then subtract 4 from 7. Visualizing from the leftmost column the difference is 3 with no borrow.
- Hence, the result of subtraction will be 377, which can also be checked for correctness by converting the numbers to equivalent decimal numbers.

2.3 HEXADECIMAL SYSTEM ARITHMETIC

The process of hexadecimal addition and subtraction is similar to that of addition and subtraction of octal numbers except that we use Table 2.2. instead of Table 2.1. As hexadecimal numbers range from 0 to F, i.e. 0, 1, 2, 3, 4, 5, 6, 7, 8, 9, A, B, C, D, E and F. The manipulation for 16 numbers is given below in Table 2.2 which will be used for addition and subtraction of hexadecimal numbers. The procedure is as follows.

For addition, locate the least significant digit of the first number (augend) in the upper row of the table, and then locate the least significant digit of the second number (addend) in the leftmost column of the Table. The point of intersection (number) of the augend with the addend gives the sum of the two numbers. Repeat the same procedure for all other digits from right to left.

TABLE 2.2 Table for addition and subtraction of hexadecimal numbers

0	1	2	3	4	5	6	7	8	9	A	B	C	D	E	F
1	2	3	4	5	6	7	8	9	A	B	C	D	E	F	10
2	3	4	5	6	7	8	9	A	B	C	D	E	F	10	11
3	4	5	6	7	8	9	A	B	C	D	E	F	10	11	12
4	5	6	7	8	9	A	B	C	D	E	F	10	11	12	13
5	6	7	8	9	A	B	C	D	E	F	10	11	12	13	14
6	7	8	9	A	B	C	D	E	F	10	11	12	13	14	15
7	8	9	A	B	C	D	E	F	10	11	12	13	14	15	16
8	9	A	B	C	D	E	F	10	11	12	13	14	15	16	17
9	A	B	C	D	E	F	10	11	12	13	14	15	16	17	18
A	B	C	D	E	F	10	11	12	13	14	15	16	17	18	19
B	C	D	E	F	10	11	12	13	14	15	16	17	18	19	1A
C	D	E	F	10	11	12	13	14	15	16	17	18	19	1A	1B
D	E	F	10	11	12	13	14	15	16	17	18	19	1A	1B	1C
E	F	10	11	12	13	14	15	16	17	18	19	1A	1B	1C	1D
F	10	11	12	13	14	15	16	17	18	19	1A	1B	1C	1D	1E

Example 2.5 Perform a hexadecimal addition of $(C629)_{16}$ and $(9A52)_{16}$.

Solution

```
  C629
  9A52  +
 -----
 1607B
```

- Start with the least significant digit column in the example, add 2 with 9 and the Table shows B as intersection point (number) with no carry.
- Now, add 5 with 2 and from the Table and the intersection point is 7 with no carry.
- Add A with 6 and the intersection point is 0 with 1 carry.
- The sum can also be checked for correctness by converting the numbers to their equivalent decimal numbers.

Example 2.6 Perform the hexadecimal subtraction of $(8B6F)_{16}$ from $(E7AC)_{16}$.

Solution

```
  E7AC
  8B6F  –
 -----
  5C3D
```

- Locate the least significant digit of the subtrahend F in the example from the first row of Table 2.2, and go downwards in the same column to find C. Number C is available as 1 C at a point (number) D in the leftmost column of the table, which is the difference D with a borrow.
- Then, because of the borrow, reduce the digit A of the minuend to 9. Now subtract 6 from 9, with the same procedure as above. The difference will be 3 with no borrow.
- Now find the difference of B from 7 with the same procedure. The difference is C with another borrow.
- Because of the previous borrow, reduce E to D. Now subtract 8 from D, which produces the difference of 5 with no borrow.
- The difference can also be checked for correctness by converting the numbers to their equivalent decimal numbers.

2.4 COMPLEMENTS OF NUMBERS

To simplify the binary subtraction operation complement of numbers are used. In digital systems for each radix or base – r, there are two types of complements, i.e. r's-complement, and the $(r – 1)$' s-complement. It simply means, for the decimal number system whose base is 10, we have 10's-complement and the 9's-complements. For octal number system with a base of 8 we have 8's-complement and 7's-complement. Similarly for binary number system with a base of 2 we have the 2's-complement and 1's-complement. And for the hexadecimal system with base of 16 we have the 16's-complement and the 15s-complement.

2.4.1 10's-Complement

The 10's-complement can be obtained by subtracting the least significant digit from 10 and all other digits from 9.

Example 2.7 Find the 10's-complement of 55274.

Solution

9 9 9 9	10
5 5 2 7	4 –
4 4 7 2	6

First subtract 4 (lsd) from 10 we obtain 6. The other digits of the given example, i.e. 7, 2, 5, and 5 will be subtracted from 9 and we obtain 2, 7, 4, and 4 respectively. Hence, the 10's-complement of 25274 is 44726.

Example 2.8 Find the 10's-complement of 0.6735.

Solution First subtract 5 (lsd) from 10 and all other digits from 9. Hence, the 10's-complement of 0.6735 is 0.3265.

Example 2.9 Find the 10's-complement of 46.653.

Solution First subtract 3 (lsd) from 10 and all other digits from 9. Hence, the 10's-complement of 46.653 is 53.347.

2.4.2 9's-Complement

The 9's-complement of a number can be obtained by subtracting every digit of the given number from 9.

Example 2.10 Find the 9's-complement of 55274.

Solution

  99999
  55274 –
  ———
  44725

Subtract every digit of the given number from 9 and the 9's-complement of 55274 is 44725. It is observed that this complement is one less than 44726 which was the 10's-complement of the same number. The 9's-complement of any number is always one less than the 10's-complement. The 10's-complement can also be obtained by adding 1 to the 9's-complement of the given number.

Example 2.11 Find the 9's-complement of 46.653.

Solution Subtract every digit of the given number from 9, the 9's-complement of 46.653 is 53.346.

2.4.3 2's-Complement

The 2's-complement can be found by unchanging all the least significant 0's and the least significant 1's and replacing all other 0's with 1's and all other 1's with 0's in the remaining number. 2's-complement of the number can also be obtained by adding 1 to the least significant digit in 1's-complement of the given number.

Example 2.12 Find the 2's-complement of 1011010.

Solution Looking from the right of the given number, the least significant one appears at 2nd digit, hence we will not alter 10 and for the remaining digits 10110, replace all ones with zeros and all zeros with ones to find the 2's-complement of a number. Hence, the 2's-complement of 1011010 is 0100110.

Example 2.13 Find the 2's-complement of 0.0101.

Solution Leave the lsd, i.e. 1, unchanged and replace all 1's with 0's and all 0's with 1's. Hence, the 2's-complement of 0.0101 is 0.1011. The first zero (0) to the left of the binary point, which separates the integer and the fractional parts remains unchanged.

Example 2.14 Find the 2's-complement of 1010.0011.

Solution Leave the lsd i.e., 1 unchanged and replace all 1's with 0's and all 0's with 1's. Hence, the 2's-complement of 1010.0011 is 0101.1101.

2.4.4 1's-Complement

To find 1's-complement of a number replace all 0's with 1's and all 1's with 0's. The 1's complement of a number is always 1 less than the 2's-complement of a number.

Example 2.15 Find the 1's-complement of 1011010.

Solution For the given example by replacing all 1's with 0's and all 0's with 1's, the 1's-complement of 1011010 is 0100101. It is also observed that 1's-complement of the given number is 1 less than 0100110 which was 2's-complement of the same number in the previous examples.

Example 2.16 Find the 1's-complement of 0.0101.

Solution Replace all 1's with 0's and all 0's with 1's, the 1's-complement of 0.0101 is 0.1010. The first zero (0) to the left of the binary point that separates the integer and the fractional part remains unchanged.

2.5 SUBTRACTION OF DIGITAL SYSTEMS

2.5.1 Subtraction with 10's-Complement and 2's-Complement

In the subtraction of digital systems it is assumed that both the numbers for the subtraction operation are positive. With the help of 10's-complement or 2's-complements the subtraction operation is performed in the following procedure:

- Obtain the 10's-complement or 2's-complement of the subtrahend and add it to the minuend.
- Verify the result (sum), and observe the carry. Discard the carry if it occurs at the end. Take the 10's-complement or 2's-complement of the result (sum) and place minus (–) sign before it if no carry occurs at the end. The examples of the both cases are given below:

Example 2.17 Find the subtraction $(51346 - 06934)_{10}$ using the 10's-complement method.

Solution

Minuend = 51346

Subtrahend = 06938

Minuend	=	51346
10's-complement of subtrahend	=	93062 +
	=	1,44408

Here, an end carry occurs, hence discard it.

The result of $(51346 - 06938)_{10}$ is $(44408)_{10}$.

Example 2.18 Find the subtraction $(06938 - 51346)_{10}$ using the 10's-complement method.

Solution

Minuend = 06938
Subtrahend = 51346

Minuend	=	06938
10's-complement of subtrahend	=	48654 +
	=	55592

In the given example it is observed that after performing the subtraction operation i.e. addition of 10's-complement of subtrahend with minuend, no carry occurs at the end.

Hence, the 10's-complement of the result 55592 is taken and we put a minus (–) sign before it resulting in – 44408. Hence, $(51346 - 06938)_{10} = (-44408)_{10}$.

Example 2.19 Find the subtraction $(1110101 - 1001101)_2$ using the 2's-complement method.

Solution

Minuend = 1110101
Subtrahend = 1001101

Minuend	=	1110101
2's-complement of subtrahend	=	0110011 +
	=	1011000

Here, an end carry occurs, hence discard it.

The result of $(1110101 - 1001101)_2$ is $(0101000)_2$.

Example 2.20 Find the subtraction $(1001101 - 1110101)_2$ using the 2's-complement method.

Solution

Minuend = 1001101
Subtrahend = 110101

Minuend	=	1001101
2's-complement of subtrahend	=	0001011 +
	=	1011000

In the given example it is observed that after performing the subtraction operation i.e. addition of 2's-complenent of subtrahend with minuend, no carry occurs at the end.

Hence, the 2's-complement of the result 1011000 is taken and we put a minus (–) sign before it resulting in – 0101000. Hence, $(1001101 - 1110101)_2 = (-0101000)_2$.

2.5.2 Subtraction with 9's-Complement and 1's-Complement

In the subtraction of digital systems it is assumed that both the numbers for the subtraction operation are positive. With the help of 9's-complment or 1's-complement the subtraction operation is performed in the following procedure:

- Obtain the 9's-complement or 1's-complement of the subtrahend and add it to the minuend.

- Verify the result (sum), and observe the carry. If there is an end carry, add 1 to lsd of the result, which is referred to as *end around carry*. If there is no end carry, obtain 9's-complement or 1's-complement of the result and put a minus (–) sign before it. The examples of both cases are given below.

Example 2.21 Find the subtraction $(51346 - 06938)_{10}$ using the 9's-complement method.

Solution

Minuend = 51346
Subtrahend = 06938

$$\begin{array}{lcr} \text{Minuend} & = & 51346 \\ \text{9's-complement of subtrahend} & = & 93061\ + \\ \hline & = & 144407 \end{array}$$

Here an end around carry occurs, hence add 1 to the lsd of the result

$$\begin{array}{lr} & 44407 \\ \text{End around carry} = & 1\ + \\ \hline & 44408 \end{array}$$

Hence, the result of subtraction of $(51346 - 06938)_{10}$ is $(44408)_{10}$.

Example 2.22 Find the subtraction $(06938 - 51346)_{10}$ using the 9's-complement method.

Solution

Minuend = 06938
Subtrahend = 51346

$$\begin{array}{lcr} \text{Minuend} & = & 03938 \\ \text{9's-complement of subtrahend} & = & 48653\ + \\ \hline & = & 55591 \end{array}$$

Here no end around carry occurs, take the 9's-complement of the result (sum) i.e. 55591 and put a minus (–) sign before it.

Hence, the result of subtraction of $(51346 - 06938)_{10}$ is $(-44408)_{10}$.

Example 2.23 Find the subtraction $(1110101 - 1001101)_2$ using the 1's-complement method.

Solution

Minuend = 1110101
Subtrahend = 1001101

$$\begin{array}{lcr} \text{Minuend} & = & 1110101 \\ \text{1's-complement of subtrahend} & = & 0110010\ + \\ \hline & = & 10100111 \end{array}$$

Here an end around carry occurs, hence add 1 to the lsd of the result.

$$\begin{array}{lr} & 0100111 \\ \text{End around carry} = & 1\ + \\ \hline & 0101000 \end{array}$$

Hence, the result of subtraction of $(1110101 - 1001101)_2$ is $(0101000)_2$.

Example 2.24 Find the subtraction of $(1001101 - 1110101)_2$ using the 2's-complement method.

Solution

Minuend = 1001101
Subtrahend = 1110101

Minuend	=	1001101
1's-complement of subtrahend	=	0001010 +
	=	1010111

Here no end around carry occurs, take the 1's-complement of the result (sum) i.e. 1010111 and put a minus (–) sign before it.

Hence, the result of subtraction of $(1001101 - 1110101)_2$ is $(-0101000)_2$.

2.6 SIGNED MAGNITUDE OF BINARY NUMBERS

Digital circuitry requires both positive and negative numbers. A signed binary number consists of a sign, either positive or negative and magnitude. In a signed magnitude representation of binary numbers, the most significant digit is zero for the representation of positive binary number and one for the representation of negative binary numbers. This msd (0 or 1) represents whether the number is positive or negative and the magnitude is the value of the numbers. There are three methods by which a signed number can be represented i.e. signed magnitude, 1's-complement and 2's-complement. Digital computers store negative binary numbers in the form of 2's-complement. Of the three methods mentioned above, for the representation of signed binary numbers, 2's-complement method is most widely used and sign magnitude is rarely used.

2.6.1 Sign Bit and Sign Magnitude

The most significant digit (msd) i.e. the leftmost bit in the signed binary number is known as signed bit. If this bit is zero (0), the number is positive and if this bit is one (1), the number is negative. When the signed binary number is represented in sign magnitude, the most significant digit, the leftmost is referred to as a sign bit and the remaining bits show the magnitude of that number. For example, the decimal number + 28 is expressed in 8-bit signed binary number as 00011100, in which the most significant zero shows that the number is positive (+ 28). On the other hand, if it is required to express –28 in 8-bit signed binary number using sign magnitude, the binary equivalent will be 10011100, in which the most significant bit, leftmost bit, represents the number as negative.

Example 2.25 Express $(+33)_{10}$ and $(-33)_{10}$ into signed magnitude binary form.

Solution The binary equivalent of 33 is 100001. Its representation in 8-bit signed binary number will be (00100001). Hence, the representation of (+33) will be (00100001) and (–33) will be (10100001). It is to be noted that there is a change in the most significant digit while representing +33 and –33 respectively.

Example 2.26 Express $(+56)_{10}$ and $(-56)_{10}$ into signed magnitude binary form.

Solution The binary equivalent of 56 is 111000. Its representation in 8-bit signed binary number will be (00111000). Hence, the representation of (+56) will be (00111000) and (–56) will be (10111000). It is to be noted that there is a change in the most significant digit while representing +56 and –56 respectively.

Example 2.27 Express $(+94)_{10}$ and $(-94)_{10}$ into signed magnitude binary form.

Solution The binary equivalent of 94 is 100001. Its representation in 8-bit signed binary number will be (01011110). Hence, the representation of (+94) will be (01011110) and (–94) will be (11011110). It is to be noted that there is a change in the most significant digit while representing +94 and –94 respectively.

2.7 NEGATIVE NUMBER REPRESENTATION USING 1's AND 2's COMPLEMENTS

The negative numbers can be represented by using 1's and 2's-complements. Digital computers generally use 2's-complement to store a negative number.

2.7.1 Using 1's-Complement

Negative numbers in 1's-complement method can be represented by taking the 1's-complement of the positive signed magnitude number. For example, (+28) is represented in positive signed magnitude number as 00011100. On the other hand (–28) is expressed as 1's-complement of (+28) i.e. 11100011.

2.7.2 Using 2's-Complement

Negative numbers in 2's-complement method can be represented by taking 2's-complement of the positive signed magnitude number. For example, (+28) is represented in positive signed magnitude number as 00011100. On the other hand (–28) is expressed as 1's-complement of (+28) i.e. 11100011.

Example 2.28 Express the decimal number (–28) in 8-bit signed magnitude, 1's-complement and 2's-complement form.

Solution The 8-bit binary representation of (+28) is 00011100.

In signed magnitude form (–28) is represented by changing the leftmost bit to 1 and the remaining bits remain unchanged, which is expressed as 10011100.

In 1's-complement form (–28) is represented by taking 1's-complement of (+28), which is expressed as 11100011.

In 2's-complement form (–28) is represented by taking 2's-complement of (+28).

$$\begin{array}{rr} \text{1's-complement of } (+28) = & 11100011 \\ & + \quad\quad 1 \\ \hline & 11100100 \end{array}$$

which is expressed as 11100100.

Example 2.29 Express the decimal number (–19) in 8-bit signed magnitude, 1's-complement and 2's-complement form.

Solution The 8-bit binary representation of (+19) is 00010011.

In signed magnitude form (–19) is represented by changing the leftmost bit to 1 and the remaining bits remain unchanged, which is expressed as 10010011.

In 1's-complement form (–19) is represented by taking the 1's-complement of (+19). which is expressed as 11101100.

In 2's-complement form (–19) is represented by taking 2's-complement of (+19).

$$\begin{array}{r} \text{1's-complement of (+19)} = 11101100 \\ + \quad\quad 1 \\ \hline 11101101 \end{array}$$

which is expressed as 11101101.

2.8 DECIMAL SIGNED NUMBERS

Decimal values of the positive and negative signed magnitude numbers can be determined by the summation of the weights of all the magnitude bits, where there are 1's and ignoring all other bits, where there are zeros (0).

Example 2.30 Express the decimal equivalent of signed binary number 10011100 expressed in its sign magnitude form.

Solution Here, there are seven magnitude bits and one sign bit. Separating sign bits and magnitude bits

sign bit = 1, which means that the magnitude of the number is negative.

magnitude bits = 0011100, assigning the weights to the bits, we get

2^6	2^5	2^4	2^3	2^2	2^1	2^0
0	0	1	1	1	0	0

Summing the weights together where 1 exists and ignoring where 0 exists,

we get $2^4 + 2^3 + 2^2 = 16 + 8 + 4 = 28.$

Adding sign magnitude bit to the solution for the signed magnitude binary number (10011100) is (–28).

Example 2.31 Express the decimal equivalent of signed binary number 01011110 expressed in its sign magnitude form.

Solution Here, there are seven magnitude bits and one sign bit. Separating sign bits and magnitude bits,

Sign bit = 0, which means that the magnitude of number is positive.

Magnitude bits = 101110, assigning the weights to the bits, we get

2^6	2^5	2^4	2^3	2^2	2^1	2^0
1	0	1	1	1	1	0

Summing the weights together where 1 exists and ignoring where 0 exists,

we get $2^6 + 2^4 + 2^3 + 2^2 + 2^1 = 64 + 16 + 8 + 4 + 2 = 94.$

Adding sign magnitude bit to the solution for the signed magnitude binary number (11011110) is (+94)

2.9 BINARY MULTIPLICATION

2.9.1 Shift and Add Multiplication (Direct Addition) Method

Multiplication of binary numbers is performed similar as in the primary school by adding a list of shifted multiplicands, according to the digits of the multiplier. This method is used to obtain the product of two unsigned binary numbers. There are three operators used in multiplication operation i.e. multiplicand, the multiplier and the product. It is illustrated by the example given below:

Example 2.32 Multiply $(10010)_2$ and $(11001)_2$.

Solution Here, the multiplication will be performed as below:

	10010	Multiplicand	18
	× 11001	Multiplier	× 25
	10010		90
	00000		36
Shift and add	00000	Shifted multiplicands	450
	10010		
	10010		
	111000010	Product	

The multiplication of $(10010)_2$ and $(11001)_2$ is $(111000010)_2$.

2.9.2 Partial Product Method

In shift and add method it is sometimes difficult to list all the shifted multiplicands and then adding them in a digital system. Multiplication of binary numbers can also be performed by partial product method in which the shifted multiplicand is added to multiply the binary number.

This technique is more reliable than the previous one.

Example 2.33 Multiply $(10010)_2$ and $(11001)_2$ using the partial product method.

Solution Here, the multiplication will be performed as below:

10010	Multiplicand	18
× 11001	Multiplier	× 25
00000	Patrial product	90
10010	Shifted multiplicand	36
010010	Patrial product	450
00000$\downarrow$	Shifted multiplicand	
0010010	Patrial product	
00000$\downarrow\downarrow$	Shifted multiplicand	
0010010	Patrial product	
10010$\downarrow\downarrow\downarrow$	Shifted multiplicand	

10100010	Patrial product
10010↓↓↓↓	Shifted multiplicand
111000010	Product

The multiplication of $(10010)_2$ and $(11001)_2$ using the partial product method is $(111000010)_2$.

2.9.3 Multiplication of the Signed Numbers

A signed number can be multiplied using unsigned multiplication and usual primary school rules by multiplying the magnitudes using unsigned multiplication methods. The product will be made to be positive if the two numbers are same in magnitude and will be negative if they have different signs. Since the sign and magnitude are separately treated, this method is very convenient for multiplication.

2.10 BINARY DIVISION

Instead of shift and add methods, shift and subtract method is used for binary division. Dividend, the divisor and the quotient are taken up while doing binary division. In this we compare the reduced dividend with multiples of the divisor to determine which multiple of the shifted divisor to subtract. The fundamental of binary division is elaborated by the below given examples.

Example 2.34 Divide $(1100)_2$ with $(0100)_2$ using partial product method.

Solution

Binary		Decimal
11	quotient	3
100) 1100	dividend	4) 12
100	shifted divisor	12
100	reduced dividend	**0 remainder**
100	shifted divisor	
000	**remainder**	

Example 2.35 Divide $(11011001)_2$ and $(1011)_2$ using partial product method.

Solution

Binary		Decimal
10011	quotient	19
1011) 11011001	dividend	11) 217
1011	shifted divisor	11
0101	reduced dividend	107
0000	shifted dividend	99
1010	reduced dividend	**8 remainder**
0000	shifted divisor	
10100	reduced dividend	
1011	shifted divisor	
10011	reduced dividend	
1011	shifted divisor	
1000	**remainder**	

2.11 SOLVED EXERCISES

Example 2.1 Add $(110100111)_2$ and $(1110101)_2$.

Solution

```
  11100111   Carries
 110100111   +
   1110101
----------
1000011100
```

Example 2.2 Add $(4712)_8$ and $(1624)_8$.

Solution

```
  101   Carries
 4715   +
 6624
-----
13541
```

Example 2.3 Subtract $(232)_8$ from $(417)_8$.

Solution

```
 417
 232   –
----
 165
```

Example 2.4 Perform a hexadecimal addition of $(B49C)_{16}$ and $(4E2F)_{16}$.

Solution

```
 B49C
 4E2F   +
-----
102CB
```

Example 2.5 Perform the hexadecimal subtraction of $(C92D)_{16}$ from $(7F9E)_{16}$.

Solution

```
C92D
7F9E   –
----
498F
```

Example 2.6 Find the 10's-complement of 63918.

Solution The 10's-complement of 63918 is 36082.

Example 2.7 Find the 10's-complement of 28.4592.

Solution The 10's-complement of 28.4592 is 71.5408.

Example 2.8 Find the 10's-complement of 0.5813.

Solution The 10's-complement of 0.5813 is 0.4187.

Example 2.9 Find the 9's-complement of 63918.

Solution The 9's-complement of 63918 is 36081.

Example 2.10 Find the 9's-complement of 28.4187.

Solution The 9's-complement of 28.4187 is 71.5812.

Example 2.11 Find the 2's-complement of 1011.11010000.

Solution The 2's-complement of 1011.11010000 is 0100.00110000.

Example 2.12 Find the 2's-complement of 111001100000.

Solution The 2's-complement of 111001100000 is 000110100000.

Example 2.13 Find the 2's-complement of 0.10001.

Solution The 2's-complement of 0.10001 is 0.01111.

Example 2.14 Find the 1's-complement of 101000011.

Solution For the given example by replacing all ones with zeros and all zeros with ones, the 1's-complement of 101000011 is 010111100.

Example 2.15 Find the 1's-complement of 1011.11010000

Solution The 1's-complement of 1011.11010000 is 0100.00101111.

Example 2.16 Find the subtraction $(96258 - 43271)_{10}$ using the 10's-complement method.

Solution

Minuend = 96258
Subtrahend = 43271

Minuend	=	96258
10's-complement of subtrahend	=	56729 +
	=	1,52987

Here, an end carry occurs, hence discard it.
The result of $(96258 - 43271)_{10}$ is $(52987)_{10}$.

Example 2.17 Find the subtraction (128722 – 439811)10 using the 10's-complement method.

Solution

Minuend = 128722
Subtrahend = 439811

Minuend	=	128722
10's-complement of subtrahend	=	560189 +
	=	688911

In the given example it is observed that after performing the subtraction operation i.e. addition of 10's-complement of subtrahend with minuend, no carry occurs at the end.

Hence, the 10's-complement of the result 688911 is taken and we put a minus (–) sign before it resulting in – 311089. Hence, $(128722 - 439811)_{10} = (-311089)_{10}$.

Example 2.18 Find the subtraction $(1011110 - 1000111)_2$ using the 2's-complement method.

Solution

Minuend = 1011110

Subtrahend = 1000111

Minuend	=	1011110
2's-complement of subtrahend	=	0111001 +
	=	10010111

Here, an end carry occurs, hence discard it.

The result of $(1011110 - 1000111)_2$ is $(0010111)_2$.

Example 2.19 Find the subtraction $(1001011 - 1011110)_2$ using the 2's-complement method.

Solution

Minuend = 1001011

Subtrahend = 1011110

Minuend	=	1001011
2's-complement of subtrahend	=	0100010 +
	=	1101101

In the given example it is observed that after performing the subtraction operation i.e. addition of 2's-complement of subtrahend with minuend, no carry occurs at the end.

Hence, the 2's-complement of the result 1101101 is taken and we put a minus (–) sign before it resulting in – 0010011. Hence, $(1001011 - 1011110)_2 = (-0010011)_2$.

Example 2.20 Find the subtraction $(439811 - 128722)_{10}$ using the 9's-complement method.

Solution

Minuend = 439811

Subtrahend = 128722

Minuend	=	439811
9's-complement of subtrahend	=	871277 +
	=	1311088

Here an end around carry occurs, hence add 1 to the lsd of the result

	311088
End around carry =	1 +
	311089

Hence, the result of subtraction of $(439811 - 128722)_{10}$ is $(311089)_{10}$.

Example 2.21 Find the subtraction $(43271 - 96258)_{10}$ using the 9's-complement method.

Solution

Minuend = 43271

Subtrahend = 96258

Minuend	=	43271
9's-complement of subtrahend	=	03741 +
	=	47012

Here no end around carry occurs, take the 9's-complement of the result (sum) i.e. 47012 and put a minus (–) sign before it.

Hence, the result of subtraction of $(51346 - 06938)_{10}$ is $(-52987)_{10}$.

Example 2.22 Find the subtraction $(1011110 - 1001011)_2$ using the 1's-complement method.

Solution

Minuend = 1011110

Subtrahend = 1001011

Minuend	=	1011110
1's-complement of subtrahend	=	0110100 +
	=	10010010

Here an end around carry occurs, hence add 1 to the lsd of the result.

$$\begin{array}{rr} & 0010010 \\ \text{End around carry} = & 1 \;+ \\ \hline & 0010011 \end{array}$$

Hence, the result of subtraction of $(1011110 - 1001011)_2$ is $(0010011)_2$.

Example 2.23 Express $(+99)_{10}$ and $(-99)_{10}$ into signed magnitude binary form.

Solution The binary equivalent of 99 is 1100011. Its representation in 8-bit signed binary number will be (01100011). Hence, the representation of (+99) will be (01100011) and of (–99) will be (11100011).

Example 2.24 Express $(+45)_{10}$ and $(-45)_{10}$ into signed magnitude binary form.

Solution The binary equivalent of 45 is 101101. Its representation in 8-bit signed binary number will be (00101101). Hence, the representation of (+56) will be (00101101) and of (–56) will be (10101101).

Example 2.25 Express the decimal number (–61) in 8-bit signed magnitu
complement form.

Solution The 8-bit binary representation of (+61) is 00111101.

In signed magnitude form (–61) is represented by changing the leftmost b
unchanged, which is expressed as 10111101.

In 1's-complement form (–61) is represented by taking 1's-complement
as 11000010.

In 2's-complement form (–61) is represented by taking 2's-complement

$$\begin{array}{rr} \text{1's-complement of } (+28) = & 11000010 \\ & +\quad 1 \\ \hline & 11000011 \end{array}$$

which is expressed as 11000011.

Example 2.26 Express the decimal equivalent of signed binary number 11001000 expressed in its sign magnitude form.

Solution Here, there are seven magnitude bits and one sign bit. Separating sign bits and magnitude bits.

Sign bit = 1, which means that the magnitude of the number is negative.

Magnitude bits = 1001000, assigning the weights to the bits, we get

2^6	2^5	2^4	2^3	2^2	2^1	2^0
1	0	0	1	0	0	0

Summing the weights together where 1 exists and ignoring where 0 exists we get

$$2^6 + 2^3 = 64 + 8 = 72.$$

Adding sign magnitude bit to the solution for the signed magnitude binary number (11001000) is (–72).

EXERCISES

1. Add $(1011010111)_2$ and $(11110010)_2$.
2. Add $(1011)_2$ and $(11)_2$ and multiply without converting it to decimal.
3. What is 01100101100 + 01101101001?
4. Obtain the 1's and 2's complements of the following binary numbers.
   (a) 11101110 (b) 00001101
   (c) 11110000 (d) 10100011
5. Find the 9's and 10's complement of the following decimal numbers.
   (a) 98137642 (b) 10000000
6. Use 10's complement arithmetic to find (42-75).
7. What number, given in 2's complement notation, is represented by 10110?
8. What number, given in 2's complement form, does 110010 represent?
9. Subtract 5 from 8 using binary notation.
10. Subtract –4 from –9 using binary notation.
11. Calculate 6.5 × 2.75 using binary arithmetic.
12. Perform the following using 2's complement arithmetic (all numbers are given in natural binary notation):
    (a) 10011 – 10101 (b) 1011 – 11011
    (c) 10101 – 1110 (d) 10111 – 10010
13. Perform the following division in binary: 1011111 ÷ 101.
14. Two possible binary representations of $(-1)_{10}$ are $(10000001)_2$ and $(11111111)_2$. One of them belongs to the sign-bit magnitude format and the other to the 2's complement format. Identify.
15. Use 10's complement arithmetic to find (68-35).
16. Perform 10.625 ÷ 2.5 using binary arithmetic.
17. Perform $76_{10} - 57_{10}$ using 10's complement arithmetic.

18. Perform $16.875_{10} \div 4.5_{10}$ using binary arithmetic.
19. Perform the following addition operations.
    (a) $(275 \cdot 75)_{10} + (37 \cdot 875)_{10}$
    (b) $(AB1 \cdot F3)_{16} + (FFF \cdot E)_{16}$.
20. Subtract $(1110 \cdot 011)_2$ from $(11011 \cdot 11)_2$ using the basic rules of binary subtraction.

3

Binary Codes

INTRODUCTION

Encoding is a process by which a message is represented in a code. In the digital systems encoding represents the decimal number system and other alphanumerics by digits of binary number system i.e. 0 and 1. In this chapter the most commonly used binary codes are discussed which are specifically Binary Coded Decimal (BCD), the Excess-3 Code, the 2*421 Code, the Gray Code, and the American Standard Code for Information Interchange (ASCII) code.

3.1 BINARY CODED DECIMAL (BCD)

The Binary Coded Decimal (BCD) is an important code used to represent the decimal numbers 0 through 9 which are shown in Table 3.1. This code is also designated as 8421 code because the digits 8, 4, 2, and 1 represent the weight i.e. 2^3, 2^2, 2^1, 2^0, which represents the weight of the bit position. It is also known as a **weighted code**. Only ten combinations i.e. 0 through 9 are possible in BCD, remaining combinations 1010, 1011, 1100, 1101, 1110, and 1111 are invalid in BCD representation. To represent the non-BCD numbers hexadecimal numbers are used which represent these numbers in alphabets A through F.

TABLE 3.1 BCD codes

Decimal	BCD
0	0000
1	0001
2	0010
3	0011
4	0100
5	0101
6	0110
7	0111
8	1000
9	1001

BCD number system requires four bits to represent a decimal number between 0 and 9; eight bits are required to represent the number between 10 and 19 and similarly to represent a decimal number between 100 and 999, 12 bits are required and so on. In general, a decimal number of n digits is represented in BCD by $4n$ bits. To avoid confusion, i.e. to differentiate a binary and a BCD number space is added after every four digits to indicate the BCD groupings.

Example 3.1 Convert $(119)_{10}$, in BCD

Solution $(119)_{10} = (0001\ 0001\ 1001)_{BCD}$

Example 3.2 Convert $(78593)_{10}$, in BCD.

Solution $(78593)_{10} = (0111\ 1000\ 0101\ 1001\ 0011)_{BCD}$

Example 3.3 Convert $(264)_{10}$, in BCD

Solution $(264)_{10} = (0010\ 0110\ 0100)_{BCD}$

3.1.1 BCD Addition

BCD is a numerical code, which is used in processor arithmetic operations. Addition of BCD numbers is the most important operation among other arithmetic operations. The addition of two BCD numbers can be performed by:

1. Adding two BCD numbers by using binary addition methods, i.e. 0 + 0 = 0, 0 + 1 = 1, 1 + 0 = 1 and 1 + 1 = 0 with carry 1.
2. After addition if a four-bit sum is equal to or less than 9, it represents a valid BCD number, which will be represented as it is.
3. Otherwise, if a four-bit sum is greater than 9 or if a carry is generated out of the four bits, which represents an invalid BCD number (because a BCD number ranges 0 through 9). In such condition, add 6 (0110) to the four-bit sum in order to make it BCD. In simple words, if we add 6 (0110) to 9(1001), the result will be 15. This 15 will be treated as 1, 5 which is represented by 0001 and 0101 in BCD form. The following example illustrates the procedure of BCD addition.

Example 3.4 Add $(599)_{10}$ and $(984)_{10}$ in BCD addition.

Solution $(599)_{10} = (0101\ 1001\ 1001)_{BCD}$

$(984)_{10} = (1001\ 1000\ 0100)_{BCD}$

Now add the above two numbers

	0101	1001	1001
	1001	1000	0100
	1110	10001	1101
	↓	↓	↓
Binary sum	(1)	(2)	(3)

Binary sums 1, 2 and 3 are greater than 9, hence add 6(0110) to the binary sums

	1110	10001	1101
+ 6	0110	0110	0110
Result	0001,0101	1000	0011

Result 0001 0101 1000 0011 $(1583)_{BCD}$

3.2 THE EXCESS-3 CODE

As the name indicates, the Excess-3 code is used to represent the decimal numbers 0 through 9 in Excess-3 form, to make it larger than the BCD code by 3.

The advantage of excess-3 code over the BCD code is that, the excess-3 code is a self-complementing code.

TABLE 3.2 Excess-3 code

Decimal	*Excess-3*
0	0011
1	0100
2	0101
3	0110
4	0111
5	1000
6	1001
7	1010
8	1011
9	1100

Hence, it is also observed that decimal numbers 0 and 9, 1 and 8, 2 and 7, 3 and 6, 4 and 5 are self-complementing as illustrated below:

$$(0)_{10} = (0011)_{\text{Excess-3}}$$

and

$$(9)_{10} = (1100)_{\text{Excess-3}}$$

$$(1)_{10} = (0100)_{\text{Excess-3}}$$

and

$$(8)10 = (1011)_{\text{Excess-3}}$$

$$(2)_{10} = (0101)_{\text{Excess-3}}$$

and

$$(7)_{10} = (1010)_{\text{Excess-3}}$$

$$(3)_{10} = (0110)_{\text{Excess-3}}$$

and

$$(6)10 = (1001)_{\text{Excess-3}}$$

$$(4)_{10} = (0111)_{\text{Excess-3}}$$

and

$$(5)10 = (1000)_{\text{Excess-3}}$$

It is also observed that the left sides are 9's-complements and the right sides are 1's-complements of each other. Therefore, the excess-3 code can be used to perform subtraction of decimal numbers using 9's-complement method. The excess-3 code is not a weighted code as BCD.

3.3 THE 2*421 CODE

The 2*421 code incorporates the advantage of both BCD and excess-3 codes as this is a self-complementing as well as a weighted code. The 2*421 code uses 4 bits to represent the decimal numbers 0 through 9 as shown in the Table 3.3.

TABLE 3.3 2*421 code

Decimal	2*421
0	0000
1	0001
2	0010
3	0011
4	0100
5	1011
6	1100
7	1101
8	1110
9	1111

In this code we again observe that the left sides are 9's-complements of each other, which represents that 2*421 code is a self-complementing code i.e. 0 and 9, 1 and 8, 2 and 7, 3 and 6, 4 and 5 are self-complementing each other and the right sides are the 1's-complements of each other.

$$(0)_{10} = (0000)_{\text{Excess-3}}$$

and

$$(9)_{10} = (1111)_{\text{Excess-3}}$$

$$(1)_{10} = (0001)_{2*421}$$

and

$$(8)_{10} = (1000)_{2*421}$$

$$(2)_{10} = (0010)_{2*421}$$

and

$$(7)_{10} = (1101)_{2*421}$$
$$(3)_{10} = (0011)_{2*421}$$

and

$$(6)_{10} = (1100)_{2*421}$$
$$(4)_{10} = (0100)_{2*421}$$

and

$$(5)_{10} = (1011)_{2*421}$$

3.4 THE GRAY CODE

In electronics circuitry digital to analog conversion is very desirable. Gray code is neither a self-complementing nor a weighted code. In gray code, only a single bit changes from one code word to the next in sequence exhibits. Due to this property, it is useful in many applications such as optical or mechanical shaft position encoders. Table 3.4 shows the binary code and the gray code for the decimal numbers 0 through 9.

TABLE 3.4 Gray code

Decimal	*Binary*	*Gray Code*
0	0000	0000
1	0001	0001
2	0010	0011
3	0011	0010
4	0100	0110
5	0101	0111
6	0110	0101
7	0111	0100
8	1000	1100
9	1001	1101
10	1010	1111
11	1011	1110
12	1100	1010
13	1101	1011
14	1110	1001
15	1111	1000

The procedure of converting a binary number into its gray code is illustrated below:

3.4.1 Binary Code to Gray Code Conversion

To convert a binary code into a gray code the following procedure is adopted:

- The msb of the gray code is same as the msb of the binary number. Going from msb to lsb add each successive binary bit with the adjacent binary bit to represent the gray code bit. The procedure is illustrated below:

Convert binary number 1101 to gray code:

1	→ + →	1	→ + →	0	→ + →	1	Binary Number
↓		↓		↓		↓	
1		0		1		1	Gray Code

Hence, the gray code for binary number 1101 is 1011.

3.4.2 Gray Code to Binary Code Conversion

To convert a gray code into a binary code the following procedure is adopted:

- The msb of the binary number is same as the msb of the gray code. Going from msb to lsb add each successive binary bit with the adjacent gray code bit to represent the binary code. The procedure is illustrated below:

Convert gray code 1101 to binary number.

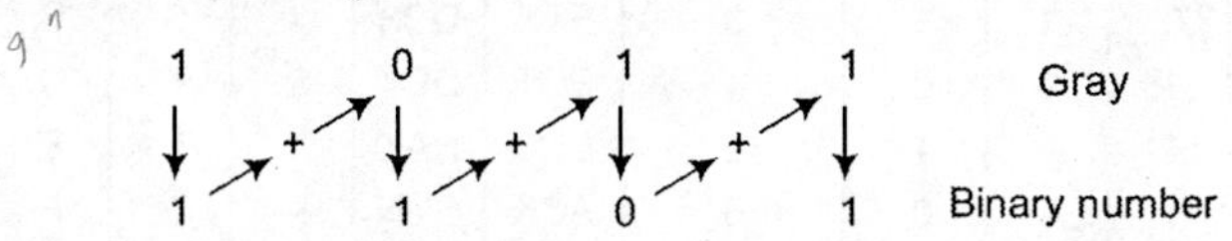

Hence, the binary number for gray code 1011 is 1101.

3.5 THE AMERICAN STANDARD CODE FOR INFORMATION INTERCHANGE (ASCII) CODE

Computers generally use alphanumeric codes. Alphanumeric codes are the codes that represent number and alphabetic characters. An alphanumeric code represents 10 decimal digits and 26 letters of the alphabet.

The American Standard Code for Information Interchange (ASCII) usually pronounced as "Askee" (ASCII), proposed by American National Standards Institute (ANSI) is the universally adopted alphanumeric code for the information interchange in digital computers and other electronic equipment. The keyboards of the computers represent ASCII code internally. When a letter, number or control command is entered through the keyboard, its corresponding ASCII code is generated in the computer through internal mechanism. In nutshell, we can say that computer understands ASCII codes.

ASCII is referred to as 7-bit ASCII and represents 128 (2^7) characters, which are numbers, letters, punctuation marks and some special characters.

The below mentioned table represents the ASCII characters. The bit representation, for example, the letter b in ASCII is as represented as follows, by using the Table 3.5 we can represent any of the symbols, alphabet, letters accordingly:

Letter b

b_7	b_6	b_5	b_4	b_3	b_2	b_1
1	1	0	0	0	1	0

Therefore, the ASCII code for the letter b is 1100010.

A description of the first 32 ASCII characters, often referred to as *control codes*, follows.

NUL A character code with a null value; literally, a character meaning "nothing." Although it is real in the sense of being recognizable, occupying space internally in the computer, and being sent

TABLE 3.5 The standard ASCII code

BIT NUMBERS															
							→	0	0	0	0	1	1	1	1
							→	0	0	1	1	0	0	1	1
							→	0	1	0	1	0	1	0	1
b_7 ↓	b_6 ↓	b_5 ↓	b_4 ↓	b_3 ↓	b_2 ↓	b_1 ↓	*COLUMN* / *ROW*	0	1	2	3	4	5	6	7
			0	0	0	0	0	NUL	DLE	SP	0	@	P	‘	p
			0	0	0	1	1	SOH	DC1	!	1	A	Q	a	q
			0	0	1	0	2	STX	DC2	"	2	B	R	b	r
			0	0	1	1	3	ETX	DC3	#	3	C	S	c	s
			0	1	0	0	4	EOT	DC4	$	4	D	T	d	t
			0	1	0	1	5	ENQ	NAK	%	5	E	U	e	u
			0	1	1	0	6	ACK	SYN	&	6	F	V	f	v
			0	1	1	1	7	BEL	ETB		7	G	W	g	w
			1	0	0	0	8	BS	CAN	(	8	H	X	h	x
			1	0	0	1	9	HT	EM	)	9	I	Y	i	y
			1	0	1	0	10	LF	SUB	*	:	J	Z	j	z
			1	0	1	1	11	VT	ESC	+	,	K	[	k	{
			1	1	0	0	12	FF	FS	.	<	L	\	l	|
			1	1	0	1	13	CR	GS	-	=	M	]	m	}
			1	1	1	0	14	SO	RS	.	>	N	^	n	~
			1	1	1	1	15	SI	US	/	?	O	_	o	DEL

or received as a character, a NUL character displays nothing, takes no space on the screen or on paper, and causes no specific action when sent to a printer. In ASCII, NUL is represented by the character code 0. It can be addressed like a physical output device (such as printer) but it discards any information sent to it.

SOH Start of Heading

STX Start of Text

TAB Horizontal Tab – Moves the cursor (or printhead) right to the next tab stop. The spacing of tab stops depends on the device, but is often either 8 or 10.

LF Linefeed – Tells a computer or printer to advance one line below the current line without moving the position of the cursor or printhead.

VT Vertical Tab

FF Form Feed – Advances paper to the top of the next page (if the output device is a printer).

CR Carriage Return – Tells a printer to return to the beginning of the current line. It is similar to the return on a typewriter but does not automatically advance to the beginning of a new line. For example, a carriage return character alone is received at the end of the words. "This is a sample line of text" would cause the cursor or printer to return to the first letter of the word "This". In the ASCII character set, the CR character has the decimal value of 13 (hexadecimal 0D).

SO Shift Out – Switches output device to alternate character set.

SI Shift In – Switches output device back to default character set.

DLE Data Link Escape

DC1 Device Control 1

DC2 Device Control 2

DC3 Device Control 3

DC4 Device Control 4

NAK Negative Acknowledgement – A control code, ASCII character 21 (hexadecimal 15H), transmitted to a sending station or computer by the receiving unit as a signal that transmitted information has arrived incorrectly.

SYN Synchronous – A character used in synchronous (timed) communications that enables the sending and receiving devices to maintain the same timing.

ETB End of Transmission Block) – Not the same as EOT

CAN Cancel

EM End of Medium

SUB Substitute

ESC Escape – Usually indicates the beginning of an escape sequence (a string of characters that gives instructions to a device such as a printer). It is represented internally as character code 27 (hexadecimal 1 B).

FS File Separator

GS Group Separator

RS Record Separator

US Unit Separator

DEL Delete

Example 3.5 Determine the ASCII codes generated by the entering RAHUL and rahul from a computer keyboard.

Solution The ASCII codes for RAHUL and rahul are given below:

Symbol/Character	*ASCII Code*
R	1010010
A	1000001
H	1001000
U	1010101
L	1001100

Symbol/ Character	*ASCII Code*
r	1110010
a	1100001
h	1101000
u	1110101
l	1101100

3.6 THE EXTENDED BINARY CODED DECIMAL INTERCHANGE CODE (EBCDIC)

The Extended Binary Coded Decimal Interchange Code (EBCDIC), also known as Extended ASCII Character Set, consists of 128 additional decimal numbers, i.e., 256 decimal numbers. The additional decimal numbers ranging from 128 through 255 represent special, mathematical, graphical and foreign characters. EBCDIC was developed by IBM to be used with IBM mainframes.

EXERCISES

1. What would be the BCD equivalent of decimal 28 in 16-bit representation?
2. Find the excess-3 equivalent of $(248 \cdot 84)_{10}$.
3. Find the gray code equivalent of decimal 13.
4. Find the gray code equivalent of binary number 1101.
5. By writing the parity code (even) and threefold repetition code for all possible 4-bit straight binary numbers, prove that the Hamming distance in the two cases is at least 2 in the case of the parity code and 3 in the case of the repetition code.
6. Briefly describe salient features of the ASCII and EBCDIC codes in terms of their capability to represent characters and suitability for their use in different platforms.
7. What is parity bit? Define even and odd parity.
8. Write the excess-3 equivalent codes of $(6)_{10}$, $(78)_{10}$ all in 16-bit format.
9. Determine the gray code equivalent of $(10011)_2$ and the binary equivalent of the gray code number 110011.
10. What is 243_{10} in BCD?
11. What is the significant feature of gray code?
12. How many binary digits are needed to represent a four-digit decimal number using BCD?
13. Write out the gray code for a three-digit system.
14. State the abbreviations of ASCII and EBCDIC code.
15. Convert decimal 9126 to both BCD and ASCII codes. For ASCII, an odd parity bit is to be appended at the left.
16. Represent the unsigned decimal numbers 965 and 672 in BCD and then show the steps necessary to form their sum.
17. Represent decimal number 5027 in
    (a) BCD (b) excess-3 code, and
    (c) 2 * 421 code
18. How many printing characters are there in ASCII? How many of them are special characters (not letters or numerals)?
19. List the ASCII code for the 10 decimal digits with an even parity bit in the leftmost position.
20. Find the 9's complement of 6027 and express it in 2 * 421 code.

4

Introduction to Logic Circuits

INTRODUCTION

We live in analog world, deal with voltage, current, temperature and physical quantities in real circuits. Real values are continuously variable, we could use a physical quantity such as voltage in a circuit to represent a real number. However, stability and accuracy of physical quantities are difficult to obtain in real circuits. Manufacturing defects, cosmic rays, power supply, temperature, etc. are several factors, which cause instability in the real circuits. Also, many mathematical and logical functions are difficult or impossible to perform with analog quantities.

On the other hand, a digital system is a combination of devices that are designed to manipulate logical information or physical quantities, which are represented in digital form. In other words, these quantities can take only discrete values. The devices are most often electronic but they can be mechanical, magnetic or pen-man. Some of the familiar digital systems include digital computers, digital cameras, calculators, digital audio and video equipment, corded, cordless and cellular phones. Telephone system is considered the world's largest digital system.

Digital computers house a number of printed circuit boards (PCBs), a power supply, storage units like a horde disk and DVD or CD-ROM drives. Each unit is plugged into a main PCB, called the *motherboard*. Motherboard holds several integrated circuit chips (ICs). A chip comprises a number of subcircuits, which are interconnected to build a complete circuit. Examples of subcircuits are those that perform arithmetic operations, store data, or control the flow of data. Each of these subcircuits is a *logic circuit*.

The study of logic circuits is motivated mostly by their use in digital computers. Such circuits also form the foundation of many other digital systems. Logic circuits perform operations on digital signals and are usually implemented as electronic circuits where the signal values are restricted to a few discrete values. In binary logic there are only two values 0 *and* 1. Since a signal value is naturally represented by a digit (0 or 1), such logic circuits are referred to as *digital circuits*. In contrast, the analog circuits exhibit a continuous range of values between minimum and maximum levels.

In digital technology, we deal with binary circuits with an understanding of how are they represented in mathematical notations, and how are they designed using modern design automation techniques.

4.1 LOGIC FUNCTIONS AND GATES

Each logical condition or element has a logic value of either 0 or 1. We also need the ways to combine different logical signals or conditions to prove a logical result.

The simplest binary element is the switch that has two states. If a switch is controlled by an input element x, then the switch will open if $x = 1$ and closed if $x = 0$ as shown in Fig. 4.1(a). Figure 4.1(b) represents the symbol of a switch.

$x = 0$ $x = 1$ S x

(a) (b)

Fig. 4.1 (a) States of a switch, (b) symbol of a switch.

Let us consider that a switch is used to turn on or off a small light bulb. Figure 4.2 shows a circuit in which a small light bulb is controlled by a switch and a battery provides the power source. The light bulb glows when the switch is closed i.e., sufficient amount of current flows through the circuit and the bulb will not glow when the switch is open i.e., when the control input $x = 0$, light will not glow and if $x = 1$, chain light will glow. It defines the two states used in binary logic.

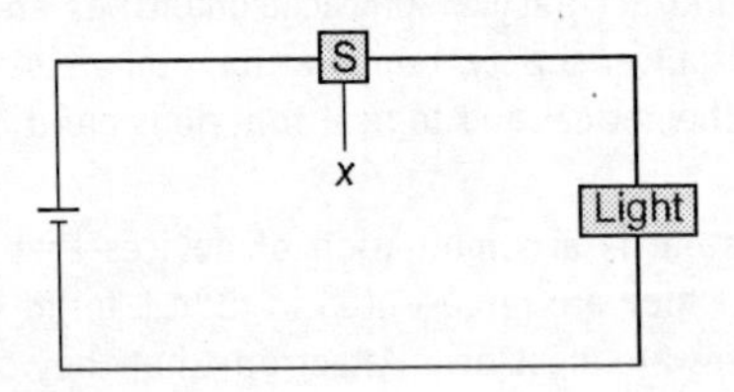

Fig. 4.2 A light controlled by a switch.

Now, consider that the light is controlled by two switches. Let x_1 and x_2 be the control inputs of these switches. The switches can be connected either in series or in parallel as shown in Fig. 4.3 and 4.4. If 'A' denotes the condition of light, the circuit's behaviour can be described by the expression of series and parallel connection.

(a) Series Connection: In series connection, light will turn on only if both the switches are closed as shown in Fig. 4.3. This behaviour can be described by the following expression:

$A(x_1, x_2) = x_1 \cdot x_2$ AND Function

where A = output of light bulb

x_1 and x_2 are control inputs.

If $x_1 = 1$ and $x_2 = 1$ then $A = 1$

If $x_1 = 0$ and $x_2 = 1$ then $A = 0$

If $x_1 = 1$ and $x_2 = 0$ then $A = 0$

If $x_1 = 0$ and $x_2 = 0$ then $A = 0$

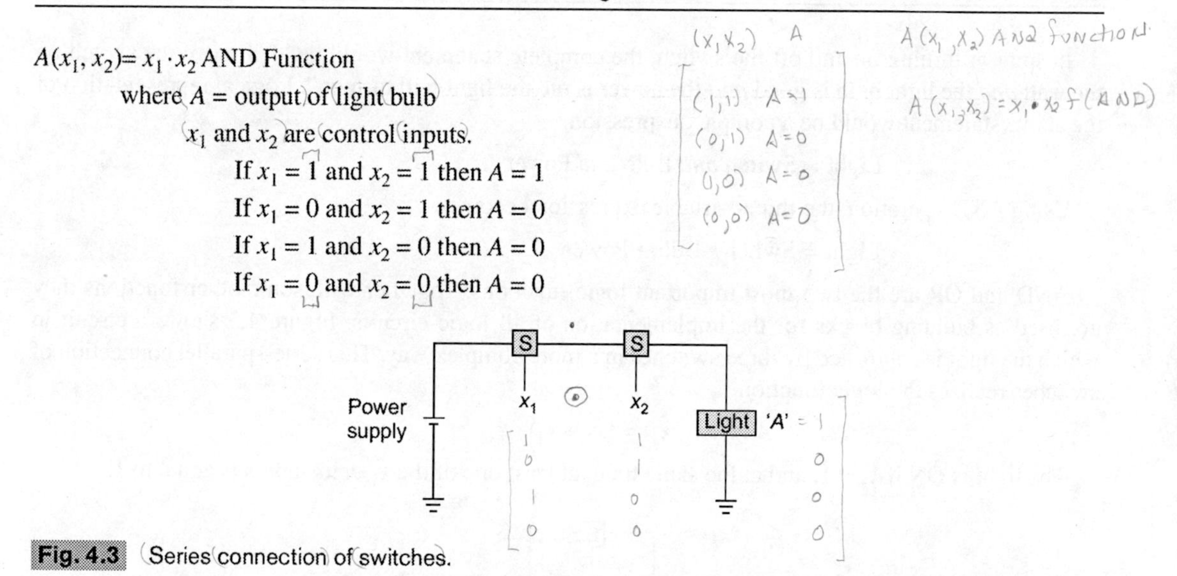

Fig. 4.3 Series connection of switches.

The above conditions can be easily understood by Fig. 4.3. The "•" symbol is the AND operator and the circuit shown in Fig. 4.3 is considered to be a *Logical AND function*.

(b) Parallel Connection: In parallel connection of two switches, light will be on if either of the two switches x_1 or x_2 is closed or both the switches are closed. Figure 4.4 shows the parallel arrangement of two switches. The behaviour can be described by the following expression.

$A(x_1, x_2) = x_1 + x_2$ 'OR' Function

where A = output of light bulb

x_1 and x_2 are control inputs

If $x_1 = 0, x_2 = 1$ then $A = 1$

If $x_1 = 1, x_2 = 0$ then $A = 1$

If $x_1 = 1, x_2 = 1$ then $A = 1$

If $x_1 = 0, x_2 = 0$ then $A = 0$

The above conditions are easily understood by Fig. 4.4. The + symbol is called OR operator and the circuit shown in Fig. 4.4 is said to implement a *logical OR function*.

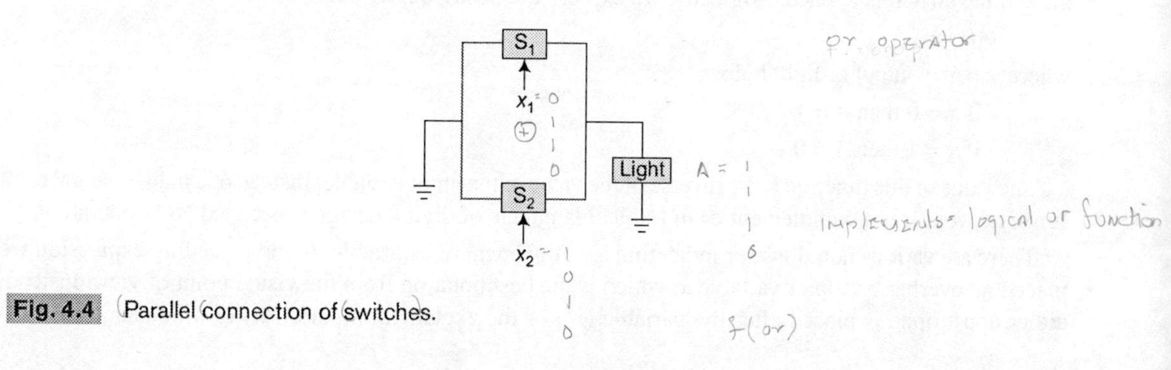

Fig. 4.4 Parallel connection of switches.

In spite of turning on and off the switch, the complete statement would be "If I move the switch on the wall *and* the light bulb is good *and* the power is on, the light will turn on". Logical representation of the above statement would be a complex expression.

Light = Switch *and* Bulb *and* Power

Using AND operation, the three-variable expression becomes

Light = Switch • Bulb • Power

AND and OR are the two most important logic functions. Together with some other functions they are used as building blocks for the implementation of all logic circuits. Figure 4.5 shows a circuit in which the light is controlled by three switches in a more complex way. This series-parallel connection of switches realizes the logic function

$$A\,(x_1, x_2, x_3) = (x_1 + x_2) \cdot x_3$$

The light is ON if $x_3 = 1$, and at the same time, at least one of the x_1 or x_2 inputs is equal to 1.

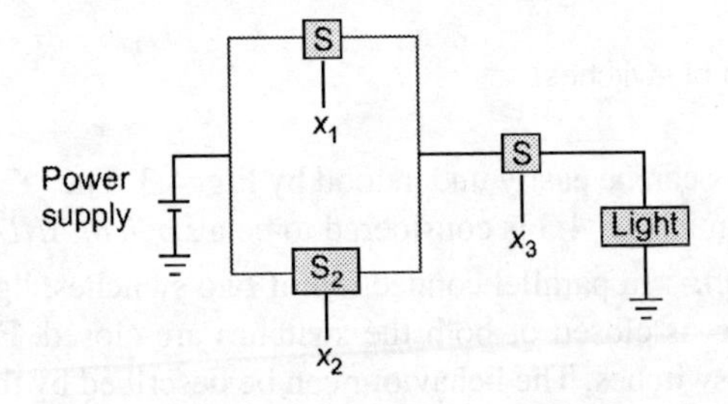

Fig. 4.5 A series-parallel connection.

4.2 INVERSION

In the previous series and parallel combination of switches, we have assumed the positive action when the switch is closed, such as turning the light on. It is equally interesting and important to consider the possibility of positive action when the switch is opened. Suppose that we connect the light bulb as shown in Fig. 4.6. In this case the switch is connected in parallel with the light, rather than in series. Consequently, a closed switch will short-circuit the light and no current will flow from it. The light will glow if the switch is opened. Logically, we express the behaviour as

$$A(x) = \overline{x}$$

where A = output of light bulb

If $x = 0$ then $A = 1$

If $x = 1$ then $A = 0$.

The value of this function is the inverse of the value of the input variable. Instead of a mathematical term 'inverse' we can use complement or in Logical language of digital design it is called NOT operation.

There are various notations for indicating a complement of a variable. In the preceding expression we placed an overbar on top of variable x, which is the best notation from the visual point of view. Instead, either apostrophe is placed after the variable (x') or the exclamation mark ($!x$) or tilde character ($\sim x$)

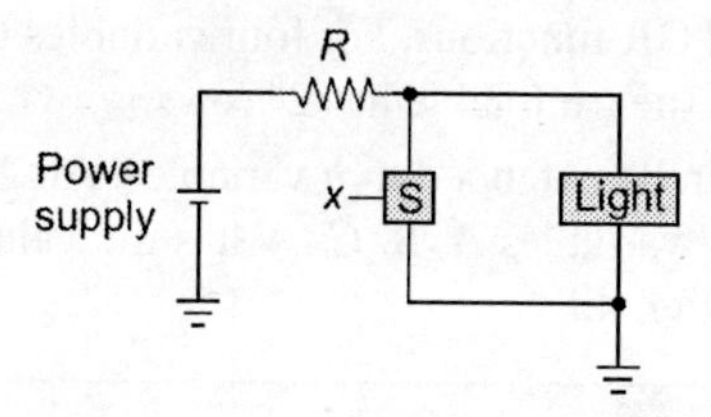

Fig. 4.6 An inverting circuit.

or the word NOT is placed before the variable to denote the complement. Different books use different notations. Thus, the following are to be observed as equivalent.

$$\bar{x} = x' = !x = \sim x = \text{NOT } x$$

The NOT operation can be applied to a single variable or to more complex variable equations. For example, if

$$F(A, B) = A + B$$

then the NOT operation of *f* is

$$\bar{f}(A, B) = \overline{A + B}$$

The above expression yields logic value 1 only when neither *A* nor *B* is equal to 1, that is, when $A = B = 0$. Again the following notations are equivalent.

$$\overline{A + B} = (A + B)' = !(A + B) = \sim (A + B) = \text{NOT } (A + B).$$

4.3 TRUTH TABLES

AND, OR and NOT logical operators have been introduced in the preceding section with an example of light bulb and switches. This approach gives these operations a certain "physical meaning". The same operation can be represented using a mathematical table known as *Truth Table* in logic ckts.

A truth table shows how a logic circuit's output responds to various combinations of inputs, using logic 1 for true and logic 0 for false. All permutations of the inputs are listed on the left, and the output of the circuit is listed on the right as shown in Fig. 4.7.

All permutations of inputs for a two-variable expression → A, B; Output → Y

A	B	Y
0	0	0
0	1	0
1	0	0
1	1	1

Fig. 4.7 A truth-table representation.

In digital electronics, truth tables can be used to reduce basic Boolean operations to simple correlations of inputs to outputs.

Figure 4.7 shows a truth table having two input variables, *A* and *B*. Small truth tables are easy to deal with. However, they grow exponentially in size with the number of variables. A truth table for three-input variables has eight rows because there are eight possible permutations. Such a table is shown in Fig. 4.3, which defines three-input AND and OR functions. For four variables the truth tables has 16 rows, and so on. In general, for n input variables the truth table has 2^n rows.

The AND and OR operations can be extended to n variables. An AND function of variables $A, B, C, D.... N$ has the value 1 only if all N variables $A, B, C...N$ has the value 1 if at least one, or more, of the variables is equal to 1 as shown in Fig. 4.8.

Inputs			*AND*	*OR*
A	B	C	A · B · C	A + B + C
0	0	0	0	0
0	0	1	0	1
0	1	0	0	1
0	1	1	0	1
1	0	0	0	1
1	0	1	0	1
1	1	0	0	1
1	1	1	1	1

Fig. 4.8 Three-input AND and OR truth table representation.

4.4 LOGIC GATES

The three basic logic operations discussed in the previous section (AND, OR, NOT) can be used to implement any logic function. Each logic function can be implemented electronically with transistors, resulting in a circuit element called a *logic gate*. Logic gates are made up of transistors, diodes and resistances which process one or more input signal in a logical fashion. Depending on the input value or voltage the logic gate will either output a value of '1' for ON or a value of '0' for OFF. Logic gates allow simplification of circuit operation. A basic understanding of logic gates is necessary to implement a logic circuit. It is often convenient to describe a logic circuit by drawing a circuit diagram, or *schematic*, consisting of graphical symbols representing logic gates. A logic gate has one or more inputs and one output that is a function of its inputs. The graphical symbols of AND, OR and NOT gates are shown in Fig. 4.9 in which the left side indicates how gates are drawn when there are only a few inputs. On the right side it shows how the symbols are augmented to accommodate a greater number of inputs.

A larger circuit is implemented by a *network* of gates. For example, a logic function of Fig. 4.5 can be implemented by the network shown in Fig. 4.10. The complexity of a network has a direct impact on its cost. It is always desired to reduce the cost of a manufactured product, it is important to implement a logic circuit as inexpensively as possible i.e. to reduce the logic function. Various techniques are there to reduce the logic function for example, Boolean Algebra, Karnaugh Maps, Quine-Mc Clussy Technique which we will discuss in later chapters.

4.5 TRUTH TABLES OF BASIC GATES

A truth table is a means for describing how a logic circuit's output depends on the logic levels at the circuit's inputs. All the logic circuits operate in the binary mode where each input and output is either 0 or 1. This characteristic of logic circuits allows us to use *Boolean algebra* as a tool for the analysis and

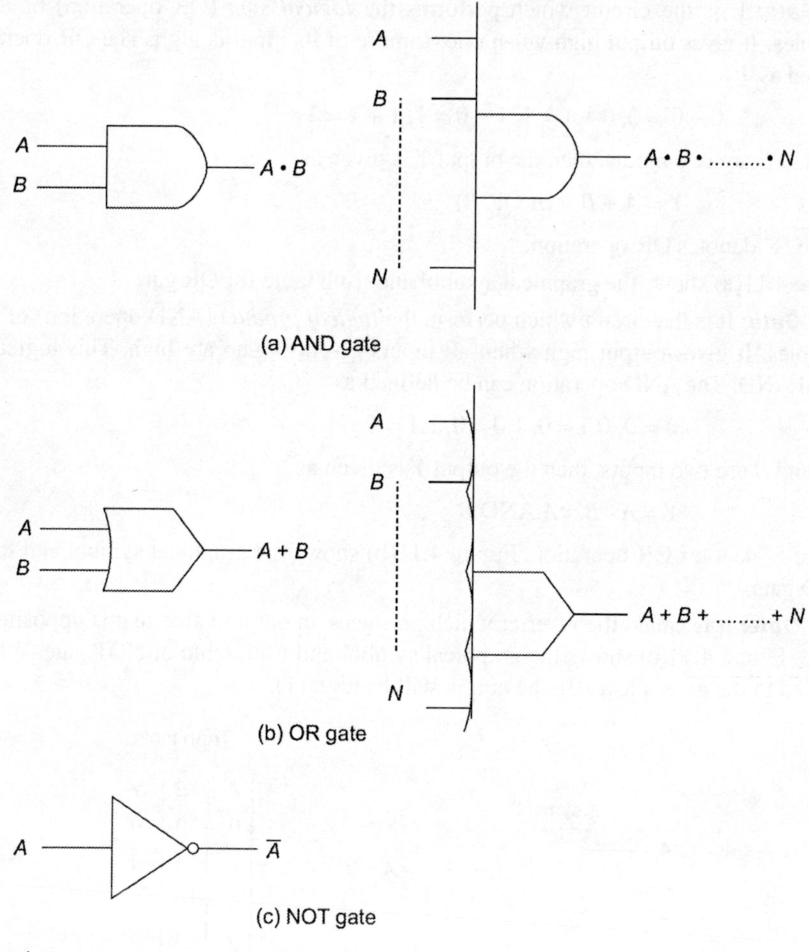

Fig. 4.9 Basic gates.

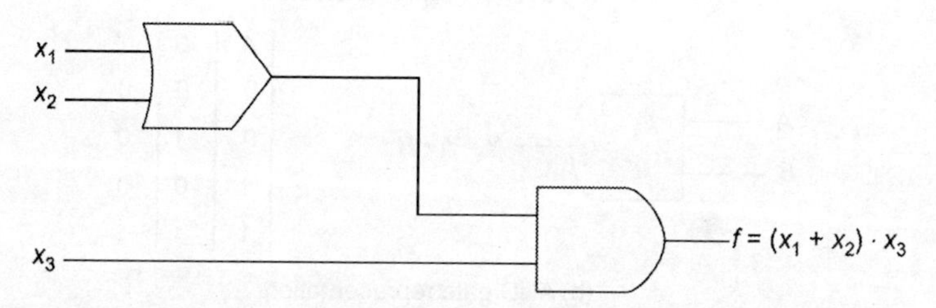

Fig. 4.10 Logic representation of Fig. 4.5.

design of digital circuits. *Boolean algebra* is a simple mathematical tool which describes the relationship between logic circuit output(s) and its input as an algebraic equation. This algebraic equation is known as *Boolean expression* or *Boolean equation*, which we will discuss in the next section. Figure 4.11 shows the truth table representation of all basic logic gates. It shows all possible relations between input and output in tabular form.

***(i)* OR Gate:** It is the circuit which performs the *logical sum* (OR operation) of two or more variables. It gives output high when one or more of its input is high. The OR operation can be defined as

$$0 + 0 = 0, 0 + 1 = 1, 1 + 0 = 1, 1 + 1 = 1$$

If A and B are two inputs, then the output Y is given as

$$Y = A + B = (A \text{ OR } B)$$

where '+' denotes OR operation.

Figure 4.11(a) shows the graphical symbol and truth table for OR gate.

***(ii)* AND Gate:** It is the circuit which perform the *logical product* (AND operation) of two or more variables. It gives output high when all inputs given to gate are high. This logical product is called AND. The AND operation can be defined as

$$0.0 = 0, 0.1 = 0, 1.0 = 0, 1.1 = 1$$

If A and B are two inputs, then the output Y is given as

$$Y = A \cdot B = A \text{ AND } B$$

where '·' denotes OR operation. Figure 4.11(b) shows the graphical symbol and truth table for AND gate.

***(iii)* NOT Gate:** It is called the inverter which produces an output value that is opposite of the input value. Figure 4.11(c) shows the graphical symbol and truth table of NOT gate. When the input applied to the gate is low (0), the output will be high (1).

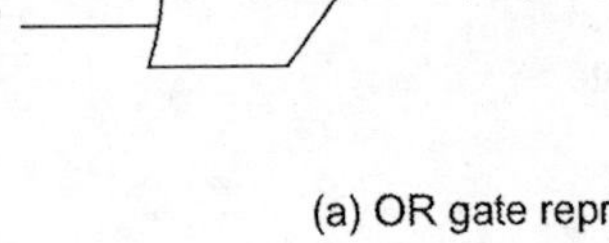

symbol

A, B → Y = A + B

Truth table

A	B	Y
0	0	0
0	1	1
1	0	1
1	1	1

(a) OR gate representation

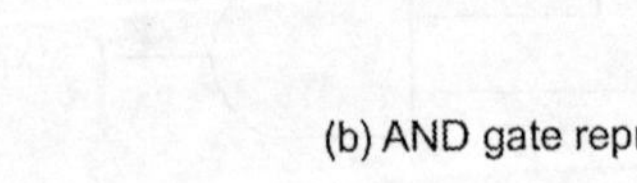

A, B → Y = A • B

A	B	Y
0	0	0
0	1	0
1	0	0
1	1	1

(b) AND gate representation

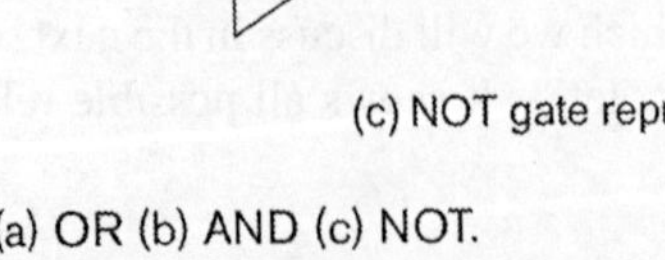

A → $Y = \overline{A}$

A	Y
0	1
1	0

(c) NOT gate representation

Fig. 4.11 Logic gates (a) OR (b) AND (c) NOT.

4.6 TIMING DIAGRAM

In majority of digital applications, the input to a logic gate (of any type) is not stationary voltage levels, but voltage waveform changes frequently between 1 and 0 logic levels. Timing diagrams are the representations of various signals in a network as a function of time. We have determined the behaviour of Fig. 4.12(a) by considering the four possible variations of inputs x_1 and x_2. The change in signals at various points would be as indicated in Fig. 4.12(d). The time runs from left to right, and each input is held for some fixed period. The figure shows the waveforms for the inputs and output of the network, as well as the internal signals labelled at points *A* & *B* (Fig. 4.12(c)). The timing diagram shows the changes in the waveforms at point *A* & *B* and the output *f* takes place instantaneously when the inputs x_1 and x_2 change their values.

The waveforms shown in Fig. 4.12(d) are based on the assumption that the logic gates respond to their changes without any delay. Such timing diagrams are useful for showing the *functional behaviour* of logic circuits. However, practical logic gates are implemented using electronic circuits which need some time to change their states. Thus, there is a delay between a change in input values and the corresponding change in the output of a gate.

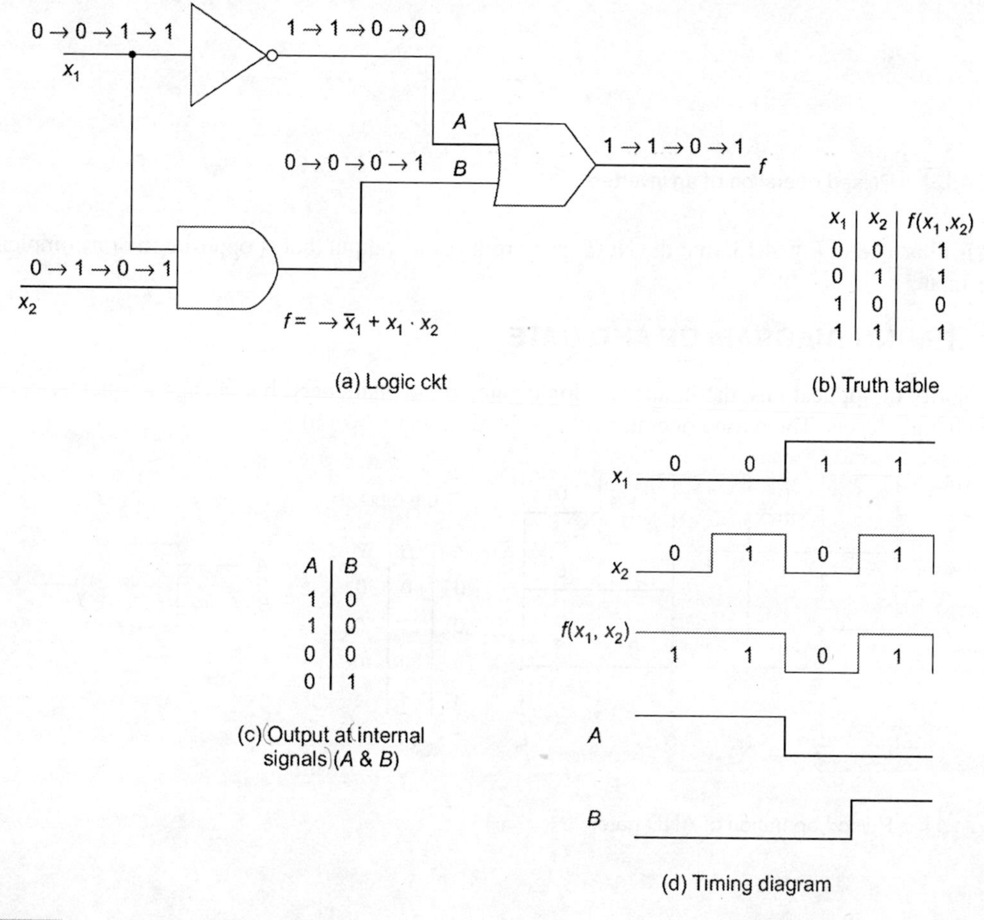

x_1	x_2	$f(x_1, x_2)$
0	0	1
0	1	1
1	0	0
1	1	1

A	B
1	0
1	0
0	0
0	1

Fig. 4.12 An example of logic network.

4.7 TIMING DIAGRAM OF NOT GATE

Consider a pulse given to a NOT gate as shown in Fig. 4.13. We will consider the time instances t_1 and t_2 to study the pulse operation.

1. At t_1 input (A) goes from low to high, therefore output (Y) goes from high to low.
2. Similarly, at t_2, input (A) goes from high to low and therefore output (Y) goes from low to high.

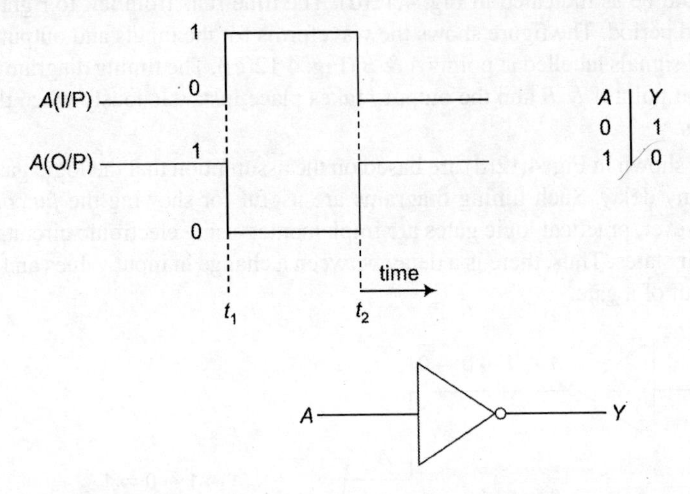

Fig. 4.13 Pulsed operation of an inverter.

It is illustated in Fig. 4.13, that the NOT gate produces an output that is opposite to or a complement of its input.

4.8 TIMING DIAGRAM OF AND GATE

In majority of applications, the inputs to a logic gate are not stationery, but change frequently between 1 and 0 logic levels. The pulsed operation obeys the truth table operation.

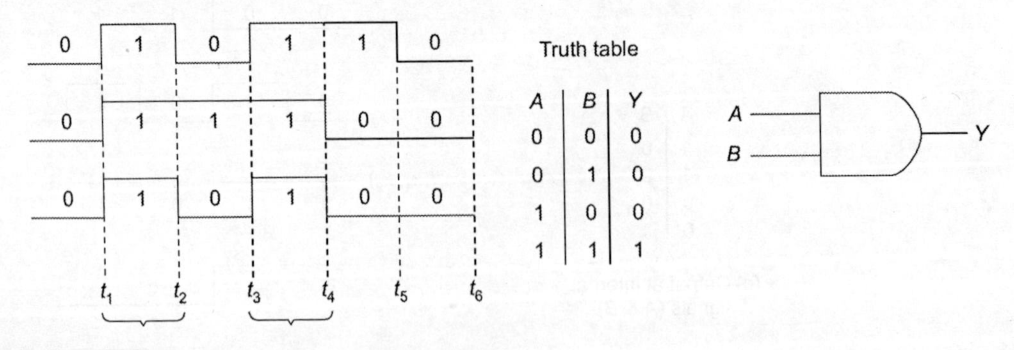

A	B	Y
0	0	0
0	1	0
1	0	0
1	1	1

Fig. 4.14 Pulsed operation of AND gate.

1. At $t = t_1$, both the inputs A and B go high, therefore output Y changes from low to high, as shown in truth table (*last column*) i.e., when both the inputs are 1 then the output is 1.
2. At $t = t_2$, input A goes 0 and B remains 1, therefore output Y changes to 0 (*second column*)
3. At $t = t_3$, input $A = 1$, $B = 1$, output $Y = 1$
4. At $t = t_4$, input $A = 1$, $B = 0$, output $Y = 0$ (third column)
5. At $t = t_5$, input $A = O$, $B = 0$, output $Y = 0$ (first column)

4.9 TIMING DIAGRAM OF OR GATE

Let us study the pulsed operation of OR gate.

1. At $t = t_1$, both input A and B are 1, then as shown in truth table, output will be 1 (last column).
2. At $t = t_2$, $A = 0$, $B = 1$ and O/P will be 1 as shown in (second column)
3. At $t = t_3$, both input $A = B = 0$, the output will be 0 (first column).
4. At $t = t_4$, input $A = 1$, $B = 0$, the output will be 1 (third column).

The resulting output wave is shown in Fig. 4.15.

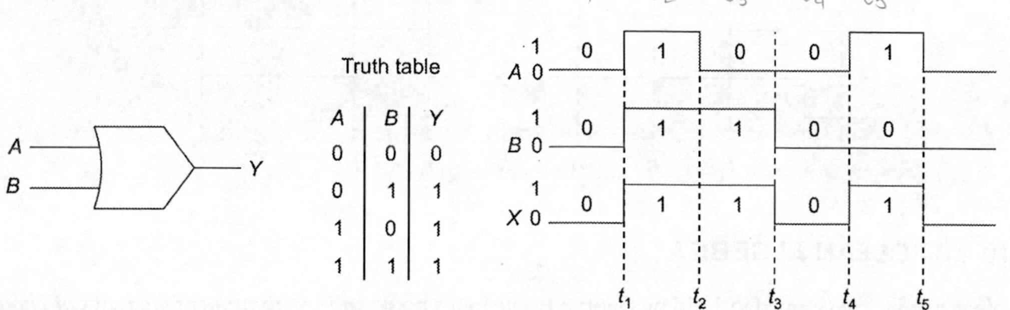

A	B	Y
0	0	0
0	1	1
1	0	1
1	1	1

Fig. 4.15 Pulsed operation of OR gate.

Example 4.1 Sketch the output wave if the input (Fig. 4.16) is applied to OR gate.

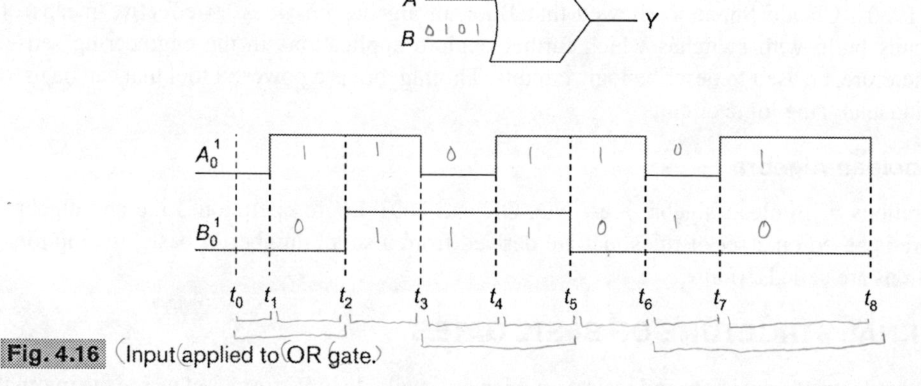

Fig. 4.16 Input applied to OR gate.

Solution Draw the truth table of OR gate. The output will be 1 when any one or both inputs are high.

1. At $t = t_0$, $A = 0$ & $B = 0$, therefore $X = 0$
2. At $t = t_1$, $A = 1$ & $B = 0$, therefore $X = 1$
3. At $t = t_2$, $A = 1$ & $B = 1$, therefore $X = 1$
4. At $t = t_3$, $A = 0$ & $B = 1$, therefore $X = 1$
5. At $t = t_4$, $A = 1$ & $B = 1$, therefore $X = 1$
6. At $t = t_5$, $A = 1$ & $B = 0$, therefore $X = 1$
7. At $t = t_6$, $A = 0$ & $B = 0$, therefore $X = 0$
8. At $t = t_7$, $A = 0$ & $B = 0$, therefore $X = 1$
9. At $t = t_8$, $A = 1$ & $B = 0$, therefore $X = 0$

Accordingly, draw the output waveform as shown below.

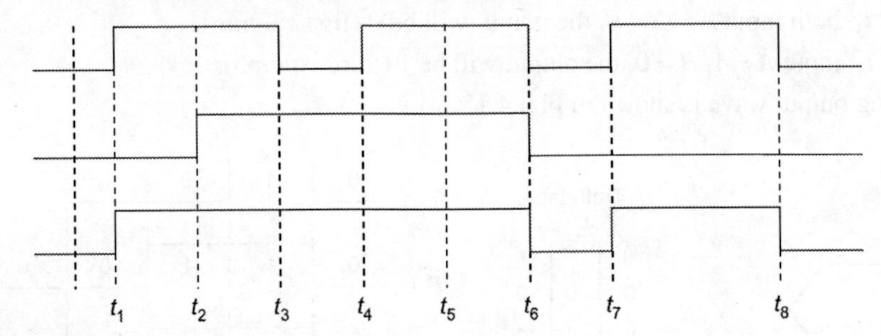

4.10 BOOLEAN ALGEBRA

Boolean algebra, developed in 1854 by George Boole in his book *An Investigation of the Laws of Though*, is a variant of ordinary algebra as taught in high school. Boolean algebra differs from ordinary algebra in three ways:

1. The variables are logical instead of a numeric character.
2. Based on assumptions, called *axioms*.
3. Operation based on laws of Boolean algebra.

In the late 1930s, Claude Shannon showed that Boolean algebra provides an effective means of describing circuits build with switches which further explore applications in the engineering sense. Algebra can, therefore, be used to describe logic circuits. This algebra is a powerful tool that can be used for designing and analyzing logic circuits.

Axioms of Boolean Algebra

The basic operations of Boolean algebra are AND, OR and NOT Logic operation. Like any algebra, Boolean algebra is based on a set of rules that are derived from a small number of basic assumptions. These assumptions are called *axioms*.

4.11 INTERNAL STRUCTURE OF BASIC GATES

Logic gates are made up of transistors and resistors which are available in IC form. Let us see the internal structure of AND and OR gates.

(a) AND Gate

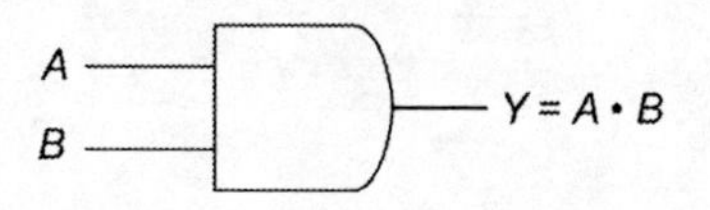

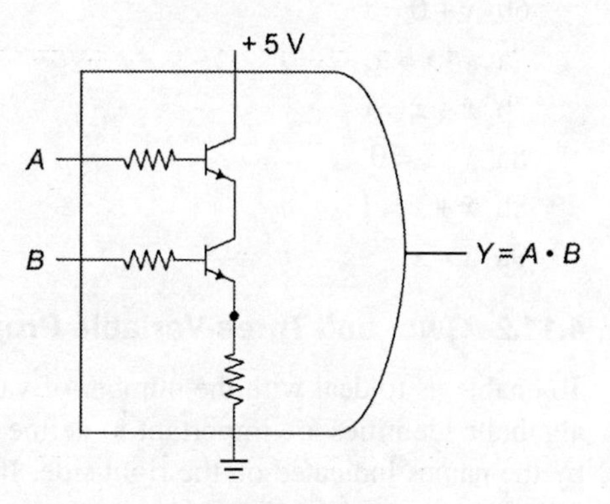

(b) OR Gate

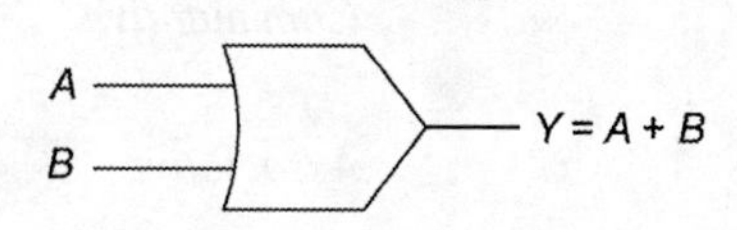

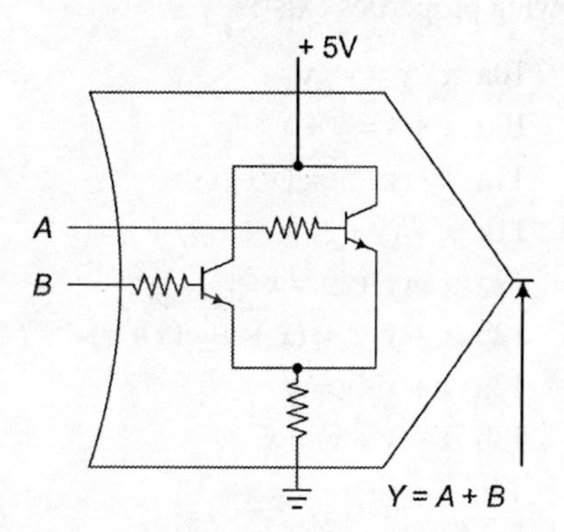

Fig. 4.17 Internal view of AND and OR gate.

Let us assume that Boolean algebra involves elements that take on one of the two values, 0 and 1. Assume that the following axioms are true:

1a. $0 \cdot 0 = 0$

1b. $1 + 1 = 1$

2a. $1 \cdot 1 = 1$

2b. $0 + 0 = 0$

3a. $0 \cdot 1 = 1 \cdot 0 = 0$

3b. $1 + 0 = 0 + 1 = 1$

4a. If $x = 0$ then $\bar{x} = 1$

4b. If $x = 1$ then $\bar{x} = 0$

4.11.1 Single Variable Theorems

From the axioms we can define some rules for dealing with single variables. These rules are often called *theorems*. If x is a variable, then the following theorems exist:

5a $x \cdot 0 = 0$

5b $x + 1 = 1$

6a $x \cdot 1 = x$

6b $x + 0 = x$

7a $x \cdot x = x$

7b $x + x = x$

8a $x \cdot \bar{x} = 0$

8b $x + \bar{x} = 1$

9a $\bar{\bar{x}} = x$

4.11.2 Two- and Three-Variable Properties

To enable us to deal with the number of variables in digital logic network, some two- or three-variable algebraic identities are important to define. These identities are referred as *properties*. They are known by the names indicated on the right side. If x, y and z are variables of any Boolean expression then the following properties exist:

10a $x \cdot y = y \cdot x$	*Commutative*
10b $x + y = y + x$	
11a $x \cdot (y \cdot z) = (x \cdot y) \cdot z$	*Associative*
11b $x + (y + z) = (x + y) + z$	
12a $x \cdot (y + z) = x \cdot y + x \cdot z$	*Distributive*
12b $x + y \cdot z = (x + y) \cdot (x + z)$	
13a $x + x \cdot y = x$	*Absorption*
13b $x \cdot (x + y) = x$	*Combining*
14a $x \cdot y + x \cdot \bar{y} = x$	
14b $(x + y) \cdot (x + \bar{y}) = x$	
15a $\overline{x \cdot y} = \bar{x} + \bar{y}$	*de Morgan's theorem*
15b $\overline{x + y} = \bar{x} \cdot \bar{y}$	
16a $x + \bar{x} \cdot y = x + y$	
16b $x \cdot (\bar{x} + y) = x \cdot y$	
17a $x \cdot y + y \cdot z + \bar{x} \cdot z = x \cdot y + \bar{x} z$	*Consensus*
17b $(x + y) \cdot (y + z) \cdot (\bar{x} + z) = (x + y) \cdot (\bar{x} + z)$	

We have listed a number of axioms, theorems and properties. These are required for performing algebraic manipulation of more complex expressions.

Example 4.2 Prove the following using Boolean axioms:

(a) $x + x \cdot y = x$

(b) $x + \bar{x} y = x + y$

(c) $(x + y)\,(x + z) = x + yz$

Solution

(a) $x + x \cdot y = x$

$x + x \cdot y = x\,(1 + y)$	(Distributive law)
$= x \cdot 1$	Rule 5(b)
$= x$	Rule 6(a)

(b) $x + \bar{x}y = x + y$

$x + \bar{x}y = (x + xy) + \bar{x}y$	(Absorption) (13a)
$= (xx + xy) + \bar{x}y$	Rule 7(a)
$= xx + xy + x\bar{y} + \bar{x}y$	Rule 8(a)
$= (x + \bar{x})\,(x + y)$	(Distributive law)
$= 1 \cdot (A + B)$	Rule 8(b)
$= A + B$	Rule 6(a).

(c) $(x + y)\,(x + z) = x + yz$

$(x + y)\,(x + z) = xx + xz + xy + yz$	(Distributive law)
$= x + xz + xy + yz$	(7a)
$= x\,(1 + z) + xy + yz$	(Distributive law)
$= x.\,1 + xy + yz$	(5b)
$= x\,(1 + y) + yz$	(Distributive law)
$= x.\,1 + yz$	(5b)
$= x + yz$	(6a)

Example 4.3 Using the truth table, show that the Boolean expression,

$$A + \bar{A}B = A + B$$

Solution There are two variables A and B in the given expression. There are four possible combinations (00, 01, 10, 11).

The first two columns (A, B) represent the input and the third column shows the complement output of A. Fourth column is obtained by ANDing the second and third columns. The fifth column is obtained by adding the first and fourth columns. The sixth column is obtained by adding the first two columns, which is required as per the expression $A + \bar{A}B = A + B$.

A	B	$\bar{A}$	$\bar{A}B$	$A + \bar{A}B$	$A + B$
0	0	1	0	0	0
0	1	1	1	1	1
1	0	0	0	1	1
1	1	0	0	1	1

The LHS and RHS of the expression is same as shown in the fifth and sixth columns. This verifies that

$$A + \bar{A}B = A + B.$$

Example 4.4 Using truth table, show the validity of the Boolean expression

$$(X + Y)\,(X + Z) = X + YZ$$

Solution The above expression has three variables X, Y, Z. These three variables can have eight possible combinations which is represented in first three columns (X, Y, Z). As per the requirement of the expression, the fourth column is obtained by OR in the first two and the fifth column is obtained by ORing the first and third columns. The sixth column is obtained by ANDing the fourth and fifth columns and the seventh column is obtained by ANDing the second and third columns. Finally the eighth column is obtained by ORing the first and seventh columns.

I	II	III	IV	V	VI	VII	VIII
X	Y	Z	$X+Y$	$X+Z$	$(X+Y)(X+Z)$	YZ	$X+YZ$
0	0	0	0	0	0	0	0
1	1	1	0	1	0	0	0
1	1	0	1	0	0	0	0
1	1	1	1	1	1	1	1
1	0	0	1	1	1	0	1
1	0	1	1	1	1	0	1
1	1	0	1	1	1	0	1
1	1	1	1	1	1	1	1

The equivalency of the sixth and eight columns $[(X+Y)(X+Z) = X+YZ]$ verifies the expression.

Example 4.5 Simplify using Boolean algebra

(a) $ABC + A\bar{B}C + AB\bar{C} = A(B+C)$

(b) $(A+B)(A+\bar{B})(\bar{A}+C) = AC$

Solution

(a) $ABC + A\bar{B}C + AB\bar{C} = A(B+C)$

LHS $ABC + A\bar{B}C + AB\bar{C}$

$= AC(B+\bar{B}) + AB\bar{C}$

$= AC + AB\bar{C}$ $\quad (B+\bar{B}=1)$

$= A(C + B\bar{C})$

$= A[(C+B)(C+\bar{C})]$ $\quad$ 16(a)

$= A(B+C)$

$=$ R.H.S.

(b) $(A+B)(A+\bar{B})(\bar{A}+C) = AC$

LHS $(A+B)(A+\bar{B})(\bar{A}+C)$

$= (AA + A\bar{B} + BA + B\bar{B})(\bar{A}+C)$ $\quad \begin{bmatrix} A\cdot A = A \\ B\cdot\bar{B} = 0 \end{bmatrix}$

$= (A + AB + \bar{A}B)(\bar{A}+C)$

$= (A(1+B) + \bar{A}B)(\bar{A}+C)$

$= A + A\bar{B}(\bar{A}+C)$ $\quad [B+1=1]$

$= A(1+\bar{B})(\bar{A}+C)$

$= A(\bar{A}+C)$ $\quad [1+\bar{B}=1]$

$= A\bar{A} + AC$

$= AC =$ R.H.S.

Example 4.6 Reduce the following expressions using Boolean algebra

(a) $Y = A\bar{B}D + A\bar{B}\,\bar{D}$

(b) $x = ACD + \bar{A}BCD$

(c) $Y = \overline{A}BC + A\overline{B}\,\overline{C} + ABC + AB\overline{C}$

(d) $Y = \overline{A}\,\overline{B}\,\overline{C} + \overline{A}B\overline{C} + A\overline{B}\,\overline{C} + AB\overline{C}$

Solution

(a) $y = A\overline{B}D + A\overline{B}\,\overline{D}$

$= A\overline{B}\,(D + \overline{D})$ $\qquad (D + \overline{D} = 1)$ (Rule 8b)

$= A\overline{B}$

(b) $x = ACD + \overline{A}BCD$

$= (A + \overline{A}B)\,CD$

$= (A + B)CD$ $\qquad (A + \overline{A}B = A + B)$ (Rule 16a)

(c) $Y = \overline{A}BC + A\overline{B}\,\overline{C} + ABC + AB\overline{C}$

$= BC(\overline{A} + A) + A\overline{C}(\overline{B} + B)$

$= BC + A\overline{C}$ $\qquad (A + \overline{A} = 1)$ (Rule 8b)

(d) $Y = \overline{A}\,\overline{B}\,\overline{C} + \overline{A}B\overline{C} + A\overline{B}\,\overline{C} + AB\overline{C}$

$= \overline{A}\,\overline{C}\,(\overline{B} + B) + A\overline{C}\,(B + \overline{B})$

$= \overline{A}\,\overline{C} + A\overline{C}$ $\qquad (B + \overline{B} = 1)$ (Rule 8b)

$= \overline{C}\,(\overline{A} + A)$

$= \overline{C}$

Example 4.7 Let us prove the validity of the logic equations:

(a) $(x_1 + x_3) \cdot (\overline{x}_1 + \overline{x}_3) = x_1\,\overline{x}_3 + \overline{x}_1\,x_3$

(b) $x_1\,\overline{x}_3 + \overline{x}_2\,\overline{x}_3 + x_1\,x_3 + \overline{x}_2\,x_3 = \overline{x}_1\,\overline{x}_2 + x_1\,x_2 + x_1\,\overline{x}_2$

Solution

(a) LHS

$(x_1 + x_3) \cdot (\overline{x}_1 + \overline{x}_3)$

$= (x_1 + x_3) \cdot \overline{x}_1 + (x_1 + x_3) \cdot \overline{x}_3$ $\qquad$ (Distributive property)

$= x_1\,\overline{x}_1 + x_3\,\overline{x}_1 + x_1\,\overline{x}_3 + x_3\,\overline{x}_3$ $\qquad$ (– do –)

$= 0 + x_3\,\overline{x}_1 + x_1\,\overline{x}_3 + 0$ $\qquad$ (6b)

$= x_3\,\overline{x}_1 + x_1\,\overline{x}_3 =$ R.H.S.

Hence, the equation is valid.

(b) LHS

$x_1\,\overline{x}_3 + \overline{x}_2\,\overline{x}_3 + x_1\,x_3 + \overline{x}_2\,x_3$

$= x_1\,(\overline{x}_3 + x_3) + \overline{x}_2\,(\overline{x}_3 + x_3)$ $\qquad$ (12a)

$= x_1\,(1) + \overline{x}_2\,(1)$ $\qquad$ (8b)

$= x_1 + \overline{x}_2$ $\qquad$ (6a)

Similarly, RHS

$\overline{x}_1\,x_2 + x_1\,x_2 + x_1\,\overline{x}_2$

$= \overline{x}_1\,\overline{x}_2 + x_1\,(x_2 + \overline{x}_2)$ $\qquad$ (12a)

$= \overline{x}_1\,\overline{x}_2 + x_1\,(1)$ $\qquad$ (8b)

$= x_1 + \bar{x}_1 \bar{x}_2$ (10b)

$= x_1 + \bar{x}_2$ (16a)

Manipulation of LHS and RHS of the above expression establishes the validity of the equation.

Example 4.8 Show that $AB + AC + B\overline{C} = AC + B\overline{C}$ by using truth table.

Solution

$AB + AC + B\overline{C} = AC + B\overline{C}$

The above expression has three variables (A, B, C). Hence the possible number of combinations would be 8.

I	II	III	IV	V	VI	VII	VIII	IX
A	B	C	$\overline{C}$	AB	AC	$B\overline{C}$	$AB + AC + B\overline{C}$	$AC + B\overline{C}$
0	0	0	1	0	0	0	0	0
0	0	1	0	0	0	0	0	0
0	1	0	1	0	0	1	1	1
0	1	1	0	0	0	0	0	0
1	0	0	1	0	0	0	0	0
1	0	1	0	0	1	0	1	1
1	1	0	1	1	0	1	1	1
1	1	1	0	1	1	0	1	1

Equivalent

The above truth table shows equivalency of the eighth and ninth columns which shows the LHS of the expression is equal to RHS. Hence, the above equation is valid.

Example 4.9 Simplify the expression using Boolean algebra

$Y = AC\,[\overline{B} + A\,(B + \overline{C})]$

Solution $Y = AC\,[\overline{B} + A(B + \overline{C})]$

$= AC\,[\overline{B} + AB + A\overline{C}]$

$= AC\overline{B} + (AC)\,(AB) + (AC)\,(A\overline{C})$

$= A\overline{B}C + ABC + AC\overline{C}\;[A \cdot A = A]$

$= A\overline{B}C + ABC + 0\;[C \cdot \overline{C} = 0]$

$= A\overline{B}C + ABC = AC\,(\overline{B} + B)$

$= AC\;[\overline{B} + B = 1]$

Example 4.10 Prove using Boolean algebra

$AB + \overline{A}B + \overline{A}\,\overline{B} = \overline{A} + B$

Solution $AB + \overline{A}B + \overline{A}\,\overline{B} = \overline{A} + B$

RHS $\overline{A} + B = \overline{A}\,(B + \overline{B}) + B\,(A + \overline{A})$

$$= \overline{A}B + \overline{A}\,\overline{B} + AB + \overline{A}B$$

$$= AB + \overline{A}B + \overline{A}B\ [\overline{A}B + \overline{A}B = \overline{A}B]$$

$$= \text{LHS}$$

Hence, proved.

The above examples illustrate the purpose of axioms, theorems and properties as a mechanism for algebraic manipulation.

4.12 NAND AND NOR LOGIC GATES

We have discussed the use of AND, OR and NOT gates in the synthesis of logic circuits. There are other basic logic functions that are also used for this purpose. These gates are NAND and NOR which are obtained by complementing the output of AND and OR gates. A bubble is placed on the output side of AND and OR gate symbols to represent the complemented output signal.

(a) NAND Gate

NAND is a contraction of NOT-AND. It implies an AND function with a complemented output. The standard symbol of NAND gate is shown in Fig. 4.18(a) which is same as the AND gate symbol except for a small bubble on its output. This small circle denotes inversion output. Thus, the NAND gate operates like an AND gate followed by inverter as shown in Fig. 4.18(b). Figure 4.18(c) shows the truth table of NAND gate, which is the inversion of AND gate.

A, B → Y

(a)

A, B → Y

(b)

A	*B*	*Y*
0	0	1
0	1	1
1	0	1
1	1	0

(c)

Fig. 4.18 (a) Logic symbol (b) NOT-AND equivalent (c) Truth table.

If A and B are two inputs applied to NAND gate, then the output will be

$$Y = \overline{A \cdot B}$$

(b) NOR Gate

The term NOR is a contraction of NOT-OR. It implies an OR function with a complemented output. The logical symbol of NOR gate is shown in Fig. 4.19(a). The symbol is like OR gate symbol with a bubble on the output, which shows inversion operation. The NOR gate operates like OR gate followed by a NOT gate.

A	B	Y
0	0	1
0	1	0
1	0	0
1	1	0

Fig. 4.19 (a) Logical symbol of NOR gate (b) Equivalent ckt. (c) Truth table.

If A and B are two inputs applied to NOR gate, then the output will be

$$Y = \overline{A + B}$$

4.13 EXCLUSIVE-OR AND EXCLUSIVE-NOR GATES

The exclusive-OR and exclusive-NOR gates are formed by a combination of other logic gates (i.e., inverter, AND & OR gates). For their wide applications in digital circuits, the exclusive-OR and exclusive-NOR gates are treated as basic logic gates.

(a) Exclusive-OR Gate

The exclusive-OR gate has only two inputs. Unlike other logic gates, the EX-OR gate never has more than two inputs. Figure 4.20 (a) shows the logic symbol and 4.20(b) shows the truth table of EX-OR gate.

A	B	Y
0	0	0
0	1	1
1	0	1
1	1	0

Fig. 4.20 (a) Logic symbol of EX-OR (b) Truth table of EX-OR gate.

If A and B are two inputs applied to EX-OR gate, then the output will be

$$Y = A \oplus B$$

$$= A \text{ EX-OR } B$$

where, $\oplus$ is the symbol for EX-OR operation.

(b) Exclusive NOR Gate

The EX-NOR gate has only two inputs like EX-OR gate. Figure 4.21(a) shows the logical symbol of EX-NOR gate and Fig. 4.21(b) shows the truth table of EX-NOR gate.

A	B	Y
0	0	1
0	1	0
1	0	0
1	1	1

(a) (b)

Fig. 4.21 (a) Logical symbol of EX-NOR (b) Truth table of EX-NOR gate.

If A and B are two inputs applied to EX-NOR gate then the output will be

$$Y = A \odot B$$
$$= A \text{ Ex-NOR } B$$

where $\odot$ is the symbol of EX-NOR operation.

4.14 TIMING DIAGRAM OF NAND GATE

Consider the two waveforms A and B are applied to NAND gate. In order to determine the pulsed operation of NAND gate, we will look at the inputs with respect to each other of the truth table.

1. At $t = t_1$, input $A = 1$, $B = 1$, therefore output $y = 0$
2. At $t = t_2$, input $A = 1$, $B = 0$, therefore output $y = 1$
3. At $t = t_3$, input $A = 0$, $B = 1$, therefore output $y = 1$
4. At $t = t_4$, input $A = 1$, $B = 0$, therefore output $y = 1$
5. At $t = t_5$, input $A = 0$, $B = 1$, therefore output $y = 1$
6. At $t = t_6$, input $A = 1$, $B = 1$, therefore output $y = 0$

The resulting output waveform (Y) is shown at waveform Fig. 4.22.

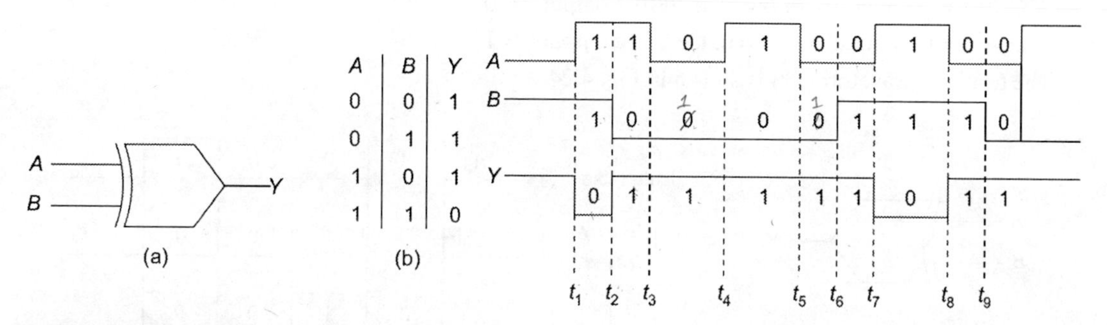

A	B	Y
0	0	1
0	1	1
1	0	1
1	1	0

Fig. 4.22 Pulsed operation of NAND gate.

4.15 TIMING DIAGRAM OF NOR GATE

Consider waveforms A and B applied to NOR gate as shown in Fig. 4.23. Look at the truth table while determining the output waveform.

1. At $t = t_1$, input $A = 1$, $B = 0$, therefore output $y = 0$
2. At $t = t_2$, input $A = 0$, $B = 0$, therefore output $y = 1$
3. At $t = t_3$, input $A = 0$, $B = 1$, therefore output $y = 0$
4. At $t = t_4$, input $A = 0$, $B = 0$, therefore output $y = 1$
5. At $t = t_5$, input $A = 1$, $B = 0$, therefore output $y = 0$

The resulting output waveform (Y) is shown in Fig. 4.23.

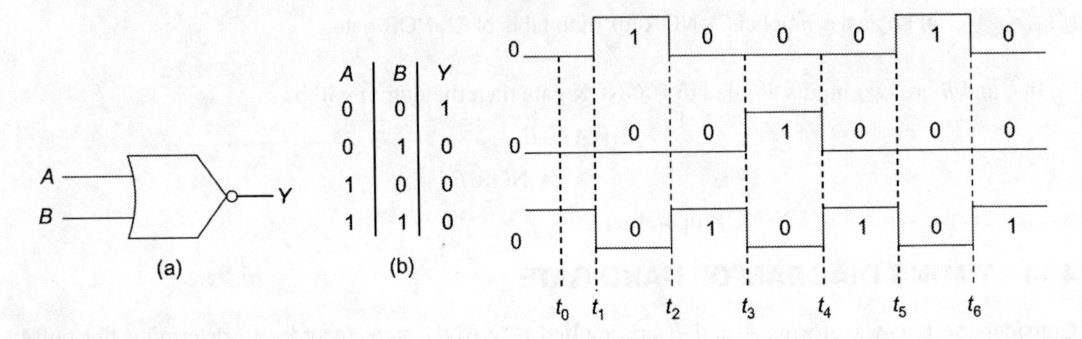

Fig. 4.23 Pulsed operation of NOR gate.

4.16 TIMING DIAGRAM OF EX-OR GATE

Consider the waveforms A and B applied at the inputs of EX-OR gate. The output pulsed waveform can be drawn while observing the behaviour of truth table.

1. At $t = t_1$, input $A = 1$, $B = 1$, therefore output $y = 0$
2. At $t = t_2$, input $A = 0$, $B = 1$, therefore output $y = 1$
3. At $t = t_3$, input $A = 0$, $B = 0$, therefore output $y = 0$.
4. At $t = t_4$, input $A = 1$, $B = 0$, therefore output $y = 1$

The resulting waveform (Y) is shown in Fig. 4.24.

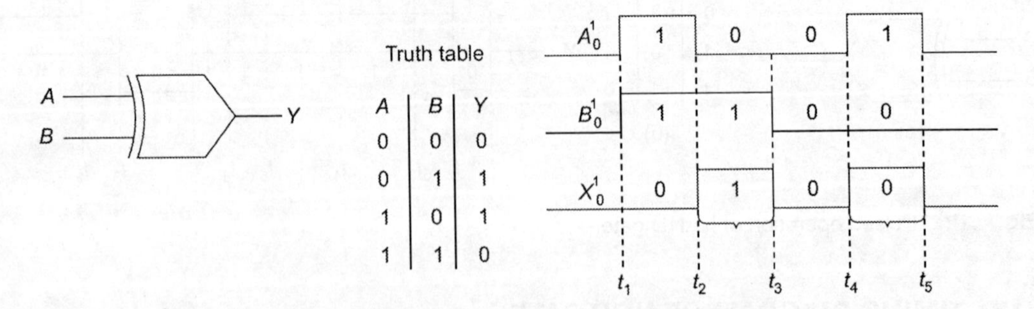

Fig. 4.24 Pulsed operation of EX-OR gate.

4.17 TIMING DIAGRAM OF EX-NOR GATE

Consider A and B are two waveforms applied at input of EX-NOR gate. Like in the previous timing diagram, see the truth table and draw the output waveform of EX-NOR gate as shown in (Y) Fig. 4.25.

1. At $t = t_1$, input $A = 1$, $B = 0$, output $y = 0$
2. At $t = t_2$, input $A = 1$, $B = 1$, output $y = 1$
3. At $t = t_3$, input $A = 0$, $B = 1$, output $y = 0$
4. At $t = t_4$, input $A = 0$, $B = 0$, output $y = 1$
5. At $t = t_5$, input $A = 1$, $B = 0$, output $y = 0$.

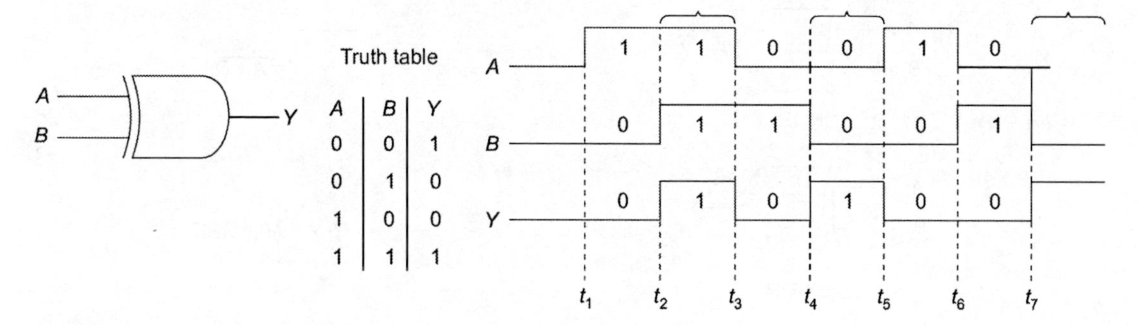

Fig. 4.25 Pulsed operation of EX-NOR gate.

4.18 BOOLEAN EXPRESSION OF A LOGIC CIRCUIT

In the previous section, we have discussed logic gates, Boolean algebra which are the basic building blocks of a logic circuit. A digital system consists of logic circuits. A logic circuit of any complexity is described by using Boolean expression. The Boolean expressions of the various logic gates are given in Fig. 4.26.

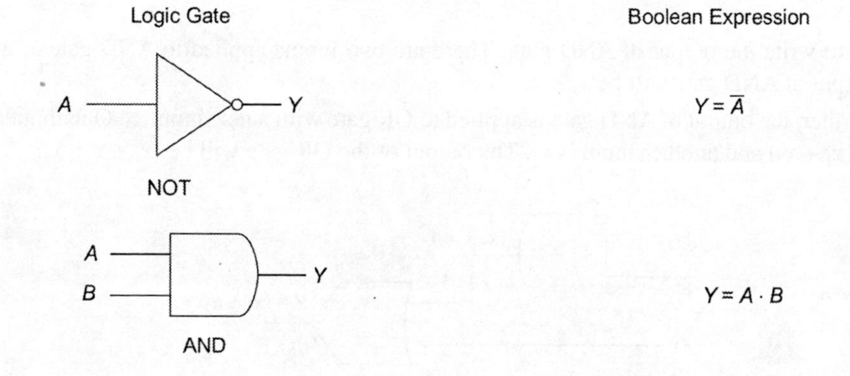

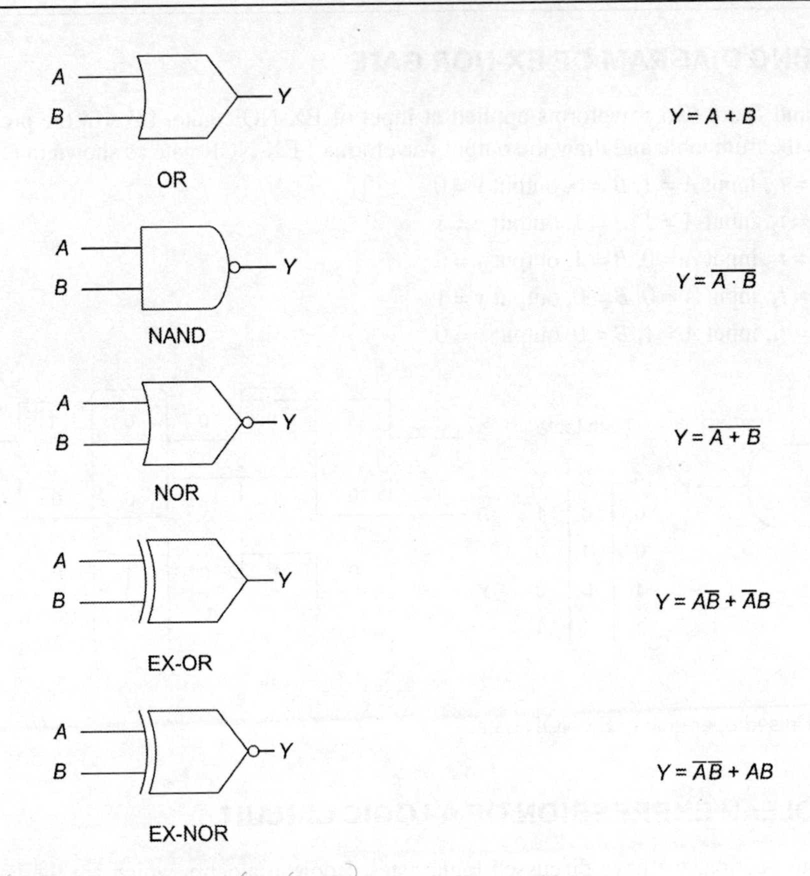

Fig. 4.26 Boolean expressions of logic gates.

Now consider an example of the circuit given in Fig. 4.27. The circuit has three inputs x_1, x_2, x_3 and a single output. We can easily determine the output expression of the circuit using Boolean expressions of logic gates.

1. First, write the output of AND gate. There are two inputs applied to AND gate x_1 and x_2. The output of AND gate will be $x_1 + x_2$.
2. Further, the output of AND gate is applied to OR gate with a new input x_3. One input of OR gate is $(x_1 + x_2)$ and another input is x_3. The output of the OR gate will be $(x_1 + x_2)$. x_3.

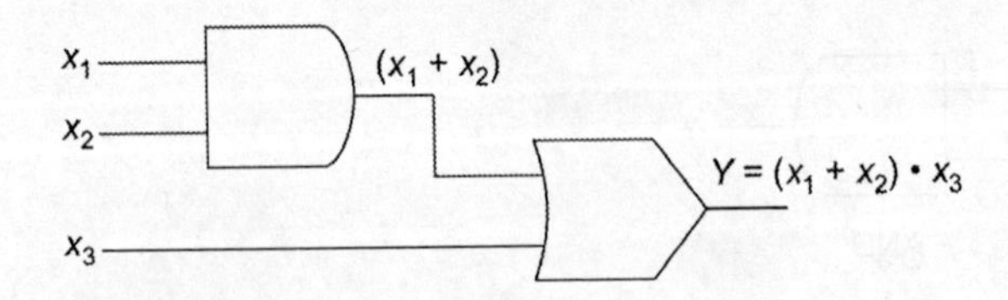

Fig. 4.27 Logic circuit with its Boolean expression.

4.19 TRUTH TABLE OF A LOGIC CIRCUIT

In the previous section, we have learnt to write a Boolean expression of a logic circuit. A truth table is a mathematical representation of a logic circuit or a Boolean expression. Once we have evaluated a Boolean expression for all the possible combinations of the input variables, we can summarize the result in the form of a truth table.

In order to understand the concept, consider the circuit shown in Fig. 4.28. The Boolean expression of this circuit can be written as,

$$Y = (A + B)\ (B + \overline{C})$$

Let us evaluate this Boolean expression for all combinations of A, B and C. Since there are three variables, the possible number of combinations would be 8.

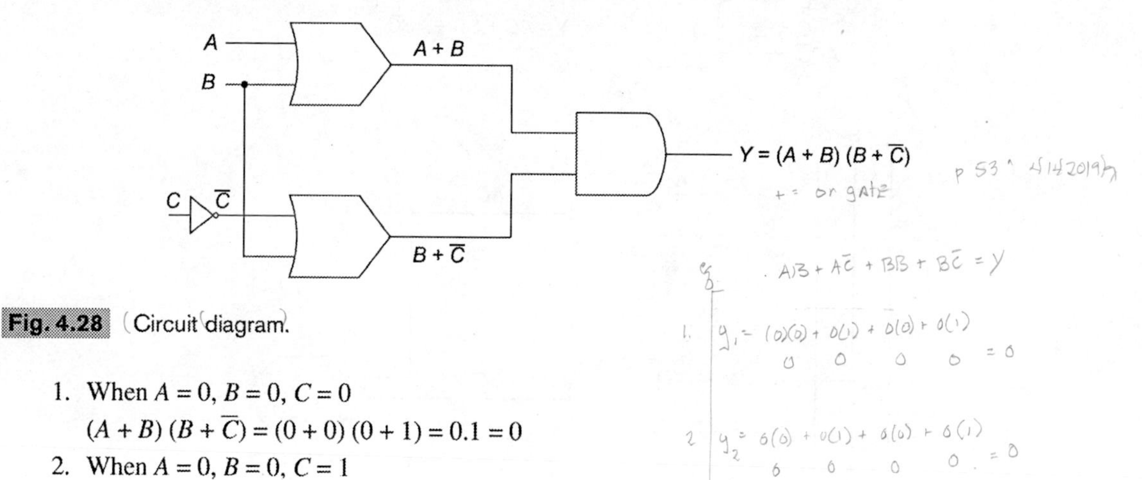

Fig. 4.28 Circuit diagram.

1. When $A = 0, B = 0, C = 0$
   $(A + B)\ (B + \overline{C}) = (0 + 0)\ (0 + 1) = 0.1 = 0$
2. When $A = 0, B = 0, C = 1$
   $(A + B)\ (B + \overline{C}) = (0 + 0)\ (0 + 0) = 0.0 = 0$
3. When $A = 0, B = 1, C = 0$
   $(A + B)\ (B + \overline{C}) = (0 + 1)\ (1 + 1) = 1.1 = 1$

Similarly, all values can be calculated, and the output values for various inputs will be shown in the truth table shown in Fig. 4.29.

Input			Output
A	B	C	Y [(A + B) (B + C̄)]
0	0	0	0
0	0	1	0
0	1	0	1
0	1	1	1
1	0	0	1
1	0	1	0
1	1	0	1
1	1	1	1

Fig. 4.29 Truth table.

4.20 IMPLEMENTATION OF A LOGIC CIRCUIT FROM BOOLEAN EXPRESSION

It will be interesting to know the behaviour of a circuit defined by a Boolean expression. Truth table, Boolean expression and a logic circuit are relevant to each other. If any one of them is given, it is easier to implement the others directly. Let us take an expression

$$Y = AC + \overline{B}C + AB\overline{C}$$

which uses AND, OR and NOT operations. The above Boolean expression contains three terms (AC, $\overline{B}C$ and $AB\overline{C}$) which are ORed together. Three input OR gates are required with inputs AC, $\overline{B}C$ and $AB\overline{C}$. This is shown in Fig. 4.30.

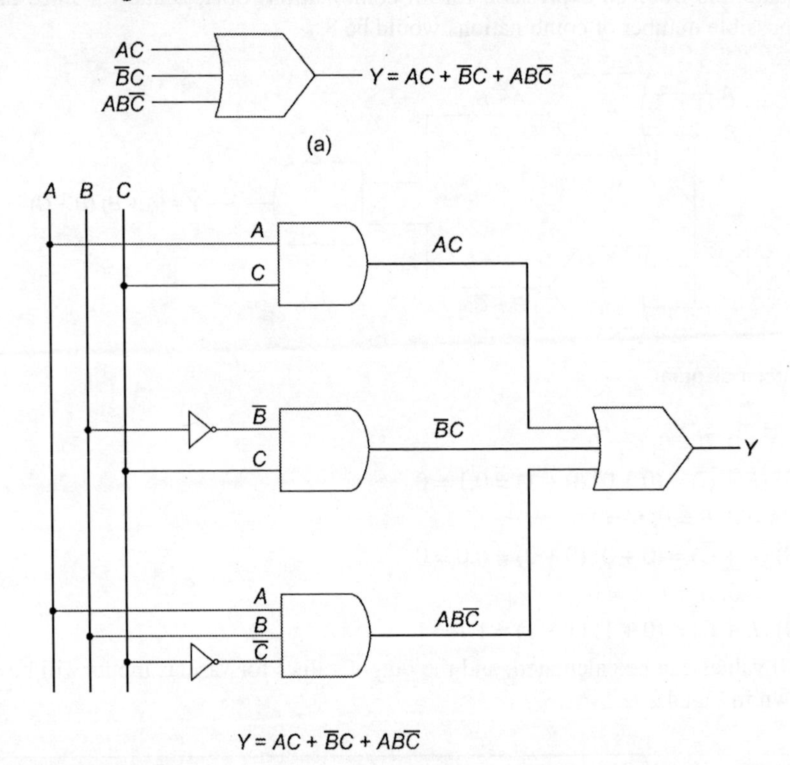

Fig. 4.30 Logic circuit implementation from a Boolean expression.

4.21 IMPLEMENTATION OF A TRUTH TABLE USING BOOLEAN EXPRESSION

In implementing a logic circuit, truth table of a Boolean expression is the heart of digital circuit. A truth table of a Boolean expression can be constructed by the following methods.

1. Observe the Boolean expression.
2. Calculate the number of variables ($2^m = n$)
3. For the valid entry of the given expression, the output will be 1, otherwise 0.

Let us see the following expression

$$Y = ABC + \overline{A}BC + A\overline{B}C + \overline{A}\,\overline{B}\,\overline{C}$$

Number of variables in the above expression is 3. Therefore, $2^m = n$ i,e. $2^3 = 8$ possible combinations are possible. But the output will be GH (1), only for the valid entries of Boolean expression, i,e. for ABC, $\overline{A}BC$, $A\overline{B}C$ & $\overline{A}\,\overline{B}\,\overline{C}$.

A	B	C	Boolean Expression of Input A, B, C	Output (1 for valid entries only)
0	0	0	$\overline{A}\overline{B}\overline{C}$	1
0	0	1	$\overline{A}\overline{B}C$	0
0	1	0	$\overline{A}B\overline{C}$	0
0	1	1	$\overline{A}BC$	1
1	0	0	$A\overline{B}\overline{C}$	0
1	0	1	$A\overline{B}C$	1
1	1	0	$AB\overline{C}$	0
1	1	1	ABC	1

Example 4.11 Design the logic circuit using AND, OR or NOT gates or combination of them for the following Boolean expression.

(a) $(A + B) \cdot \overline{C}$

(b) $A\overline{B} + C$

(c) $Y = A\overline{B} + \overline{A}B$

(d) $Y = \overline{P}\,\overline{Q} + PQ$

(e) $Y = AB + B\overline{C}$

(f) $Y = \overline{A + P\overline{Q}}$

(g) $Y = \overline{PQ\,(A + \overline{Q})}$

(h) $Y = \overline{PQ\,(A + B)}$

(i) $Y = \overline{(A + B + \overline{P}Q\overline{R})} + \overline{B}P\overline{Q}$

Solution

(a) $(A + B) \cdot \overline{C}$

Inputs are A, B, C

We use gates: OR, AND and NOT

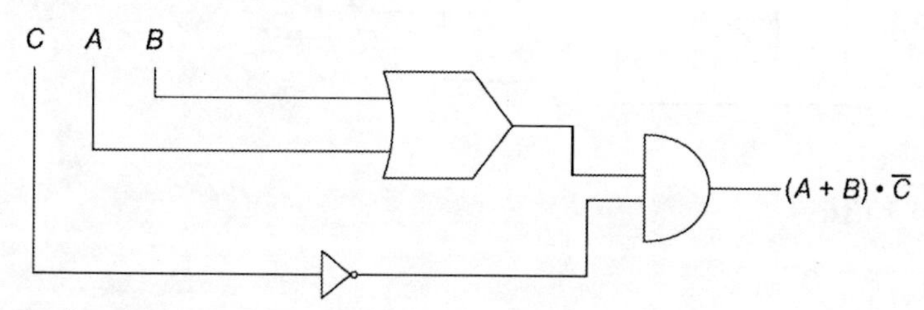

(b) $A\overline{B} + C$

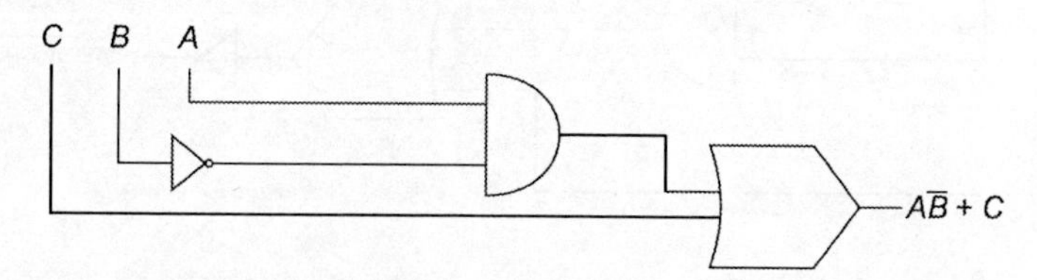

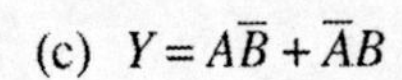

(c) $Y = A\overline{B} + \overline{A}B$

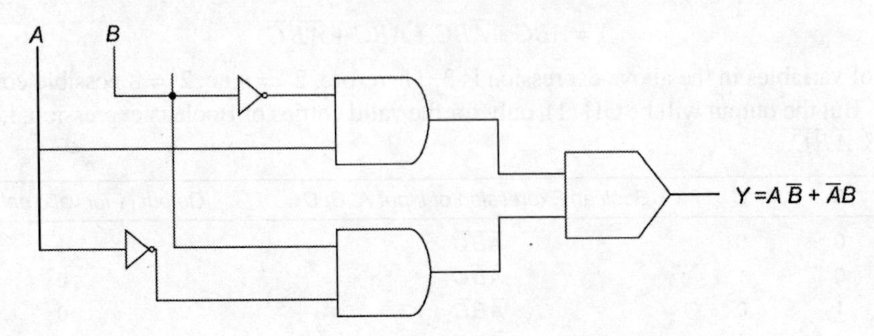

(d) $Y = \overline{P}\,\overline{Q} + PQ$

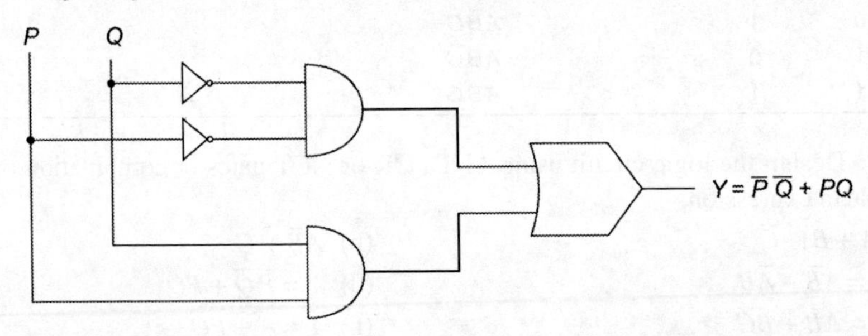

(e) $Y = AB + B\overline{C}$

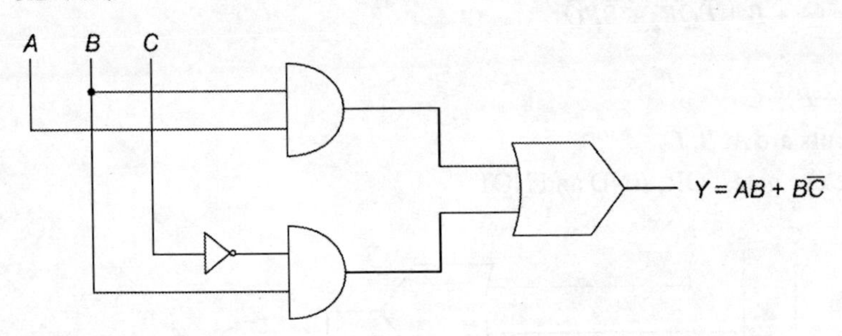

(f) $Y = \overline{A + P\overline{Q}}$

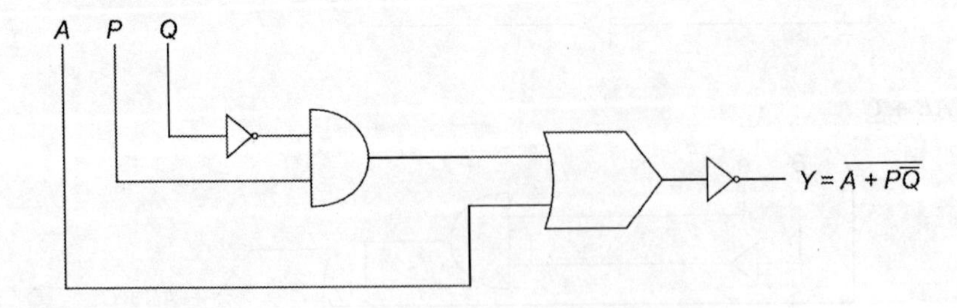

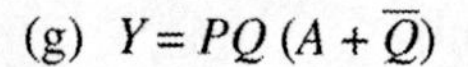

(g) $Y = PQ\,(A + \overline{Q})$

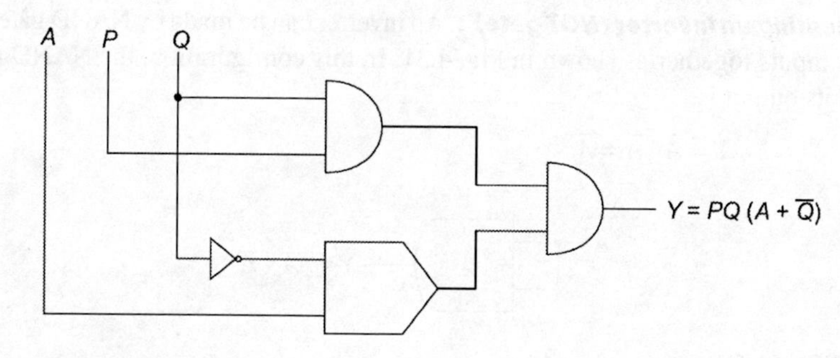

(h) $Y = \overline{PQ\,(A + B)}$

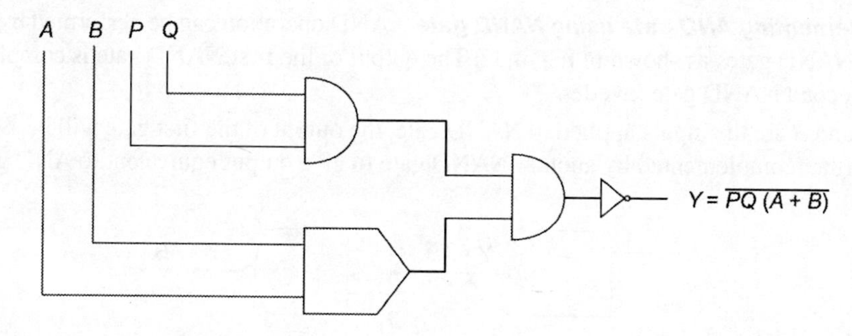

(i) $Y = \overline{(A + B + \overline{P}Q\overline{R})} + \overline{B}P\overline{Q}$

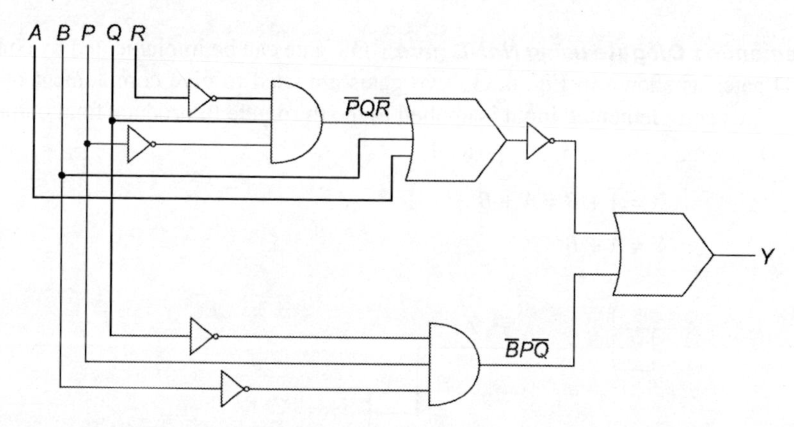

4.22 UNIVERSAL GATES (NAND AND NOR)

In digital circuitry, it is always important to reduce the complexity of the circuit. NAND and NOR gates are designated as universal gates so that any logic function can be implemented by these gates without the need to use any other gate. In practice, this is advantageous since NAND and NOR gates are economical and easier to fabricate and are the basic gates used in all IC digital logic families. It is an interesting task to see how NAND and NOR gates are used to implement all other types of logic gates.

Implementation of basic gates (AND, OR, NOT) using NAND gates only:

(a) *Implementing an Inverter (NOT gate)* An inverter can be made by NAND gate by connecting all of its inputs together as shown in Fig. 4.31. In this configuration, the NAND acts as inverter because its output is

$$Y = \overline{A} \cdot \overline{A} = \overline{A}$$

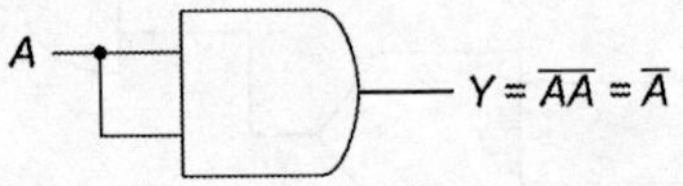

A

$Y = \overline{AA} = \overline{A}$

Fig. 4.31 NAND as NOT gate.

(b) *Implementing AND gate using NAND gate* AND operation can be performed by connecting two NAND gates as shown in Fig. 4.32. The output of the first NAND gate is complemented by the second NAND gate inverter.

If A and B are the inputs applied to NAND gate, the output of the first gate will be $Y = \overline{AB}$ which is further complemented by another NAND gate to give output equivalent to AND gate $Y = AB$.

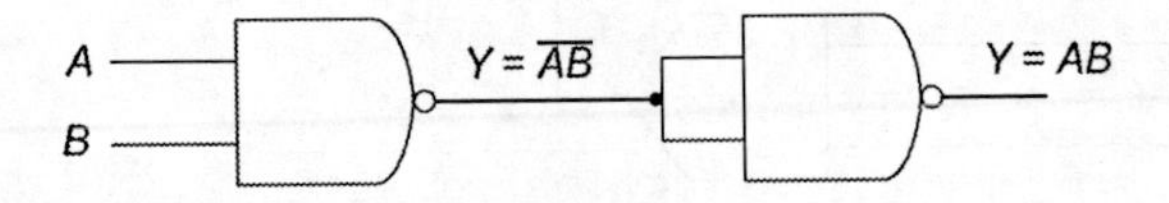

Fig. 4.32 NAND gate as AND gate.

(c) *Implementing OR gate using NAND gate* OR gate can be implemented by connecting three NAND gates as shown in Fig. 4.33. Two gates are used to give complement output $Y = \overline{A}\,\overline{B}$. Further, this complemented input is applied to the third gate to produce final output $Y = A + B$.

Using De Morgan Theorem, it is possible

$$\overline{\overline{A}\,\overline{B}} = \overline{\overline{A}} + \overline{\overline{B}} = A + B$$

$$Y = A + B$$

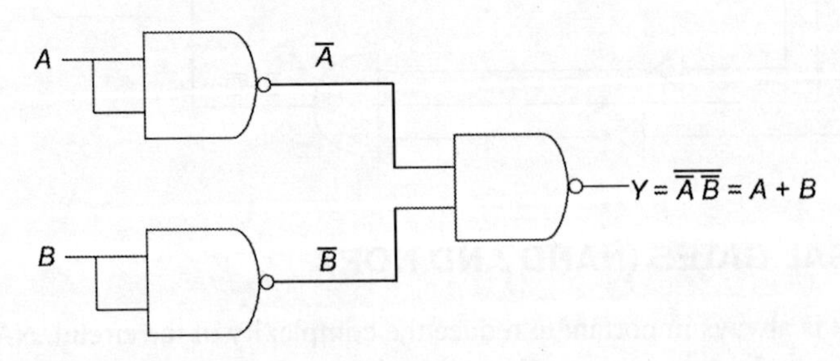

Fig. 4.33 NAND as OR gate.

(a) Implementing NOT gate using NAND gate

NOT gate operation can be performed by two methods shown in Fig. 4.34 (a & b). If all the NAND gate input pins connect to the input signal A, i.e. if all the inputs are short, the NAND gate gives an output $\overline{A}$ the [Fig. 4.34 (a)]. In the other method, if the one input of the NAND gate is connected to input signal $\overline{A}$ and all other input pins, are connected to logic 1, the output will be $\overline{A}$ as shown in Fig. 4.34 (b).

A — $(\overline{A{\cdot}A}) = \overline{A}$

(a)

A, 1 — $(\overline{A{\cdot}1}) = \overline{A}$

(b)

Fig. 4.34 NAND as NOT gate.

Implementation of basic gates using NOR gate only:

(a) Implementing NOT gate using NOR gate Similar to NAND realization of NOT gate, NOR gate also implements NOT gate operation by two methods shown in Fig. 4.35 (a & b). If all the inputs of NOR gate are connected together and a single input 'A' is given, the NOR gate simply acts as a NOT gate as shown in Fig. 4.35 (a). In the second method if an input signal 'A' is applied to one of the inputs of NOR gate and all other inputs are connected to logic 0, the output will be $\overline{A}$ as shown in Fig. 4.35 (b).

A — $\overline{A + A} = A$

(a)

A, 1 — $\overline{A + 0} = \overline{A}$

(b)

Fig. 4.35 NOR gate as NOT gate.

Example 4.12 Implement the following Boolean expression using NAND gates only.

$$Y = A\overline{B}C + ABC + \overline{A}BC$$

Solution

Step I While implementing the above expression using AND-OR-NOT Logic, we have,

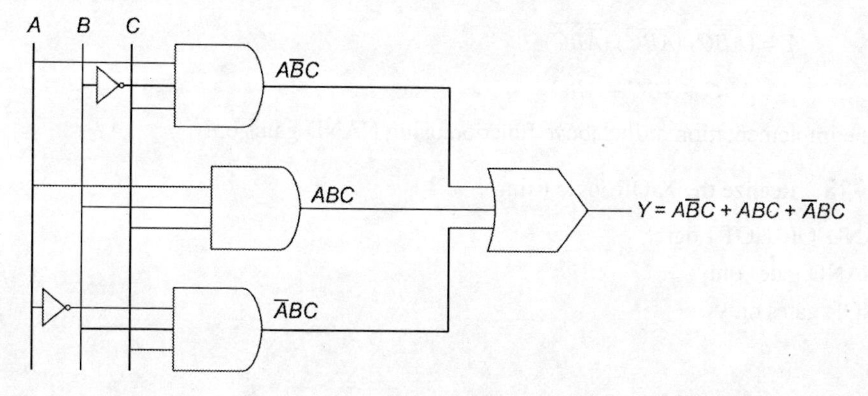

Step II Replace AND gate by NAND gate, OR gate by bubbled OR gate and NOT gate by NAND inverter.

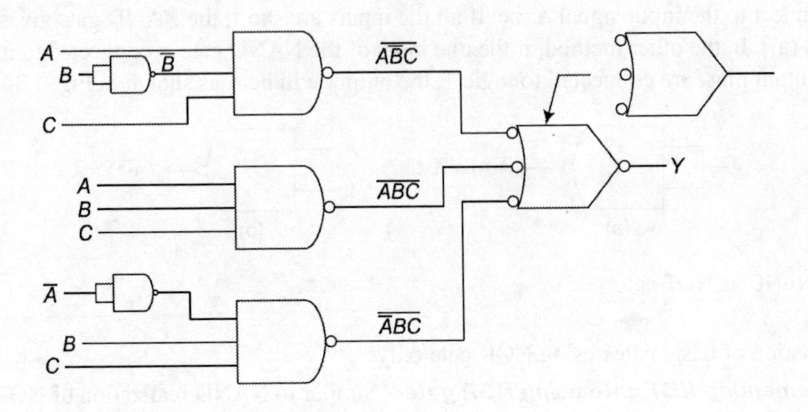

Step III Draw the circuit using NAND gates only.

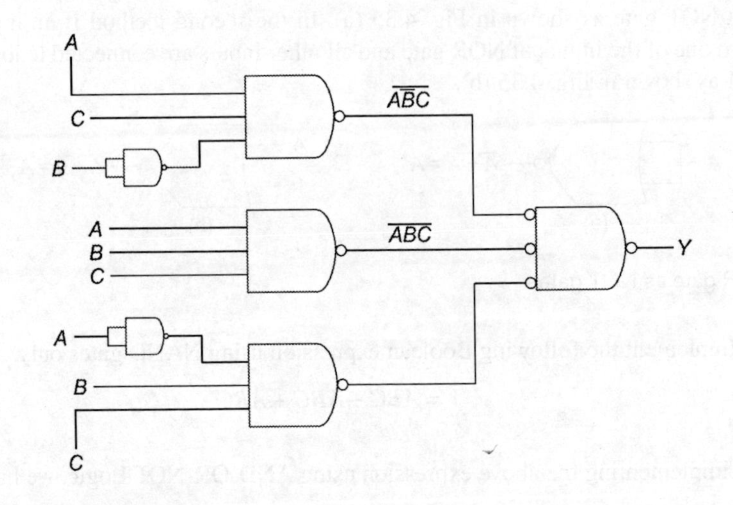

$$Y = \overline{(A\bar{B}C)}\ \overline{(ABC)}\ \overline{(\bar{A}BC)}$$

$$= A\bar{B}C + ABC + \bar{A}BC$$

which is the implementation of the above function using NAND gates only.

Example 4.13 Realize the X-OR logic using

(a) AND-OR-NOT Logic

(b) NAND gates only

(c) NOR gates only.

Solution The Ex-OR gate representation is

$$Y = A\overline{B} + \overline{A}B = A \oplus B$$

(a) Realization by AND-OR-NOT Logic

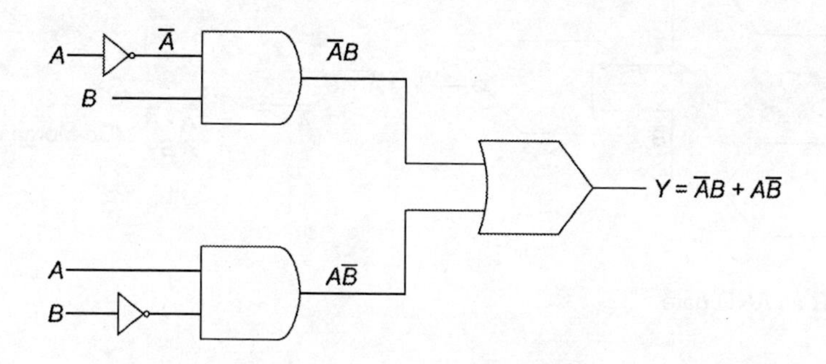

(b) Realization by NAND Logic

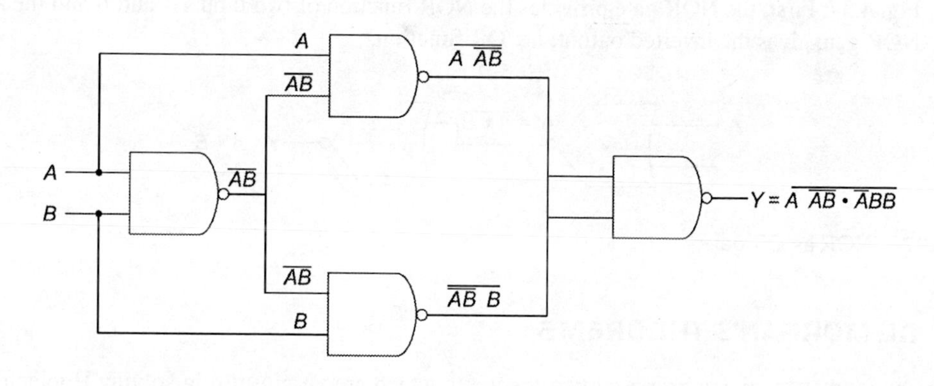

$$Y = A\overline{B} + \overline{A}B$$

$$= A\overline{A} + A\overline{B} + \overline{A}B + B\overline{B}$$

$$= A\,(\overline{A} + \overline{B}) + B\,(\overline{A} + \overline{B})$$

$$= A\,\overline{AB} + B\,\overline{AB}$$

$$= \overline{\overline{A\,\overline{AB} + B\,\overline{AB}}}$$

$$= \overline{\overline{A\,\overline{AB}} \cdot \overline{B\,\overline{AB}}}$$

(b) **Implementing AND gate using NOR gate**

AND gate can be performed by using NOR gate by connecting two NOR gates as shown in Fig. 4.36. The figure shows three NOR gates. The first and second NOR gates are used as a NOT gate to give complemented outputs of the respective inputs. Third NOR gate gives the complemented output (De Morgan function) as shown in the figure.

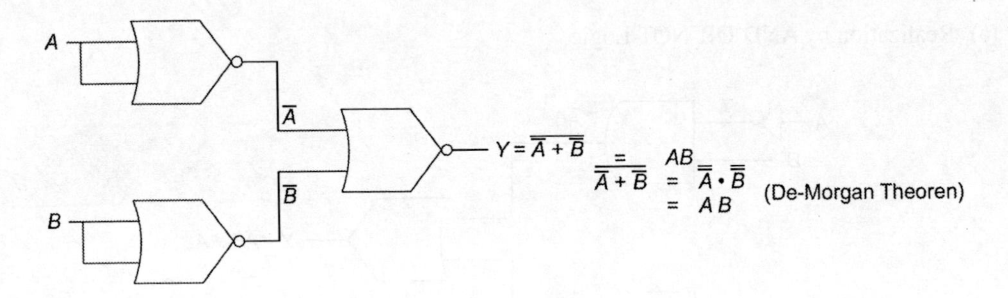

Fig. 4.36 NOR as AND gate.

(c) **Implementing OR gate using NOR gate**

OR gate operation can be performed by NOR gate by connecting two NOR gates as shown in Fig. 4.37. First, the NOR gate provides the NOR function of two inputs A and B and the second NOR gate gives the inverted output, i.e. OR function.

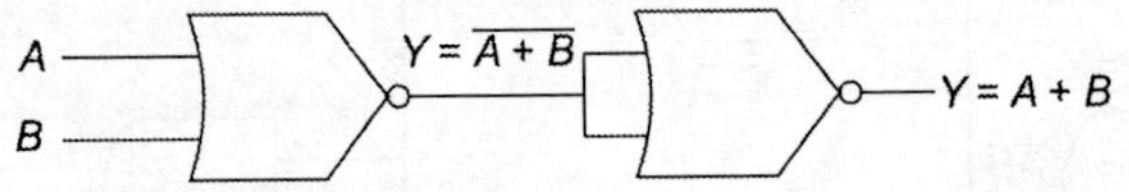

Fig. 4.37 NOR as OR gate.

4.23 DE MORGAN'S THEOREMS

De Morgan, a mathematician proposed two theorems which are very useful in solving Boolean expressions:

Theorem 4.1 It states that the complement of a product of variables is equal to the sum of the complements of individual variables. That is, the complement of two or more variables ANDed together is equal to the sum of complements of the individual variables, i.e.

$$\overline{AB} = \overline{A} + \overline{B}$$

Figure 4.38 shows the schematic representation of the above theorem.

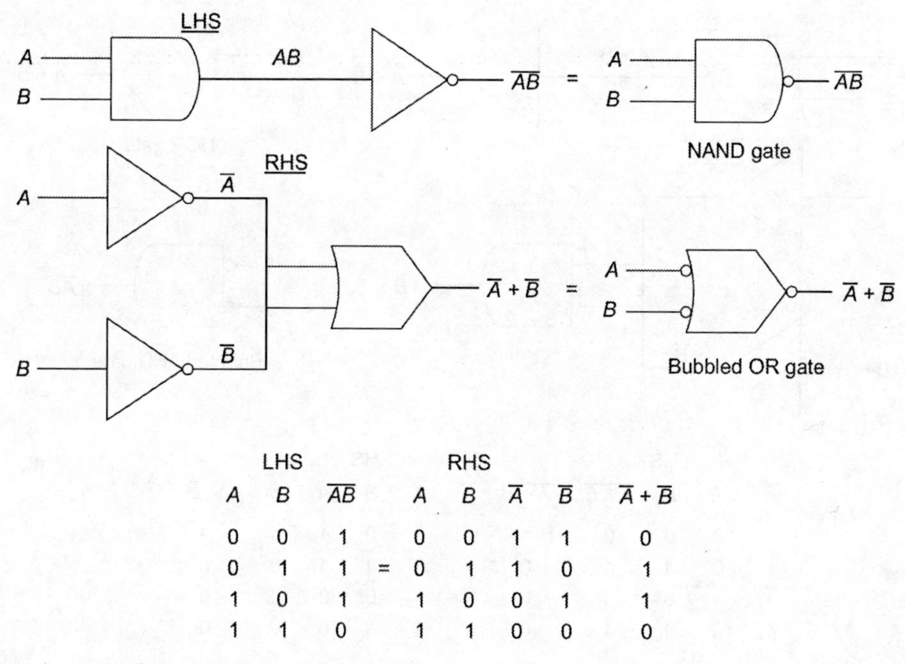

LHS				RHS				
A	B	$\overline{AB}$		A	B	$\overline{A}$	$\overline{B}$	$\overline{A} + \overline{B}$
0	0	1		0	0	1	1	0
0	1	1	=	0	1	1	0	1
1	0	1		1	0	0	1	1
1	1	0		1	1	0	0	0

Fig. 4.38 Schematic representation of theorem 4.1.

It is shown in Fig. 4.38 that the NAND gate is equivalent to bubbled OR gate as per Theorem 4.1 ($\overline{AB} = \overline{A} + \overline{B}$). This has also been proven simply by truth table in Fig. 4.38. This theorem can be extended to any number of variables or combination of variables which is very useful in Boolean solving

$$\overline{ABCD} = \overline{A} + \overline{B} + \overline{C} + \overline{D} + ...$$

$$\overline{(ABC)\,(DE)\,(FGH)\,....} = \overline{ABC} + \overline{DE} + \overline{FGH}\,....$$

The above theorem permits the removal of individual variables, and transformation from a product-of-sum to sum-of-product form.

Theorem 4.2 It states that the complement of sum of variables is equal to the product of individual variables. It simply means that the complement of two or more variables ORed together, is same as the AND of complements of each of the individual variables.

$$\overline{A + B} = \overline{A}\,\overline{B}$$

Schematic diagram of the above function is represented in Fig. 4.39, which shows that NOR gate is equivalent to a bubbled AND gate, which has also been proven using truth table.

This theorem can also be extended to any number of variables or combination of variables. For example,

$$\overline{ABCDE\,...} = \overline{A} + \overline{B} + \overline{C} + \overline{D} + \overline{E} + ...$$

$$\overline{(ABC)\,(DEF)\,(GHI)\,...} = \overline{ABC} + \overline{DEF} + \overline{GHI} + ...$$

The above theorem permits the removal of individual variables and transformation of sum-of-product to product-of-sum form.

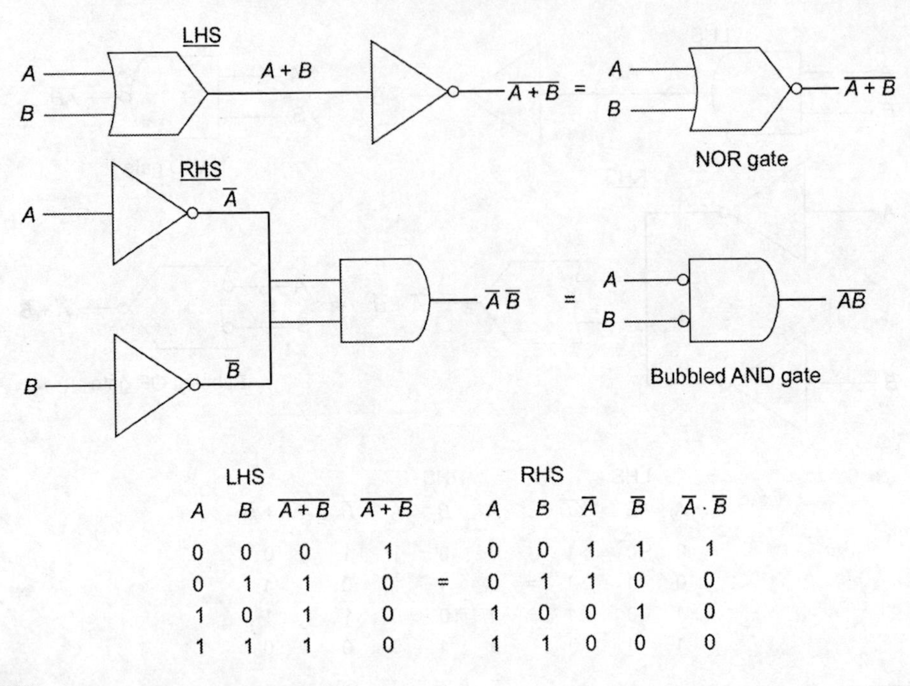

LHS					RHS				
A	B	$\overline{A+B}$	$\overline{A+B}$		A	B	$\overline{A}$	$\overline{B}$	$\overline{A}\cdot\overline{B}$
0	0	0	1		0	0	1	1	1
0	1	1	0	=	0	1	1	0	0
1	0	1	0		1	0	0	1	0
1	1	1	0		1	1	0	0	0

Fig. 4.39 Schematic representation of Theorem 4.2.

Demorganisation Process

It is described in the above two theorems that the transformations

$$\overline{AB} = \overline{A} + \overline{B}$$

$$\overline{A + B} = \overline{A}\,\overline{B}$$

can be extended to any number of variables or expressions. The complex Boolean expression can be simplified by using Demorganisation process by the following three steps.

1. Complement the entire expression.
2. Change all ANDs to ORs and all ORs to ANDs.
3. Complement each of the individual variables.

4.24 "BREAK LINE-CHANGE SIGN" DE MORGAN'S METHOD

"Break line-change sign" is the second method to perform de Morgan's function to any Boolean expression. For example, $\overline{ABC} + \overline{DE}$ is to Demorganise then we can break the line between A, B and the line between B, C and that between D and E and change the sign from ANDing to ORing. This yields

$$\overline{A} + \overline{B} + \overline{C} + \overline{D} + \overline{E}.$$

Example 4.14 Reduce the given Boolean expression using break line-change sign De Morgan's method.

Solution

$$Y = \overline{\overline{AB} + \overline{A} + AB}$$

Break the upper line between $\overline{AB}$ and $\overline{A}$ and that between $\overline{AB}$, $\overline{A}$ and AB.

$$= \overline{\overline{AB}}.\, \overline{\overline{A}}\, .\, \overline{AB}$$

$$= AB.\, A.\, \overline{AB}$$

$$= AB.\, \overline{AB}$$

$$= 0$$

Alternatively, break the lower line between A and B and change the sign between them.

$$= \overline{\overline{A} + B + \overline{A} + AB}$$

$$= \overline{\overline{A} + \overline{B} + AB}$$

Similarly, break the line between $\overline{A}$ and $\overline{B}$ and that between $\overline{B}$ and AB and change the sign between them.

$$= \overline{\overline{A}}.\, \overline{\overline{B}}.\, \overline{AB} = AB.\, \overline{AB} = 0.$$

4.25 IMPLEMENTING A BOOLEAN EXPRESSION USING NAND AND NOR GATES ONLY

In a digital complex circuit, the minimum solution of a Boolean expression is usually obtained by (sum-of-product) SOP or (product-of-sum) POS form. Sometimes SOP form is implemented by using AND gates whose output is ORed, whereas POS form is implemented using OR gates whose output are ANDed. This is known as a hybrid combination of SOP and POS forms. Hybrid logic reduces the number of gate inputs required for realization, results in multilevel logic. Hybrid logic results in propagation delay between inputs and output as the different inputs pass through different number of gates to reach the output.

Since NAND and NOR gates are universal gates, they perform better than basic gates. The digital circuits which are first minimised and implemented using basic gates (AND, OR, NOT) may then be converted into NAND or NOR gate as per the requirements in the following procedure.

(a) Realization Using NAND Gates Only

According to De Morgan's theorem NAND gate is equivalent to bubbled OR gate. Any digital circuit may be realized by NAND gates only by the following steps:

Step I Simplify the given Boolean expression using Boolean laws, K-maps or by any another method to a reduced expression in SOP form.

Step II Implement the reduced expression using basic gates—AND, OR, NOT.

Step III Replace every AND gate by NAND gate, every OR gate by a bubbled OR gate and NOT gate by a NAND inverter.

Step IV Draw the circuit using NAND gates only. (Bubbled-OR gate is a NAND gate.)

(b) Realization Using NOR Gates Only

According to De-Morgan's theorem, NOR gate is equivalent to bubbled AND gate. An alternate method to realize any digital circuit is NOR-gate realisation, with the following steps:

Step I Simplify the Boolean expression using laws, K-maps and represent the reduced expression in POS form.

Step II Implement the reduced expression using basic gates i.e., AND, OR, NOT.

Step III Replace every OR gate by NOR gate, every AND gate by a bubbled AND gate, and every inverter by a NOR inverter.

Step IV Draw the circuit using NOR gates only.

Example 4.15 Realize Ex-NOR gate using

(a) Only NOR gates
(b) Only NAND gates

Solution

Step I Simplify the given expression

EX-NOR gate

$$Y = \overline{A \oplus B}$$
$$= \overline{A}\,\overline{B} + A\,B$$

Step II Realize the given/simplified expression using basic AND-OR-NOT Logic.

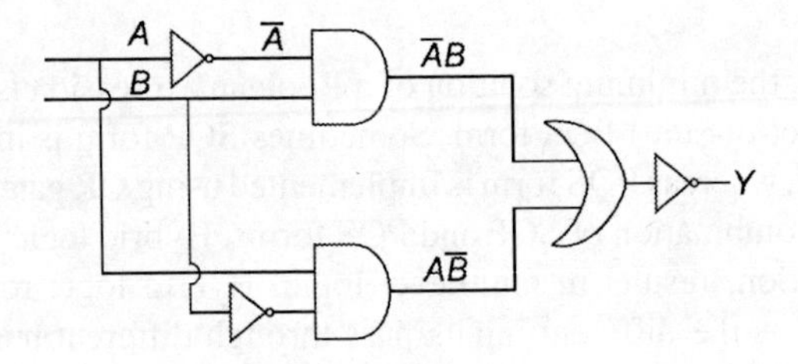

Step III

Case 4

(a) FOR realizing the circuit with only NOR gates:

(i) Replace AND gate with bubbled AND gate.
(ii) Add bubble on the output of each OR gate
(iii) Add an inverter ((NOT) gate) where bubbles were added.

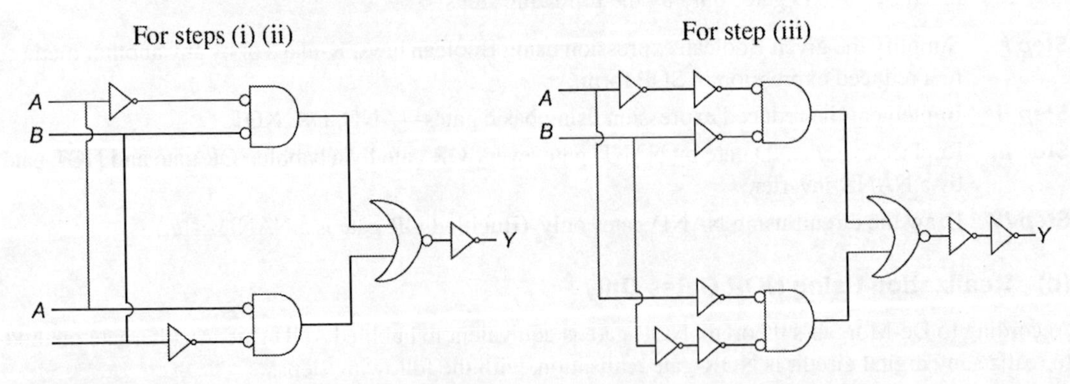

Simplified circuit for step (iii)

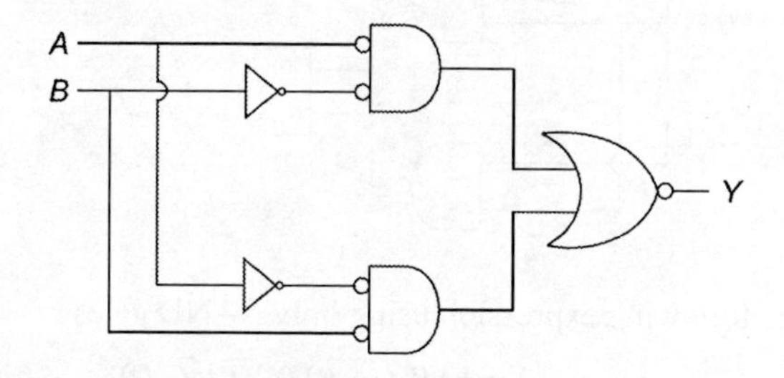

Note Two indenters appearing consecutively cancel out.

Step IV Replace bubbled AND by NOR & inverter by NOR inverter.

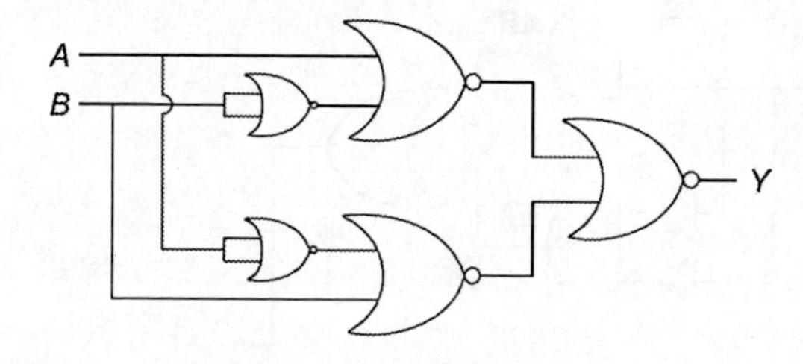

Case 2 FOR realizing the circuit using NAND gates only:

Step I Replace OR gate by bubbled OR gate

Step II Put bubbles on the output of each AND gate

Step III Add inverter where bubbles were added.

For step (i) For steps (ii) & (iii)

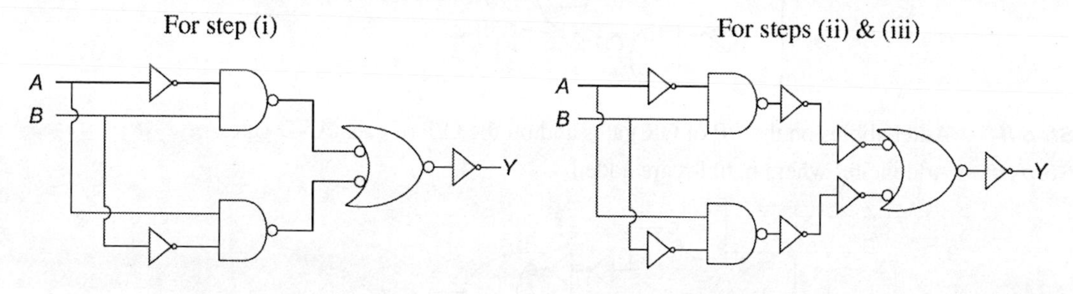

Simplified circuit for steps (ii) (iii)

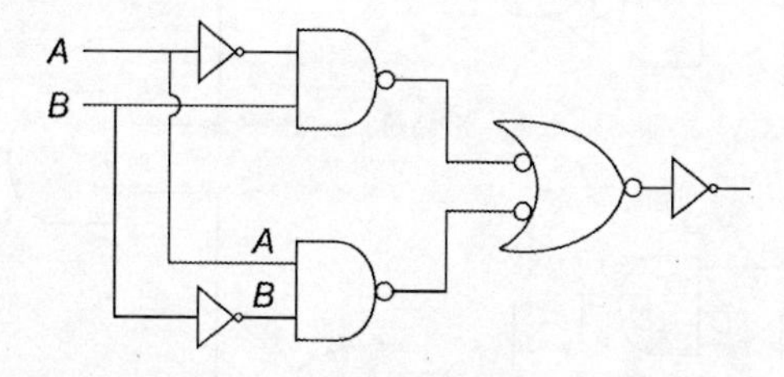

Step IV Replace bubbled OR by NAND gate AND INVERTER with NAND inverter.

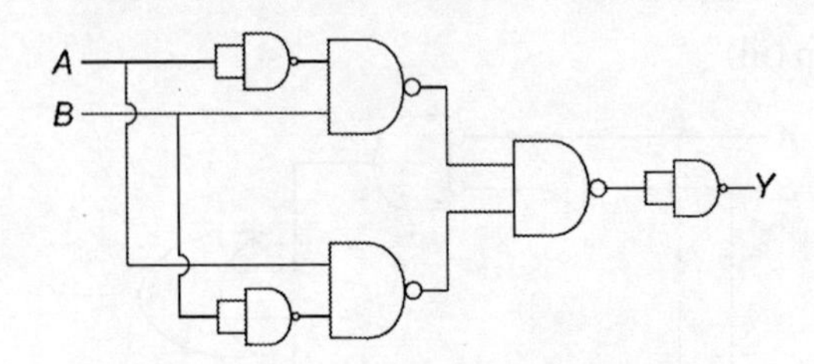

Example 4.16 Realize the following expression using only NAND gates

$$Y = (AB + \overline{A}\,\overline{B})\,(C\overline{D} + \overline{C}\,D)$$

Solution

Step I Realize the expression using AOI Logic.

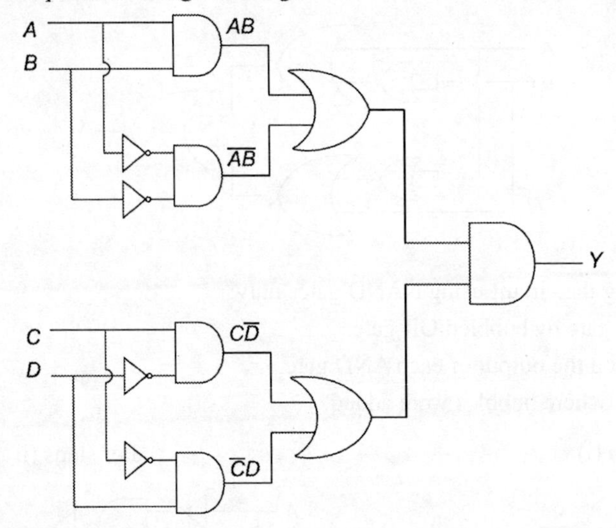

Step II Add bubbles on the I/P of OR gates and on the O/P of each AND gate.

Step III Add inciter where bubbles are added.

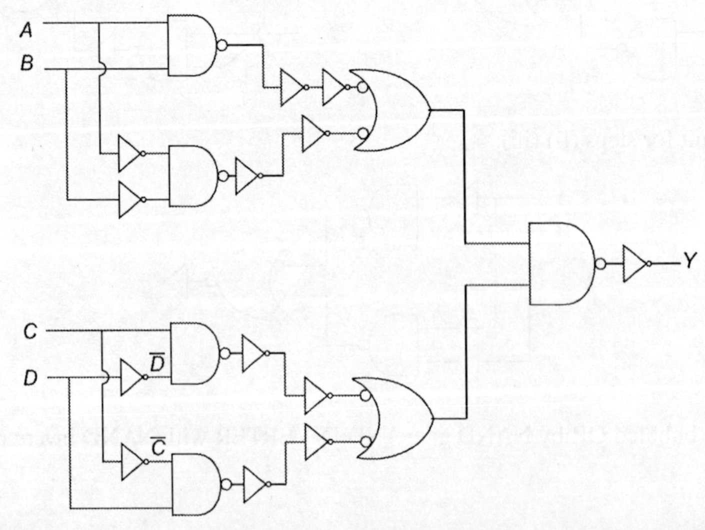

Step IV Simplify the circuit and replaced bubbled OR by NAND gates and inciters with NAND inverters.

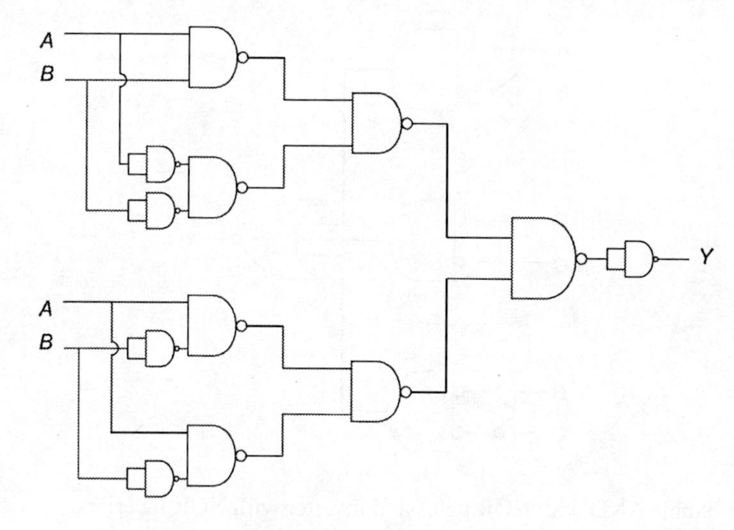

Example 4.17 Realize the following expression in SOP form using only NOR gates

$$Y = \Sigma m\ (0, 3, 4, 5, 7).$$

Solution

Step I Simplify the given SOP form expression:

$$\begin{aligned} Y &= \Sigma m\ (0, 3, 4, 5, 7) \\ &= \overline{A}\,\overline{B}\,\overline{C} + \overline{A}\,B\,C + A\,\overline{B}\,\overline{C} + A\,\overline{B}\,C + ABC. \\ &= A\,\overline{B}\,(C + \overline{C}) + \overline{B}\,\overline{C}\,(A + \overline{A}) + BC\,(A + \overline{A}) \\ &= A\overline{B} + BC + \overline{B}\,\overline{C} \end{aligned}$$

Step II Realize the expression using AOI Logic.

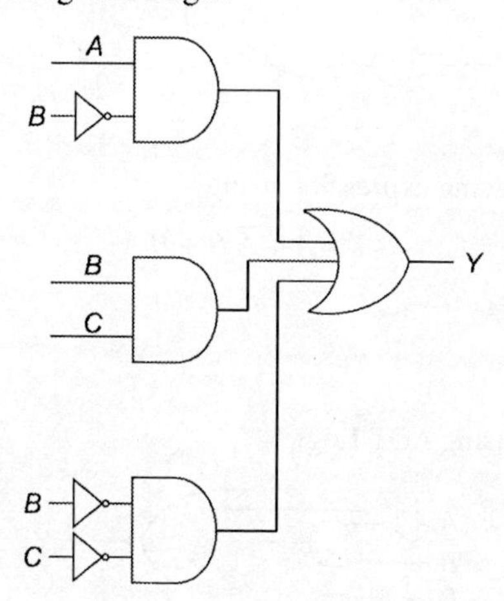

Step III Replace AND gates by bubbled AND add bubble at the O/P of OR gate.

Step IV Add inverter where bubbles are added.

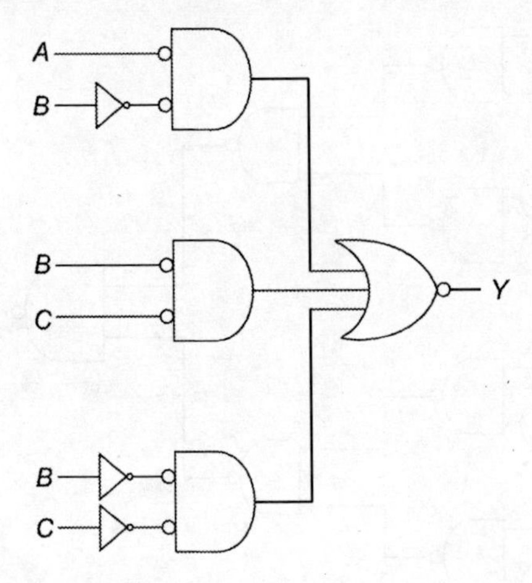

Step V Replace bubble AND with NOR gates and inverter with NOR inverters.

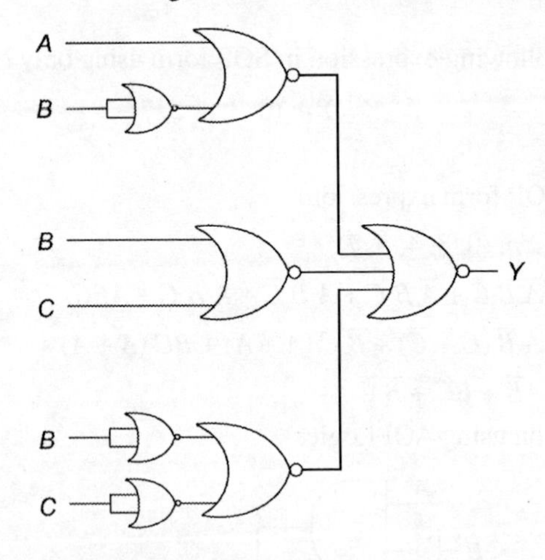

Example 4.18 Realize the following expression using

$$Y = A + \overline{A}\,B + \overline{B}\,A$$

(a) NAND gates only

(b) NOR gates only.

Solution

Step I Realize the expression using AOT Logic

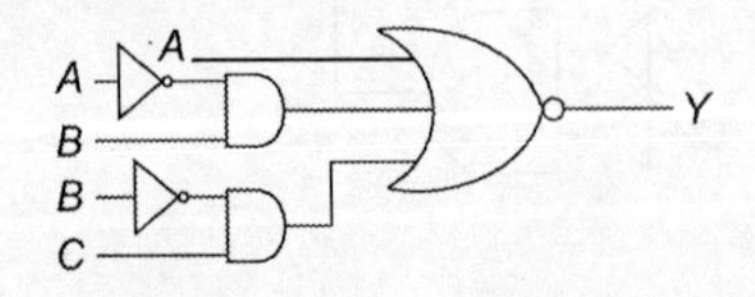

Step II

Case 4 For realization with NAND gates only.

(a) Add bubbles inverter on the I/P of the OR gates and on the O/P of the AND gates.

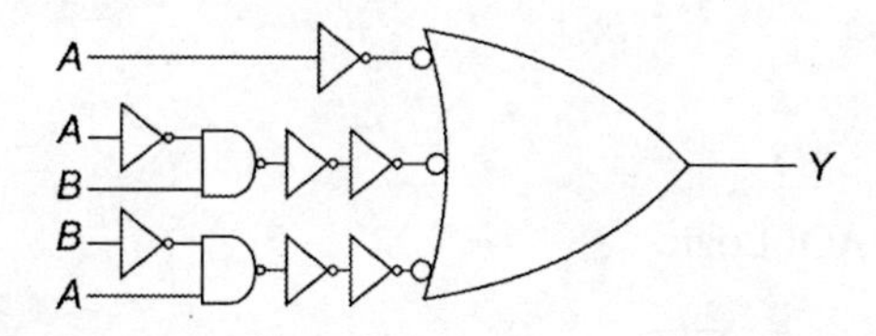

Step III Replace bubbled OR by NAND gate and inverter with the NAND inverter.

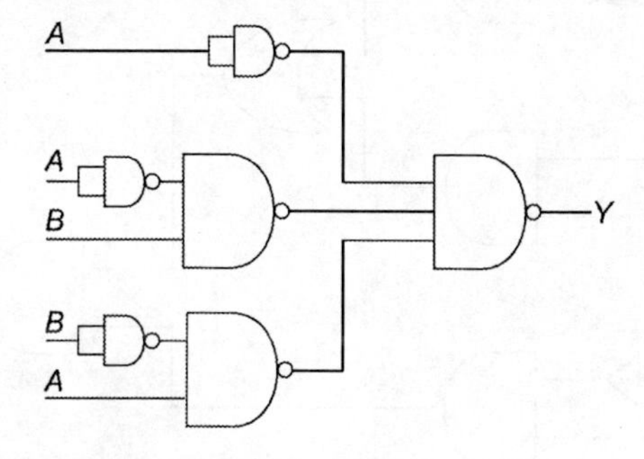

Case 2 For realization with only NOP gates

(a) Add bubbles inverter on the O/P AND gates and bubble inverter on the O/P of OR gates.

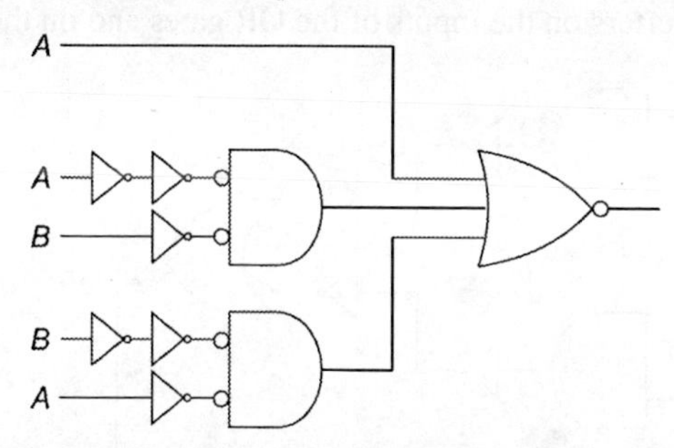

(b) Replace bubble AND with NOR gate AND inverter with NOR inverters.

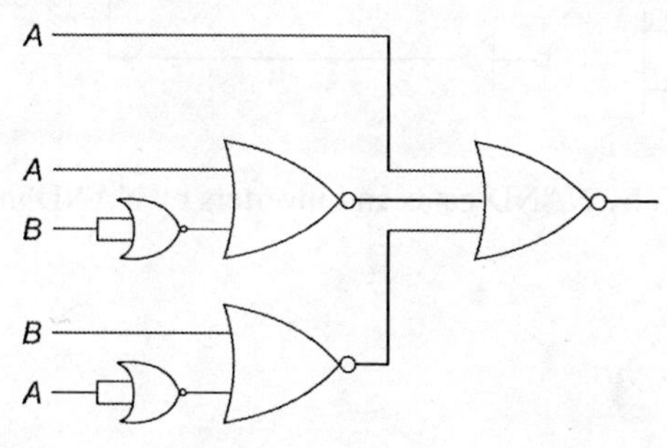

Example 4.19 Realize the following expression

$$Y = (\bar{A}\,B + CD)\,\bar{A}\,C$$

using

(a) Only NAND gates

(b) Only NOR gates.

Solution

(a) Realize the circuit using AOI Logic

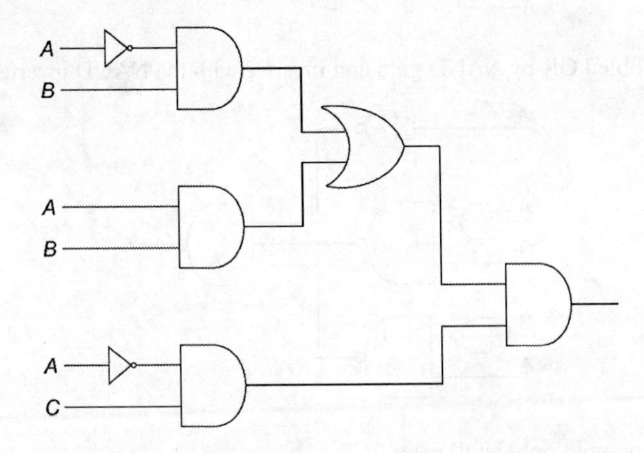

Case 4 Realize circuit using only NAND gates.

Step I Add bubbles and inverters on the inputs of the OR gates and on the O/P of the AND gates.

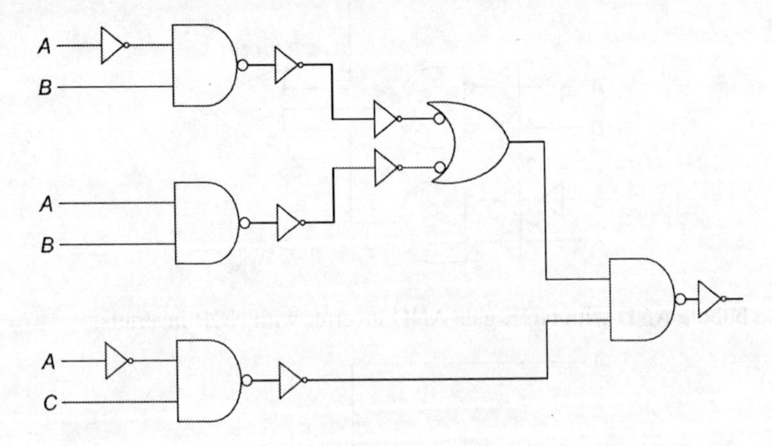

Step II Replace bubbled OR by NAND gates and inverters by NAND inverters.

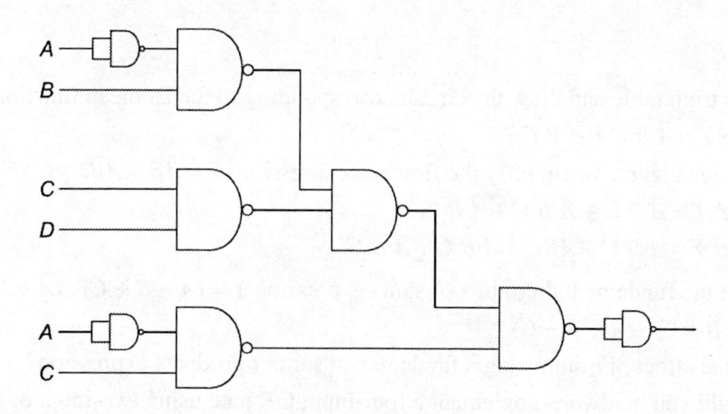

Case 2 Realizing the circuit using only NOR gates.

Step I Add bubbles and inverters on the I/P of AND gates and on the O/P of the OR gates.

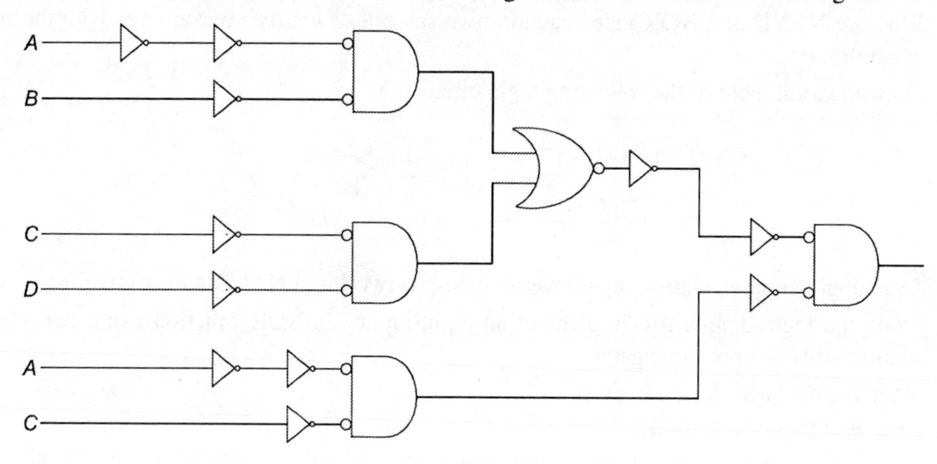

Step II Replace bubbled AND with NOR gates and inverters with NOR inverters.

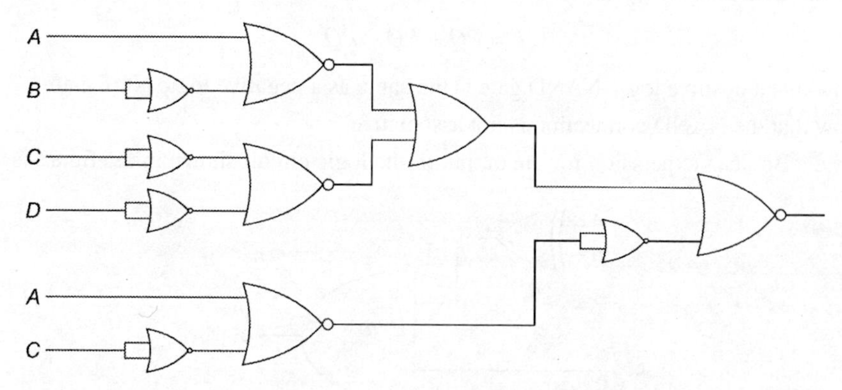

EXERCISES

1. Write the truth table and draw the circuit corresponding to the Boolean function $Y = \overline{A}C + AB$.
2. Minimise $Y = \overline{A}\,B\,\overline{C} + \overline{A}\,B\,C$.
3. Use Boolean algebra to simplify the Boolean expression: $Y = AB + A\overline{B}$.
4. Minimise: $Y = \overline{A}\,B\,\overline{C} + \overline{A}\,B\,C + A\,B\,C$.
5. Minimise: $Y = A\,B\,\overline{C} + ABC + A\,\overline{B}\,\overline{C} + A\,\overline{B}\,C$.
6. Minimise the fundamental product of sums expression $Y = (A + \overline{B} + C) \cdot (\overline{A} + B + C) \cdot (\overline{A} + B + \overline{C}) \cdot (\overline{A} + \overline{B} + C) \cdot (\overline{A} + \overline{B} + \overline{C})$.
7. What is the effect of minimising a fundamental sum of products expression?
8. How would you hardware-implement a four-input OR gate using two-input or gates only?
9. How can you implement a NOT circuit using a two-input EX-OR gate?
10. Show the logic arrangement for implementing a NOT circuit using a two-input NOR gate.
11. Why are NAND and NOR gates called universal gates? Justify your answer with the help of examples.
12. Draw the truth table of the following logic circuit.

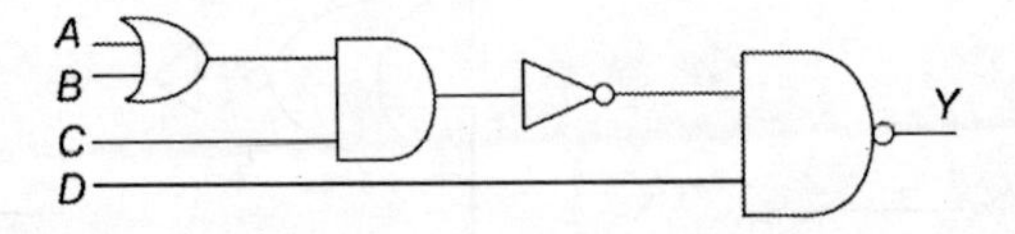

13. Draw logic implementation of an inverter using (i) two-input NAND and (ii) two-input NOR.
14. Draw the logic diagram to implement an eight-input EX-NOR function using the minimum number of two-input logic gates.
15. What are the basic digital logic gates?
16. State de Morgan's theorem.
17. State the associative property of Boolean algebra.
18. List the truth table of the function

$$F = PQ + P\overline{Q} + \overline{P}Q$$

19. Show that a positive logic NAND gate is the same as a negative logic NOT gate.
20. Show that the NAND connection is not associative.
21. Write a Boolean experssion for the output of the logic circuit shown in the figure is

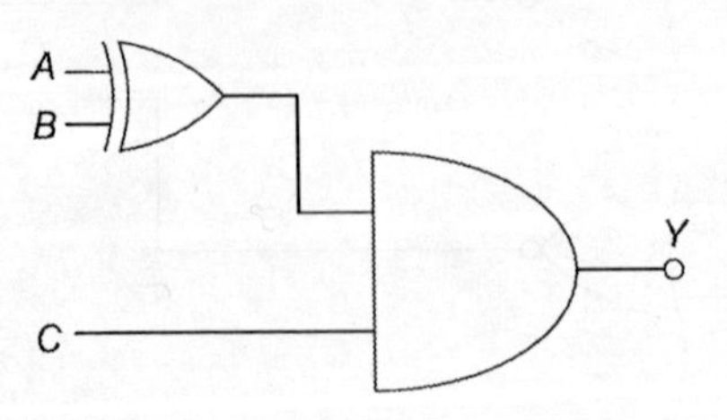

22. The dual of the Boolean theorem is

$$A \cdot (B + C) = A \cdot B + A \cdot C$$

23. Complement of the Boolean expression is

$$AB \cdot (C + AC)$$

24. Simplify the Boolean expression

$$\overline{X}Y\overline{Z} + \overline{X}\,\overline{Y}Z + XY\overline{Z} + X\overline{Y}Z + XYZ$$

25. For the logic circuit shown in the figure below, the output Z is equal to:

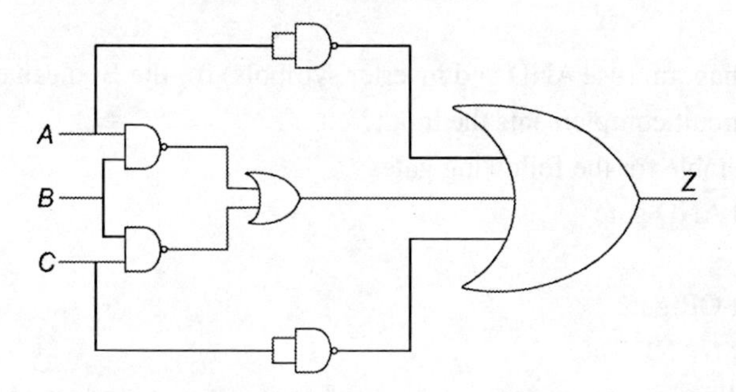

26. The minimum number of NAND gates required to implement $A + A\overline{B} + A\overline{B}C$ is equal to:
27. Write the truth table of the given logic

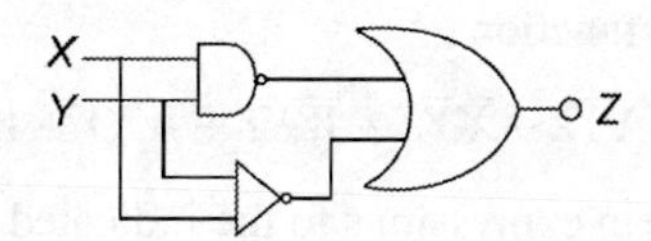

28. Simplify the following Boolean expressions to a minimum number of literals.
    (a) $ABC + \overline{A}B + AB\overline{C}$
    (b) $PQ + P(RS + R\overline{S})$
    (c) $(B\overline{C} + \overline{A}D)\,(A\overline{B} + C\overline{D})$
    (d) $XYZ + \overline{X}Y + XY\overline{Z}$
29. Find the complement of the following expressions:
    (a) $P\overline{Q} + \overline{P}Q$
    (b) $(A + \overline{B} + C)\,(\overline{A} + \overline{C})\,(A + B)$
    (c) $(X\overline{Y} + Z)\,\overline{P} + Q$
30. Draw the logic diagrams for the following Boolean expressions:
    (a) $Y = \overline{X}\,\overline{Y} + Y(X + Z)$
    (b) $Z = BC + A\overline{C} + \overline{B}\,\overline{C}$
    (c) $X = (A + B)\,(\overline{C} + D)\,(\overline{A} + D)$
31. Find the complement of $F = x + yz$; then show that $FF' = 0$ and $F + F = 1$:

32. Draw a waveform to represent the logic levels (H and L) at output Y of the AND gate in Fig. (1).

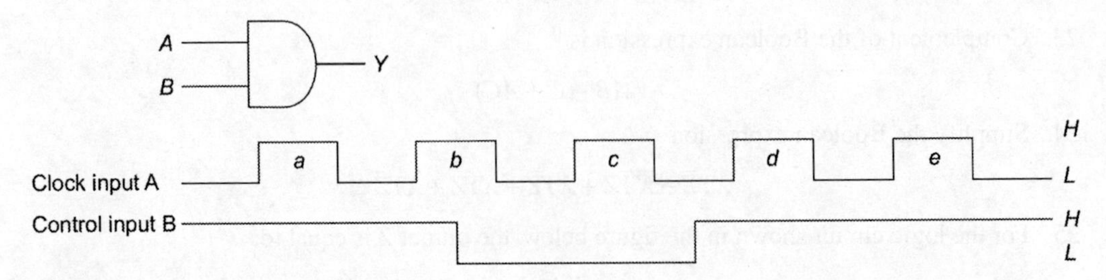

33. Draw a logic diagram (use AND and inverter symbols) for the Boolean expression $C\overline{A}B = Y$.
34. Which logic circuit complements the input?
35. Draw the truth table for the following gates:
    (a) Two-input AND gate
    (b) Inverter
    (c) Two-input OR gate
    (d) Two-input NAND gate
36. Convert the following expressions into sum-of-products and product-of-sums:
    (a) $(AB + C)(B + \overline{C}D)$
    (b) $\overline{P} + P(P + \overline{Q})(Q + \overline{R})$
37. Obtain the truth table of the function

$$F = X\overline{Y}Z + \overline{X}\overline{Y}Z + \overline{W}XY + W\overline{X}Y + WXY$$

38. Reduce the following Boolean expressions to the indicated number of literals:
    (a) $\overline{P}\overline{Q} + PRQ + P\overline{Q}$ to three literals
    (b) $\overline{(\overline{A}\overline{B} + C)} + C + AB + DC$ to three literals

5

Logic Families

INTRODUCTION

There are many ways to design an electronic logic circuit. In 1930s Bell Laboratories developed the first electrically controlled logic circuit, which was based on relays. In mid-1940s the first electronic digital computer, the Eniac, used logic circuits based on vacuum tubes. The inventions of *semiconductor diode* and the *bipolar junction transistor* allowed the development of smaller, faster and more capable computers in late 1950s. In the 1960s the invention of the *integrated circuits* allowed multiple diodes, transistors and other components to be fabricated on a single chip, and computers go still better.

5.1 LOGIC FAMILIES

A logic family is a collection of various integrated circuit chips that have similar input, output and internal circuit characteristics, but perform different logic functions. To perform any logic function, chips from the same logic family can be interconnected. Chips from the different logic family may use different power supply voltages or may use different input and output conditions to represent logic values.

Various integrated circuits are available which can be used as logic circuits. The ICs can be classified into three main categories:

(i) ***Small scale integrated circuits (SSI)*** Those ICs, which have less than 12 gates integrated on a single chip.

(ii) ***Medium scale integrated circuits (MSI)*** Those ICs which have 12 to 100 gates integrated on a single chip.

(iii) Large scale integrated circuits (LSI) Those ICs which have more than 100 gates integrated on a single chip.

ICs are manufactured by using two basic techniques, namely, bipolar and metal oxide semiconductor (MOS) technologies. Bipolar technique is preferred in SSI and MSI because it is faster and MOS technology is preferred due to increased density of MOSFET in the same chip area. Logic families can be generally classified into two types.

1. Bipolar Logic Families

Bipolar logic families are the families using bipolar junction semiconductor devices such as BJT. Bipolar logic families can be further divided into two groups:

(a) Saturated logic family: In which transistor goes into saturation. Members belonging to this family are:

1. RTL	Resistor-transistor logic
2. DTL	Diode-transistor logic
3. TTL	Transistor-transistor logic
4. I^2L	Integrated-injection logic

(b) Non-saturated bipolar logic family: In which the transistor cannot go into saturation, members, belonging to this family are:

1. ECL	Emitter coupled logic
2. TTL	Schottky-transistor-transistor logic

2. Unipolar Logic Families

Unipolar logic families are those in which unipolar junction semiconductors are used such as MOS. Members belonging to this family are:

1. MOS	Metal oxide semiconductor
2. CMOS	Complementary metal-oxide semiconductor

The most successful *bipolar logic family* (based on bipolar junction transistors) was *transistor-transistor logic (TTL)*, which was introduced in 1960s, evolved into a family of logic families that were compatible with each other but different in speed, power consumption, and cost. Digital systems could mix components from several different TTL families, according to design goals and constraints in different parts of the system.

MOS (metal oxide semiconductor) is a type of semiconductor device fabricated with a conducting layer and a semiconductor layer separated by an insulation layer. MOSs are unipolar devices that use either holes or electrons for conduction but not both at once, i.e., they are not bipolar devices.

MOS circuits lagged bipolar circuits considerably in speed. They were suitable only in a few applications because of lower power consumption and higher levels of integration.

CMOS (complementary MOS) circuits, tremendously increased their performance and popularity. Almost all new large-scale integrated circuits, such as memories, microprocessors, use CMOS. Likewise, small-to-medium-scale applications, for which TTL was used, are now much more likely to use CMOS devices with equivalent functionality or better, and higher speed with low power consumption. CMOS circuits now account for the vast majority of the worldwide integrated circuit market.

All logic families have certain advantages and disadvantages. The disadvantages of one logic family have been overcome by others. There are many design parameters that can affect the behaviour of logic family. The performance of the various logic families is usually evaluated by comparing the characteristics of the basic gate in each family. The reason is that sometimes speed may be the main requirement whereas another time minimum power dissipation might be the main consideration. The important parameters that can affect the behaviour of logic family are fan-out, power dissipation, propagation delay and noise margin. Before going into the designing details we need to understand what are these parameters and how they affect the functioning.

5.2 FACTORS AFFECTING PERFORMANCE OF A LOGIC FAMILY

(a) *Fan-in (Input Load Factor)* It is the number of input signals that can be connected to a gate without causing it to operate outside its intended operating range. It is expressed in terms of standard inputs or unit loads (UL).

(b) *Fan-out (Output Load Factor)* It is the maximum number of inputs that a logic gate can drive. If a gate has a fan-out of 8, then it means that 10 unit loads can be driven by the gate.

Consider an example to explain fan-in and fan-out. A unit load for a logic family is as follows:

$$IUL = 50\ \mu A \quad \text{high state}$$
$$1\ mA \quad \text{low state}$$

Determine the fan-in and fan-out in this family that has the following parameters.

$$I_{OH} = 400\ \mu A$$
$$I_{OL} = 10\ mA$$
$$I_{IH} = 150\ \mu A$$
$$I_{IL} = 4\ mA$$

$$\text{Fan-in} = \frac{\text{High-state current}}{\text{High-state unit load}} \quad \text{and} \quad \frac{\text{Low-state current}}{\text{Low-state unit load}}$$

$$= \frac{150}{50} = 3\ UL \quad \text{and} \quad \frac{4}{1} = 4\ UL$$

$$\text{Fan-out} = \frac{400}{50} = 8\ UL \quad \text{or} \quad \frac{10}{1} = 10\ UL$$

therefore, fan-out = 8UL.

In the previous example, we have seen the terminology related to digital logic. Let us discuss it before discussing other parameters.

Digital IC Terminology

Although there are many manufacturers of digital ICs and logic families, a fair amount of terminology associated with digital ICs is somewhat standardized among the various manufacturers and logic families.

Symbol	Definition
V_{IH}	High-state input voltage, corresponding to logic 1 at input.
V_{IL}	Low-state input voltage corresponding to logic 0 at input.
V_{OH}	High-state output voltage, corresponding to logic 0 at output.
V_{OL}	Low-state output voltage, corresponding to logic 0 at output.
I_{IH}	High-state input current; current flowing from input when the input voltage corresponds to logic 1.

I_{IL} Low-state input current; current flowing from an input when the input voltage corresponds to logic 0.

I_{OH} High-state output current; current flowing from an output when the output voltage corresponds to logic 1.

I_{OL} Low-state output current; current flowing from an output when the output voltage corresponds to logic 0.

(c) Propagation Delay: Propagation delay of a logic gate is the average transition delay time for the signal to propagate from input to output when the binary signal changes in value, which is measured in nanoseconds (ns).

$$1 \text{ ns} = 10^{-9} \text{ of a second.}$$

The signals that travel from the inputs of a digital circuit to its outputs pass through a series of gates. The sum of the propagation delays through the gates is the total delay of the circuit. When speed of operation is the important factor for any logic family, then each gate must have a short propagation delay and the digital circuit must have a minimum number of gates between inputs and outputs.

Figure 5.1 shows the gate delay offered by a gate for the signal appearing at its input, before it reaches the gate output. It shows a NOT gate with a delay of Δ, where *X* changes only after a delay of Δ. Propagation delay can be referred to as *gate delay*.

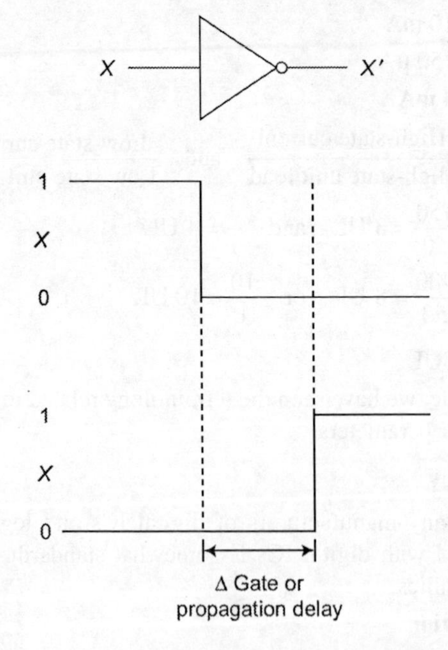

Fig. 5.1 Gate delay.

Further, gates are connected with wires and these wires delay the signal they carry, these delays become very significant when frequency increases. These wire delays are also referred to as *flight time* (i.e., signal flight time from point A to B). Wire delay is also known as *transport delay*. Figure 5.2 shows the transport delay between two gates.

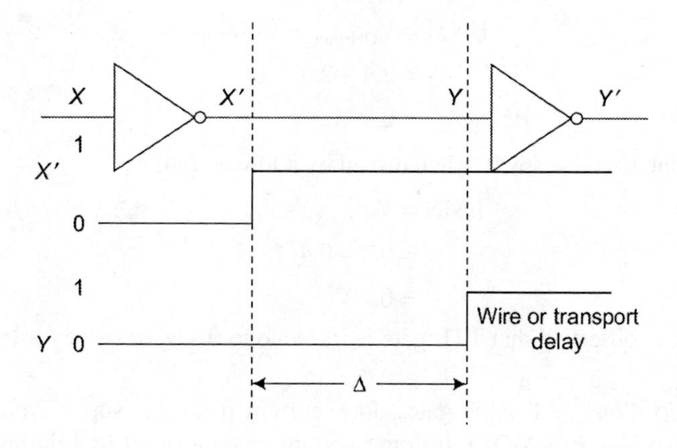

Fig. 5.2 Transport delay.

(d) Noise Immunity: Noise immunity of a logic circuit refers to the circuit's ability to tolerate noise without causing a false change in its output voltage. Due to stray capacitance in the connecting wires of logic circuits, noise voltage is produced. Too much noise voltage causes the voltage at input of the logic circuit to drop below $V_{IH(min)}$ or rise above $V_{IL(max)}$. This could produce unpredictable operation in the logic circuit.

(e) Noise Margin: A quantitative measure of a circuit's noise immunity is called noise margin. It is expressed in volts. It is also defined as the maximum noise voltage level that can be tolerated by a gate. It's normaly supplied by the manufacturer in the gate documentation.

1. LNM (Low noise margin)
2. HNM (High noise margin)

- LNM is the largest noise amplitude that is guaranteed not to change the output voltage level when superimposed on the input voltage of logic gate, when this voltage is *low*.

$$LNM = V_{IL(max)} - V_{OL(max)}$$

- HNM is the largest noise amplitude that is guaranteed not to change the output voltage level if superimposed on the input voltage of the gate, when this voltage is *high*.

$$HNM = V_{OH(min)} - V_{IH(min)}$$

Example 5.1: For the different parameters of TTL family shown below, find: (a) the maximum value of noise spike that can be tolerated when a HIGH output is driving an input and (b) the maximum value of noise spice when a LOW output is driving an input.

Parameter	*Min(V)*	*Typical (V)*	*Max (V)*
V_{OH}	2.4	3.4	–
V_{OL}	–	0.2	0.4
V_{IH}	2.0	–	–
V_{IL}	–	–	0.8

Solution: The maximum value of noise spike, when driven by a HIGH output,

$$\text{HNM} = V_{OH(min)} - V_{IH\,(min)}$$
$$= 2.4 - 2.0$$
$$= 0.4\text{ V}$$

and the maximum value of noise spike, when driven by a low output,

$$\text{LMN} = V_{IL\,(max)} - V_{OL\,(max)}$$
$$= 0.8 - 0.4$$
$$= 0.4\text{ V}$$

From the above, it is observed that TTL gate is immune to 0.4 V of poise for both the high and low inputs.

(f) Power Dissipation: All logic gates draw current from d.c. supply voltage (V_{CC} in case of logic gates and V_{DD} in CMOS). It draws certain amount of current during its operation. Each gate can be in high transition or low transition states. There are three different currents drawn from the power supply.

- IccH Current drawn during high state.
- IccT Current drawn during high to low or low to high transition.
- IccL Current drawn during low state.

For TTL, IccT the transition current is negligible, in comparison to IccH and IccL. Assume IccH and IccL are equal then,

Average power dissipation $$\text{Pav} = V_{CC} \times \frac{(\text{IccH} + \text{IccL})}{2}$$

The power dissipation in logic gates can also be specified as the product of the d.c. supply voltage (V_{CC}) and the amount of current drawn from the supply (i.e., IccH or IccL). Thus, power dissipation is given by.

$$\text{PD} = V_{CC}\ \text{IccH}$$

or $$V_{CC}\ \text{IccL}.$$

For example, if IccH is 2.5 mA when Vcc is 5V, the power dissipation

$$\text{PD} = V_{CC}\ \text{IccH}$$
$$= 5 * 2.5\text{ mA} = 12.5\text{ mW}.$$

Figure 5.3 shows IccH and IccL for a logic gate.

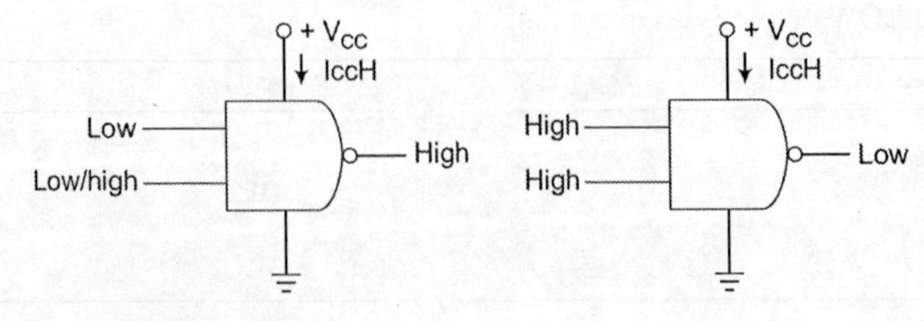

Fig. 5.3 IccH and IccL in a logic gate.

Example 5.2: A TTL logic gate draws 2 mA when its output is in high state and draws 3.5 mA when its output is low. If the supply voltage is 5 V and logic gate is operated on 50% duty cycle, calculate the average power dissipation.

Solution: The average supply current

$$Icc = \frac{IccH + IccL}{2}$$

$$= \frac{2\text{ mA} + 3.5\text{ mA}}{2} = 2.75\text{ mA}$$

∴ The average power dissipation,

$$PD = V_{CC}.I_{CC} = 5 \times 2.75\text{ mA}$$
$$= 13.7\text{ mW}.$$

5.3 RESISTOR TRANSISTOR LOGIC

Resistor transistor logic (RTL) is a class of digital circuitry which uses resistors and transistors as the input network and switching elements respectively. RTL is the earliest logic family and is not available in the monolithic I_C form. The schematic diagram of two-input RTL NOR gate is shown in Fig. 5.4.

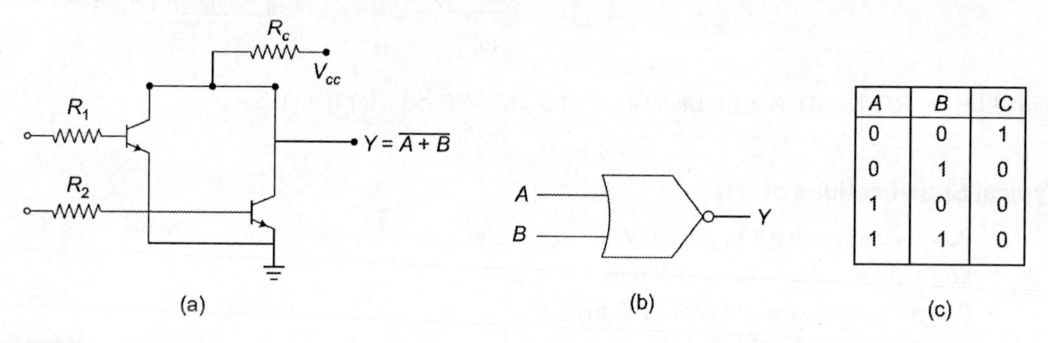

A	B	C
0	0	1
0	1	0
1	0	0
1	1	0

Fig. 5.4 (a) RTL-NOR gate, (b) Standard symbol of NOR gate, (c) Truth table.

- *A* and *B* are two inputs connected to base Q_1 and Q_2 through resistors R_1 and R_2 respectively
- The collectors of two transistors Q_1 and Q_2 are connected together for output.
- The collectors of both transistors are connected together and voltage V_C is applied through resistor R_C.
- Voltage level of RTL varies from 0.2 V to 3.6 V corresponding to 0 to 1 or low to high.

Working of RTL Circuit

The RTL NOR gate works as NOR gate, as is evident from the truth table:

- If any of the inputs out of *A* and *B* goes high, the corresponding transistor out of Q_1 and Q_2 becomes ON. The concerned transistor goes into saturation.
- If both the inputs go high, then both the transistors will be driven into saturation.
- Conversely, if both the inputs are at logic 0, both transistors will be cut off, which causes the HIGH output at common collector point.

RTL circuit can also be designed to work as NAND gate. Figure 5.5 shows RTL as NAND gate.

- When any of two inputs *A* or *B* is low or both inputs *A* and *B* are low, the transistors Q_1 and Q_2 are cut off, making output voltage high.
- When both the inputs *A* and *B* are high, both the transistors will be ON and driven into saturation, making output voltage low.

The above operation satisfies NAND gate operation.

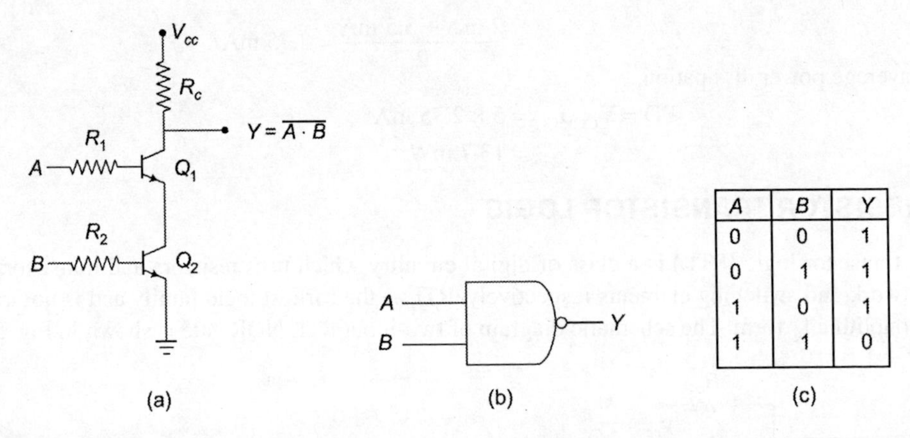

A	B	Y
0	0	1
0	1	1
1	0	1
1	1	0

Fig. 5.5 (a) RTL NAND gate, (b) Standard symbol of NAND gate, (c) Truth table.

Typical Specifications of RTL

- Power supply voltage $V_{CC} = 3.6$ V
- Propagation delay = 12 ns to 300 ns
- Power dissipation = 30 mW to 100 mW
- Noise margin = 0.2 V to 0.4 V
- Fan-out = 4

5.4 DIODE TRANSISTOR LOGIC

Diode transistor logic (DTL) was the first commercially available IC logic family. DTL logic circuits built from bipolar junction transistors (BJT), diodes and resistors.

A basic DTL gate consists of AND or OR gate followed by NOT gate. It is used to obtain NAND or NOR gates. Figure 5.6 shows the basic DTL gate behaving like two-input NAND gate.

Operation of DTL

- The diodes D_1 D_2 and resistor R_1 form AND gate. If any input out of *A* and *B* goes Low, the corresponding diode D_1 or D_2 conducts through Vcc and R_1.
- Now, the potential at point *c* is the sum of low input voltage level and forward voltage drop across the diode (0.7 V). It is insufficient to drive the current through D_3 and base emitter junction of transistor Q_1, keeps the transistor Q_1 in cut-off. As a result, high output voltage at collector which is Vcc 5 V appears.

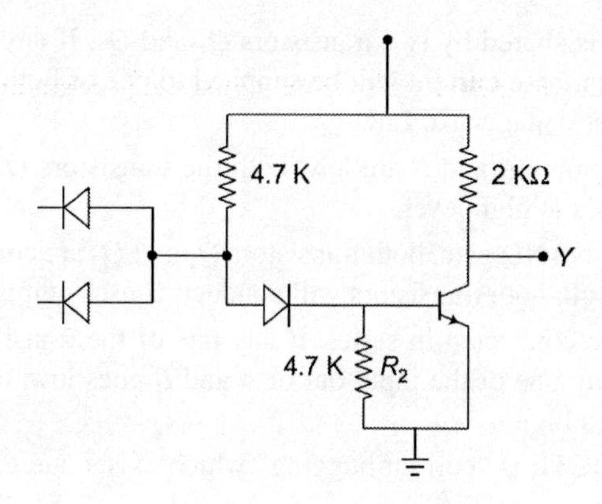

Fig. 5.6 DTL-NAND gate.

- Consider both inputs A and B are high, diodes D_1 and D_2 do not pass any current. The diode D_3 conducts and current starts passing through it and holds transistor Q_1 into saturation. The input diodes D_1 and D_2 do not conduct, base current still drives the transistor into saturation due to which the output at Y drops to low level.

Typical Specifications of DTL

- Power supply voltage : Vcc + 5 V
- Propagation delay : 25 ns
- Noise margin : 0.7 V
- Fan-out and in : 8

5.5 DIRECT-COUPLED TRANSISTOR LOGIC (DCTL)

DCTL is a simple logic family which behaves as NOR or NAND gates as shown in Fig. 5.7(a) and (b) respectively. It is readily available in IC form.

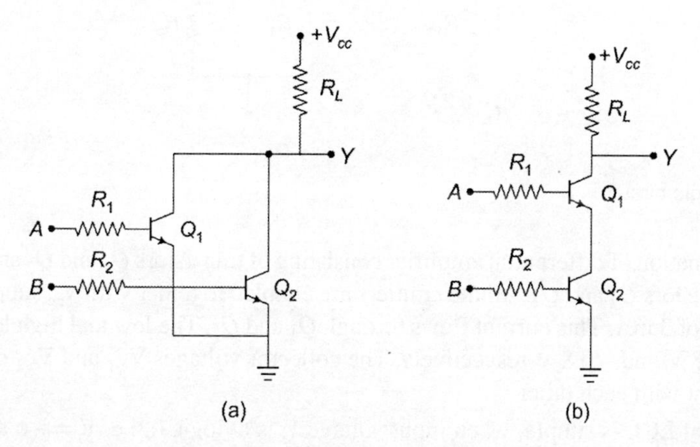

Fig. 5.7 (a) DCTL-NOR gate, (b) DCTL-NAND gate.

- A common load RL is shared by two transistors Q_1 and Q_2. If any one or both inputs *A* and *B* of Fig. 5.6(a) are high, base current will be supplied to one or both transistors, causing them to conduct makes output voltage at Y Low.
- Conversely, if both inputs *A* and *B* are low, both the transistors Q_1 and Q_2 remain cut-off and output at Y approaches at high level.
- In Fig. 5.6(b) DCTL-NAND gate, both transistors Q_1 and Q_2 are connected in series. When both inputs *A* and *B* are high, both transistors will conduct, causes output voltage low.
- As the transistors are connected in series, if any one of the transistors out of Q_1 and Q_2 does not conduct i.e., if any one of the input out of *A* and *B* goes low, then due to no conduction of transistor, output goes high.

One serious problem of DCTL is "current hogging" which exists due to variation of the transistor's parameters used in the circuit.

5.6 EMITTER COUPLED LOGIC (ECL)

Emitter coupled logic (ECL) is the fastest logic family using non-saturating transistors. All the logic families, studied earlier use transistors in saturation mode, which affects switching speeds due to delay of transistors operated in saturation. This is the reason, why ECL is called the fastest logic family.

Figure 5.8 shows an ECL circuit.

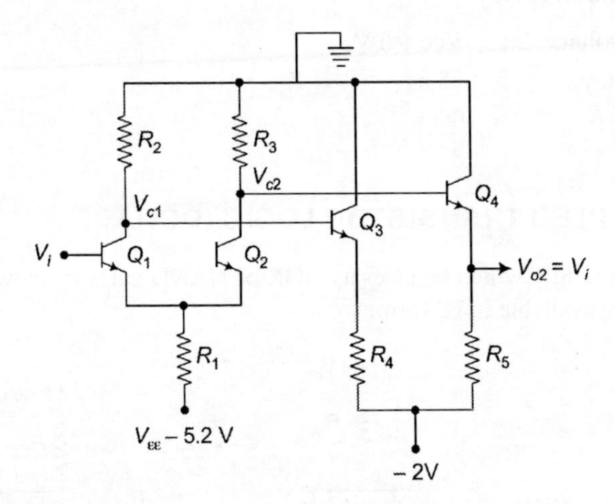

Fig. 5.8 ECL logic circuit.

ECL is a combination of differential amplifier consisting of transistors Q_1 and Q_2 and emitter followers consisting of transistors Q_3 and Q_4. All the emitters are coupled together with V_{ee} supply which produces a constant current of 3 mA. This current flows through Q_1 and Q_2. The low and high logic levels for ECL is considered – 0.8 V and – 0.8 V respectively. The collector voltages Vc_1 and Vc_2 of transistor Q_1 and Q_2 are complement with each other.

The operation of ECL is simple, when input voltage V_i is at logic 0, i.e. $V_i = -\phi$ and V, then Q_1 does not conduct and Q_2 conducts. Thus, $Vc_1 = 0$ V and $Vc_2 = -1.0$ V. Also, when input voltage V_i is at logic 1 i.e., $V_i = -0.8$ V, then Q_1 conducts and Q_2 does not conduct resulting $V_{C1} = -1.0$ V and $V_{C2} = 0$ V. This shows that the emitter follower subtracts 0.8 V from V_{C1} and V_{C2}, thus changing the output levels V_{C1} and V_{C2} to the desired ECL input logic levels.

Typical specifications

Power supply $V_{EE} = -5.2$ V

Power dissipation = 50 mW

Fan-out = 16 to 20

ECL does not have a wide range of applications in digital circuits due to high power dissipation and low noise margins.

5.7 INTEGRATED INJECTION LOGIC (I^2L)

Integrated injection logic (I^2L) is a family of digital circuits built with multiple collector BJJ's. Figure 5.9 shows the logic diagram of I^2L NOR gate and its circuit diagram. This family is considerably small in size as it does not contain resistors. It is most widely used in VLSI circuits due to miniaturization.

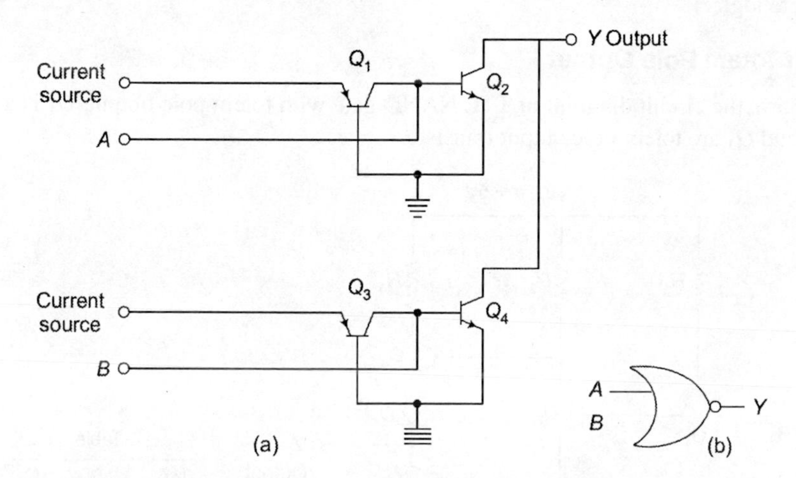

Fig. 5.9 (a) Integrated-injection logic NOR gate, (b) Logic symbol of NOR gate.

- Q_1 and Q_2 are two transistors, acting as a current source which feed current to the bases of transistors Q_1 and Q_4.
- When any of the inputs is low, the corresponding output transistor does not conduct and the output will be low. For example, when input *A* is low, the current coming from Q_1 to base of Q_2 is shorted to ground, Q_2 does not conduct and the output at *Y* will be low. In the same manner, *B* input controls transistor Q_4.
- Now consider the input *B* is high, the transistor Q_4 will conduct results the output low at *Y* terminal. The above condition shows that if any of the inputs or both inputs are high, the output is low.
- The output at terminal *Y* will be high only when both the inputs are low which shows their I^2L performs the function of NOR gate.

Typical specifications

- Propagation delay : Sns
- Power dissipation : 300 MHz
- Noise margin : 0.35 V
- Fan-in : 5
- Fan-out : 8

5.8 TTL LOGIC FAMILY

Transistor: Transistor logic performs many digital functions and has achieved the most widespread popularity. TTL 5400 and 7400 are numerical designations of TTL family. TTL versions are available in SSI package and the complex forms are available in MSI and LSI packages. The TTL gates have three different types of output configurations:

(i) TTL NAND gate with totem pole output
(ii) TTL with open collector output
(iii) Tri-state logic.

(i) TTL with Totem Pole Output

Figure 5.10 shows the circuit diagram of TTL NAND gate with totem pole output. Q_1 is a multiemitter transistor, Q_3 and Q_4 are totem pole output transistors.

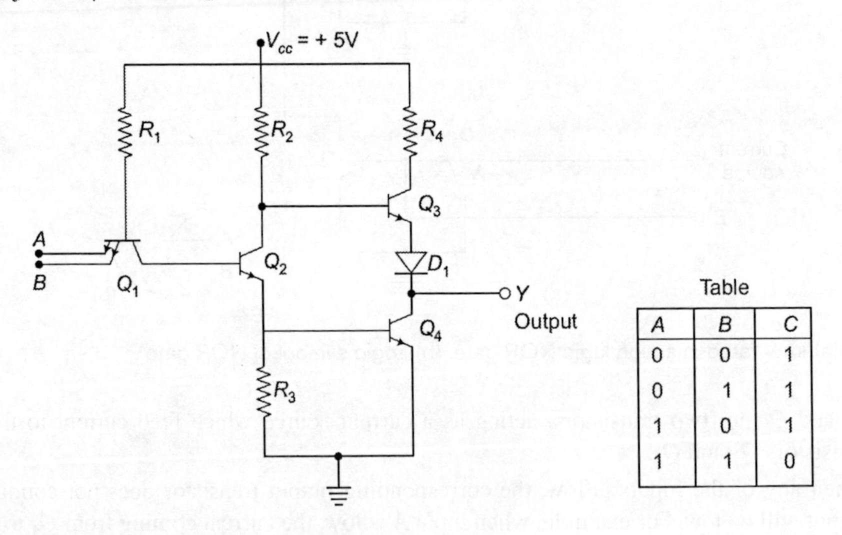

A	B	C
0	0	1
0	1	1
1	0	1
1	1	0

Fig. 5.10 TTL NAND gate (Totem pole output).

- If any of the inputs out of A or B is low, the transistor will be driven into saturation due to base-emitter current of transistor Q_1. Due to saturation, transistor Q_1 pulls the base of transistor Q_2 to ground and cutting off the transistor. This causes Q_3 to conduct and Q_4 to cut off.

This causes Q_3 to conduct and Q_4 to cut off. Q_3 will work as an emitter follower and will give high voltage at terminal Y. It obeys the truth table of NAND gate i.e., if any of the inputs or both the inputs are at logic low, the output will be high.

- Consider the case, when both the inputs *A* and *B* are high. Due to no emitter current in transistor Q_1, its base-collector junction is forward biased. This supplies base current to Q_2, which further feeds base current to Q_4, causing it to conduct. But simultaneously, due to cutting-off of Q_3 (collector of Q_2 goes low) gives low voltage at output terminal *Y*.

This obeys the truth table of NAND gate.

Advantages of TTL

1. The multiemitter of a transistor Q_1 can have upto eight emitters, which cannot be achieved otherwise.
2. The combination of resistor, diodes, transistors is replaced by multiemitter transistor, which reduces the geometrical size and yields lower cost.

Disadvantages of TTL

1. Wired-logic connection is not possible in TTL totem pole circuits. The wired-AND operation of two different gates is not possible.
2. It produces large current spikes during switching.

(ii) TTL with Open Collector Output

Wire ANDing operation is not possible in TTL with totem pole output. To allow wire-ANDing, the diode D_1, transistor Q_3 and resistance R_4 of Fig. 5.9 are removed from totem pole pair of TTL-NAND gate. This is known as open collector, because the collector of Q_4 is open. Figure 5.11 shows TTL with open collector output.

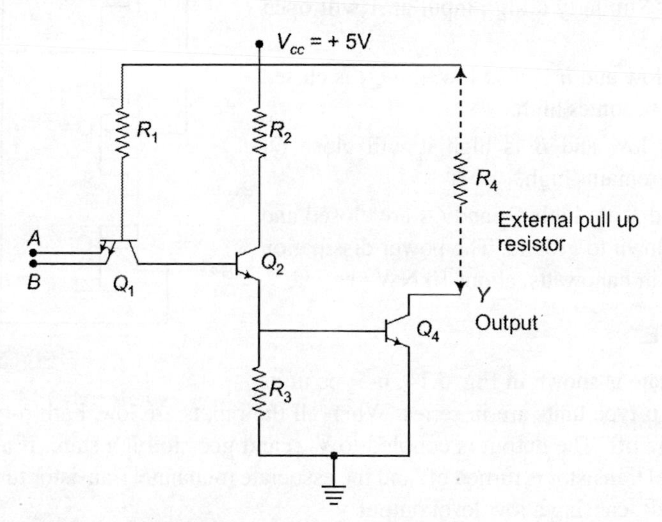

Fig. 5.11 TTL with open collector output.

5.9 COMPLEMENTARY MOS FAMILY

Complementary MOS circuits make the advantage of the fact that both n-channel and p-channel devices can be fabricated on the same substrate. To understand the behaviour of CMOS circuit, we must review the enhancement type MOSFET transistor. The n-channel MOSFET conducts only when its gate to source voltage is positive. The p-channel MOSFET conducts only when its gate to source voltage is negative. The output of MOS is low when the voltage applied to gate-source is zero.

CMOS INVERTER

The CMOS inverter circuits are shown in Fig. 5.12. Consider the state when the input is low, both gates are at zero potential. The input is at $-V_{DD}$ relative to the source of p-channel device and to 0 volt relative to the source of n-channel device. The result is that p-channel device is turned on and n-channel device is turned off. Now there is a low impedance path from V_{DD} to the output and a very high impedance path from output to ground. So the output voltage is high.

When the input is high both the gates are at V_{DD} and the situation is reversed. The p-channel device acts as off and n-channel as on. This results in high output.

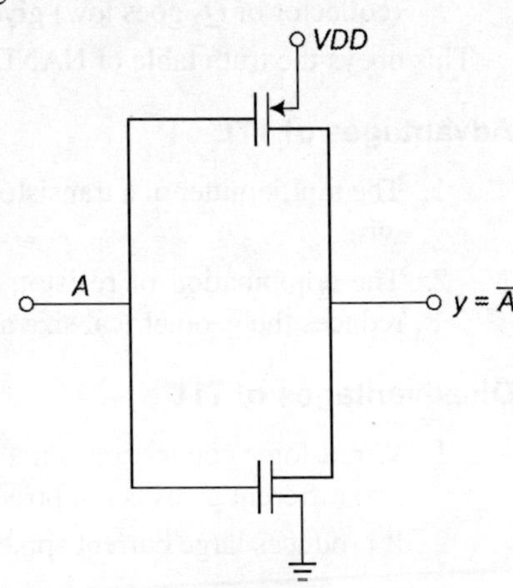

Fig. 5.12

CMOS NAND GATE

The CMOS NAND gate is shown in Fig. 5.13. Q_1 and Q_2 form one complementary stage and Q_3, Q_4 form another one and assume these as switches. A low input will close Q_1 and conduct Q_2. Similarly a high input at A will open Q_1 and close Q_2.

- When A is low and B is also low, the Q_1 is closed therefore Y becomes high.
- When A is low and B is high it will close Q_1 therefore Y remains high.

When both A and B are high Q_2 and Q_3 are closed and pulling the output down to ground. The power dissipation of CMOS family is in nanowatts, about 10 NW per gate.

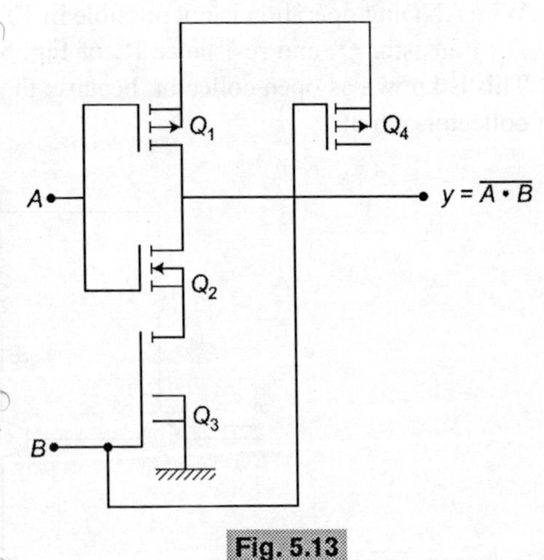

Fig. 5.13

CMOS NOR GATE

The CMOS NOR gate is shown in Fig. 5.14. n-Type units in parallel and two p-type units are in series. When all the inputs are low, both p-type units are on and both n-type units are off. The output is coupled to V_{DD} and goes to high state. If any input is high, the associated p-channel transistor is turned off and the associate n-channel transistor turns on. This connects the output to grounds, causing a low level output.

The explanation or analysis of CMOS NOR gate can also be carried out this way. The truth table of CMOS NOR gate is shown in Fig. 5.15.

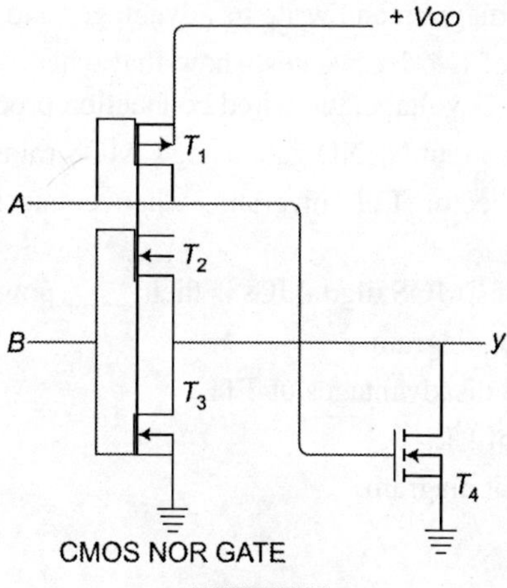

Fig. 5.14

A	B	$Y = \overline{A + B}$
0	0	1
0	1	0
1	0	0
1	1	0

Fig. 5.15

(i) When both the inputs are low Q_1 and Q_2 are closed. Therefore Y is pulled high through the small series resistance of Q_4 and Q_2.

(ii) When A is low and B is high. As B is high, Q_3 is closed pulling output down to ground.

(iii) When A is high, B is low, with A as high Q_4 is closed. The closed Q_4 turns the output low.

(iv) When A is high, B is high. Since A is still high, Q_4 is still closed and output remains low.

The output of NOR gate is high when all the inputs to the gate are high.

EXERCISES

1. What do you understand by the term logic family? What is the significance of the logic family with reference to digital integrated circuits (ICs)?
2. Briefly describe fan-in, fan-out, propagation delay, noise immunity, noise margin and power dissipation parameters.
3. What is the totem-pole output stage? What are its advantages?
4. Why is ECL called a nonsaturating logic? What is its main advantage?
5. What in a logic family decides the fan-out, speed of operation, noise immunity or power dissipation?
6. Explain ECL with a neat diagram.
7. Give the comparison between TTL and CMOS families.

8. Explain TTL with neat a diagram and write its advantages and disadvantages.
9. Using the NOR outputs of two ECL gates, show that when connected together to an external resistor and negative supply voltage, the wired connection produces an OR function.
10. Show the circuit of a four-input NAND gate using CMOS transistors.
11. Prove that two open-collector TTL inverters, when connected together, produce the NOR function.
12. The primary advantage of CMOS digital ICs is their ____ power consumption.
13. Explain a RTS with a neat diagram.
14. Write the advantages and disadvantages of TTL.
15. Write the specifications of I^2L.
16. Explain DCTL with a neat diagram.

6

Karnaugh Maps

INTRODUCTION

Karnaugh map (K-map) is a pictorial method used to minimise Boolean expression without having to use Boolean algebraic theorems and equation manipulations. Using K-map, expressions of two to four variables are easily minimised. Expressions with five to six variables are more difficult but achievable, and expressions with seven or more variables are extremely difficult to minimise using K-map.

In the previous chapter we have seen how Boolean algebra is useful to simplify a Boolean equation algebraically. Two problems arise generally when Boolean algebra is used to minimise a Boolean expression. First, the procedure is very difficult to apply in systematic way. The effectiveness of algebraic simplification depends on our familiarity with, and ability to apply Boolean algebraic rules, laws and theorems. Secondly, using Boolean algebra theorems it is difficult to be sure that we have arrived at minimal solution.

On the other hand, K-map is a systematic method to solve a Boolean expression at its minimal solution. The Karnaugh map was invented in 1952 by E.W. Veitch. It was further developed in 1953 by M. Karnaugh, a physicist at Bell Laboratories, to simplify digital electronic circuits. In K-maps the Boolean variables are transferred and ordered according to the principles of Gray code in which only one variable changes in between the squares.

Any Boolean expression can be expressed in Sum of Product (SOP) or Product of Sum (POS) form. A sum of product (SOP) is one in which a number of product terms, each one of which contains all variables of the function either in complemented or non-complemented form are summed together, whereas in product of sum (POS) a number of sum terms, each one of which contains all variables of the function either in complemented or non-complemented form are multiplied together. Each product term in standard SOP form is called minterm and each sum term in standard POS form is called maxterm. For simplicity, each minterm is represented by '0' and maxterm by '1'. For minterms, the binary words are formed by representing each non-complemented variable by 1 and each complemented variable by 0. The decimal equivalent of the binary is expressed as a subscript of lower case m, i.e., m_0, m_1, m_3, m_5 etc. For maxterm, the binary words are formed by representing each non-complemented variable by a_0 and each complemented variable by a_1, and the decimal equivalent of this binary word is expressed as a subscript of upper case, letter M i.e., M_0, M_1, M_3. etc.

Any given function which is not in standard form can always be converted in standard form by expansion of the function.

6.1 TERMINOLOGY

A huge amount of work has gone into the development of techniques for digital hardware. To facilitate the presentation of the results, certain terminology has evolved which is useful for describing the minimization process.

(a) *Literal:* A product term consisting of a number of variables in complemented or uncomplemented form is called *literal*. For example, $A\,\overline{B}\,C + A\,B\,C$ has three literals A, B and C.

(b) *Implicant:* Any single '1' or group of 1's which can be combined together on a map of function F represents a product term which is called an *implicant* of F. The K-map shown in Fig. 6.1 shows several implicants of F.

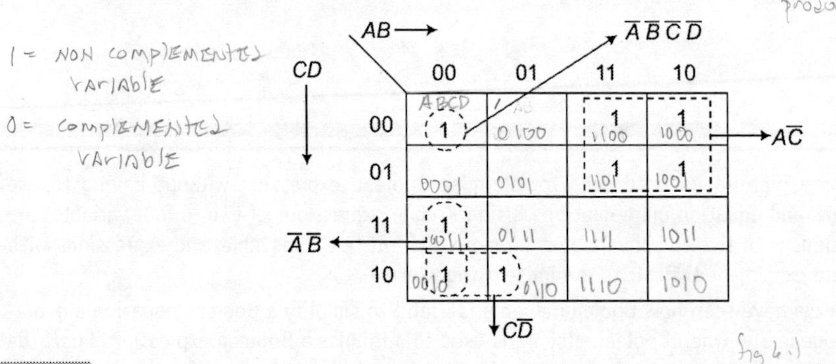

Fig. 6.1

A product term $A\overline{C}$, $\overline{A}\,\overline{B}$, $C\overline{D}$ are the *prime implicants* of function F, whereas $\overline{A}\,\overline{B}\,\overline{C}\,\overline{D}$ is not a *prime implicant* because it can be combined with $\overline{A}\,\overline{B}\,\overline{C}\,\overline{D}$ to form a group.

(c) *Cover:* A collection of implicants which accounts for all valuations for which a given function is equal to '1' is called a cover of that function. It is also apparent that the set of all prime implicants is a cover. A cover defines a particular implementation of the function. For example, a cover of a function 'F' consisting of the minterms leads to the expression:

$$f = \overline{A}\,B\,C + A\,B\,C + \overline{A}\,\overline{B}\,C$$

(d) *Cost:* Cost of a logic circuit is the number of gates plus the total number of inputs to all the gates in the circuit.

STANDARD SOP FORM

A standard SOP form is that in which each term contains all literals present in a given expression. For example,

$$Y = AB + BC + AC$$

has three literals *A*, *B* and *C*. It is not in *standard SOP form*, because all the terms contain two literals, while the expression

$$Y = ABC + A\overline{B}C + \overline{A}\,\overline{B}\,C$$

is in standard SOP form because each term contains three literals, *A*, *B* and *C*. We can standardise a given expression by using Boolean algebra. The terms in standard SOP forms are called *minterms* and denoted by '*m*'.

STANDARD POS FORM

If a given expression is in POS form, it is called standard POS if each term contains all literals present. For example

$$Y = (A + \overline{B} + C).\ (A + B + \overline{C}).\ (\overline{A} + \overline{B} + \overline{C})$$

is a standard POS form and each term in standard POS form is a '*maxterm*' and denoted by 'M'.

6.2 EXPANDING A BOOLEAN EXPRESSION TO STANDARD SOP FORM

The expansion of a Boolean expression to SOP form can be done by:

1. Write all the terms.
2. If any variable is missing in any term, expand that term by multiplying it with the sum of each missing variable and its complement.
3. Drop out the redundant terms.

Example 6.1 Find the minterms and maxterms of the following expression

$$Y = \overline{A} + \overline{B}$$

Solution Given expression $Y = \overline{A} + \overline{B}$ is a two-variable function. In the first term $\overline{A}$, the variable *B* is missing, so multiply it by $(B + \overline{B})$. In the second term $\overline{B}$, variable *A* is missing, so multiply it by $(A + \overline{A})$. Therefore,

$$\begin{aligned}\overline{A} + \overline{B} &= \overline{A}\,(B + \overline{B}) + \overline{B}\,(A + \overline{A})\\ &= \overline{A}\,B + \overline{A}\,\overline{B} + A\,\overline{B} + \overline{A}\,\overline{B}\\ &= \overline{A}\,B + A\,\overline{B} + \overline{A}\,\overline{B}\\ &= 01 + 10 + 00\\ &= m_1 + m_2 + m_0.\\ &= \Sigma m\ (0, 1, 2)\end{aligned}$$

The term m_3 is missing in SOP form. Therefore, the maxterm M_3 will be present in POS form. Hence, the POS form is ΠTM_3.

6.3 EXPANDING A BOOLEAN EXPRESSION TO STANDARD POS FORM

To expand a Boolean expression to a standard POS form the following steps are to be conducted:

1. If one or more variables are missing in any sum term, expand that term by adding the products of each missing term and its complement.
2. Drop the redundant terms.

6.4 KARNAUGH MAP

A Karnaugh map (K-map) is similar to truth table because it shows all the possible values (inputs) and the resulting output for each value. However, instead of organised into columns and rows like a truth table, K-map is an array of squares (or cells) in which each square represents a binary value.

The number of squares or cells in a K-map depends upon the total number of possible input variable combination. For a two-variable K-map, number of squares is $2^2 = 4$. Similarly, for a four-variable K-map the number of squares will be $2^4 = 16$. K-maps are used to minimise a Boolean expression into SOP or POS form into the minimum numbers of literals. Sometimes we deal with very complicated Boolean expressions for implementing a logic function. The K-map provides the simplest and systematic method for minimizing the Boolean expression.

6.5 TWO-VARIABLE KARNAUGH MAP

A two-variable Karnaugh map is shown in Fig. 6.2. If we consider two variables, ($2^2 = 4$) four cells are required, for which four terms are possible. If A and B are two variables, four combinations of variables are possible i.e., AB, $\overline{A}\,B$, $A\overline{B}$, $\overline{A}\,\overline{B}$ in SOP form. In POS form, these combinations are written as $A + B$, $\overline{A} + B$, $A + \overline{B}$, $\overline{A} + \overline{B}$. K-map for two variables is an array of four squares as shown in Fig. 6.2.

The value of a given square is the value of A at the left in the same row combined with the value of B at the top in the same column. For example, the square of upper left corner has a value of $\overline{A}\,\overline{B}$, the square of lower right corner has a value of AB.

B \ A	0	1
0	0: $\overline{A}\,\overline{B}$ 0 0	2: $A\,\overline{B}$ 1 0
1	1: $\overline{A}\,B$ 0 1	3: $A\,B$ 1 1

Fig. 6.2 A two variable K-map in SOP form.

Minterms	*Inputs*		*Boolean representation (SOP)*
	A	*B*	
0 (m_0)	0	0	$\overline{A}\,\overline{B}$
1 (m_1)	0	1	$\overline{A}\,B$
2 (m_2)	1	0	$A\,\overline{B}$
3 (m_3)	1	1	$A\,B$

In POS form, the representation of a two-variable K-map is shown in Fig. 6.3.

B \ A	0	1
0	$A + B$ (0)	$\overline{A} + B$ (2)
1	$A + \overline{B}$ (1)	$\overline{A} + \overline{B}$ (3)

Fig. 6.3 A two-variable K-map in POS form.

Maxterms	*Inputs*		*Boolean representation (POS)*
	A	*B*	
0 (M_0)	0	0	$A + B$
1 (M_1)	0	1	$A + \overline{B}$
2 (M_2)	1	0	$\overline{A} + B$
3 (M_3)	1	1	$\overline{A} + \overline{B}$

6.6 THREE-VARIABLE KARNAUGH MAP

A three-variable K-map is constructed by placing 2 two-variable maps side by side. Figure 6.4 (a) shows three-variable K-map in SOP form indicating the minterms in it. Figure 6.4 (b) shows a three-variable K-map in POS form indicating maxterm in it. If we consider three variables ($2^3 = 8$), eight cells are required, for which eight terms are possible. If *A*, *B* and *C* are three variables then, eight combinations of variables are possible as shown in Fig. 6.4 for which possible Boolean representation is also given in Fig. 6.5.

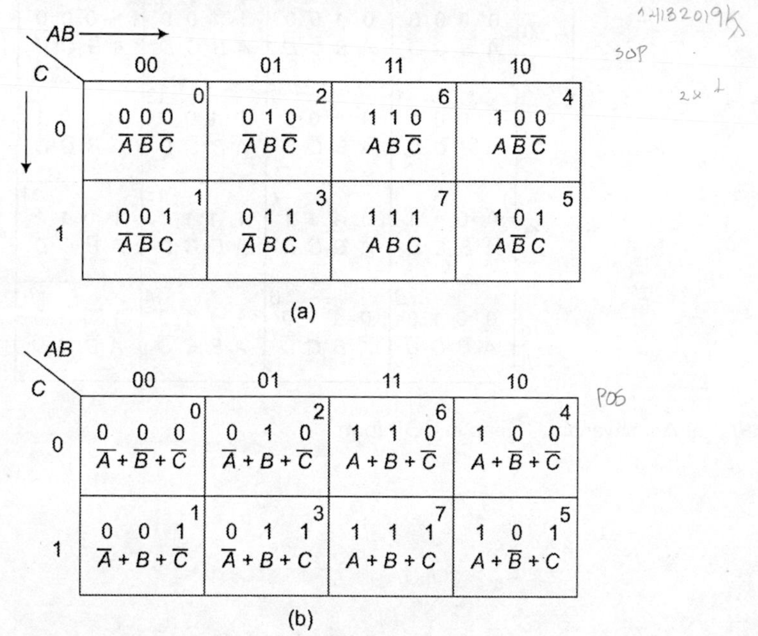

Fig. 6.4 (a) A three-variable K-map in SOP form, (b) A three-variable K-map in POS form.

Variable			Boolean representation in			
A	B	C	SOP form	(minterm)	POS form	(maxterm)
0	0	0	$\bar{A}\bar{B}\bar{C}$	(m_0)	$\bar{A}+\bar{B}+\bar{C}$	(M_0)
0	0	1	$\bar{A}\bar{B}C$	(m_1)	$\bar{A}+\bar{B}+C$	(M_1)
0	1	0	$\bar{A}B\bar{C}$	(m_2)	$\bar{A}+B+\bar{C}$	(M_2)
0	1	1	$\bar{A}BC$	(m_3)	$\bar{A}+B+C$	(M_3)
1	0	0	$A\bar{B}\bar{C}$	(m_4)	$A+\bar{B}+\bar{C}$	(M_4)
1	0	1	$A\bar{B}C$	(m_5)	$A+\bar{B}+C$	(M_5)
1	1	0	$AB\bar{C}$	(m_6)	$A+B+\bar{C}$	(M_6)
1	1	1	ABC	(m_7)	$A+B+C$	(M_7)

Fig. 6.5 Tabular representation of possible maxterms and minterms in a three-variable K-map.

6.7 FOUR-VARIABLE K-MAP

A four-variable K-map is an array of sixteen cells, as shown in Fig. 6.6. Binary values of *A* and *B* are along the left side and the values of *C* and *D* are across the top. It is constructed by placing 2 three-variable K-maps together to create four rows in the same fashion as we used 2 two-variable K-maps to form four columns in a three-variable K-map. If *A*, *B*, *C* and *D* are four variables, then sixteen combinations of variables are possible as shown in Fig. 6.6(a) and (b) for SOP and POS forms respectively. The possible 16 combinations with their Boolean representation is given in Fig. 6.7.

CD \ AB	00	01	11	10
00	0 0 0 0 0 $\bar{A}\bar{B}\bar{C}\bar{D}$	4 0 1 0 0 $\bar{A}B\bar{C}\bar{D}$	12 1 1 0 0 $AB\bar{C}\bar{D}$	5 1 0 0 0 $A\bar{B}\bar{C}\bar{D}$
01	1 0 0 0 1 $\bar{A}\bar{B}\bar{C}D$	5 0 1 0 1 $\bar{A}B\bar{C}D$	13 1 1 0 1 $AB\bar{C}D$	9 1 0 0 1 $A\bar{B}\bar{C}D$
11	3 0 0 1 1 $\bar{A}\bar{B}CD$	7 0 1 1 1 $\bar{A}BCD$	15 1 1 1 1 $ABCD$	11 1 0 1 1 $A\bar{B}CD$
10	2 0 0 1 0 $\bar{A}\bar{B}C\bar{D}$	6 0 1 1 0 $\bar{A}BC\bar{D}$	14 1 1 1 0 $ABC\bar{D}$	10 1 0 1 0 $A\bar{B}C\bar{D}$

Fig. 6.6 (a) A four-variable K-map in SOP form.

CD \ AB →	00	01	11	10
00	(0) 0 0 0 0 $\overline{A}+\overline{B}+\overline{C}+\overline{D}$	(4) 0 1 0 0 $\overline{A}+B+\overline{C}+\overline{D}$	(12) 1 1 0 0 $A+B+\overline{C}+\overline{D}$	(8) 1 0 0 0 $A+\overline{B}+\overline{C}+\overline{D}$
01	(1) 0 0 0 1 $\overline{A}+\overline{B}+\overline{C}+D$	(5) 0 1 0 1 $\overline{A}+B+\overline{C}+D$	(13) 1 1 0 1 $A+B+\overline{C}+D$	(9) 1 0 0 1 $A+\overline{B}+\overline{C}+D$
11	(3) 0 0 1 1 $\overline{A}+\overline{B}+C+D$	(7) 0 1 1 1 $\overline{A}+B+C+D$	(15) 1 1 1 1 $A+B+C+D$	(11) 1 0 1 1 $A+\overline{B}+C+D$
10	(2) 0 0 1 0 $\overline{A}+\overline{B}+C+\overline{D}$	(6) 0 1 1 0 $\overline{A}+B+C+\overline{D}$	(14) 1 1 1 0 $A+B+C+\overline{D}$	(10) 1 1 1 0 $A+\overline{B}+C+\overline{C}$

Fig. 6.6 (b) A four-variable K-map in POS form.

Variable				*Boolean representation in*			
A	*B*	*C*	*D*	*SOP form*	*(minterm)*	*POS form*	*(maxterm)*
0	0	0	0	$\overline{A}\,\overline{B}\,\overline{C}\,\overline{D}$	m_0	$\overline{A}+\overline{B}+\overline{C}+\overline{D}$	M_0
0	0	0	1	$\overline{A}\,\overline{B}\,\overline{C}\,D$	m_1	$\overline{A}+\overline{B}+\overline{C}+D$	M_1
0	0	1	0	$\overline{A}\,\overline{B}\,C\,\overline{D}$	m_2	$\overline{A}+\overline{B}+C+\overline{D}$	M_2
0	0	1	1	$\overline{A}\,\overline{B}\,C\,D$	m_3	$\overline{A}+\overline{B}+C+D$	M_3
0	1	0	0	$\overline{A}\,B\,\overline{C}\,\overline{D}$	m_4	$\overline{A}+B+\overline{C}+\overline{D}$	M_4
0	1	0	1	$\overline{A}\,B\,\overline{C}\,D$	m_5	$\overline{A}+B+\overline{C}+D$	M_5
0	1	1	0	$\overline{A}\,B\,C\,\overline{D}$	m_6	$\overline{A}+B+C+\overline{D}$	M_6
0	1	1	1	$\overline{A}\,B\,C\,D$	m_7	$\overline{A}+B+C+D$	M_7
1	0	0	0	$A\,\overline{B}\,\overline{C}\,\overline{D}$	m_8	$A+\overline{B}+\overline{C}+\overline{D}$	M_8
1	0	0	1	$A\,\overline{B}\,\overline{C}\,D$	m_9	$A+\overline{B}+\overline{C}+D$	M_9
1	0	1	0	$A\,\overline{B}\,C\,\overline{D}$	m_{10}	$A+\overline{B}+C+\overline{D}$	M_{10}
1	0	1	1	$A\,\overline{B}\,C\,D$	m_{11}	$A+\overline{B}+C+D$	M_{11}
1	1	0	0	$A\,B\,\overline{C}\,\overline{D}$	m_{12}	$A+B+\overline{C}+\overline{D}$	M_{12}
1	1	0	1	$A\,B\,\overline{C}\,D$	m_{13}	$A+B+\overline{C}+D$	M_{13}
1	1	1	0	$A\,B\,C\,\overline{D}$	m_{14}	$A+B+C+\overline{D}$	M_{14}
1	1	1	1	$A\,B\,C\,D$	m_{15}	$A+B+C+D$	M_{15}

Fig. 6.7 Tabular representation of possible maxterms and minterms in a four variable K-map.

6.8 LOOPS IN K-MAP

The output expression of any digital circuit can be simplified by properly combining the cells of K-map which contains 1's in SOP form and 0's in POS form. This process of combining cells is called *looping*.

(a) Looping of Two

Figure 6.8(a) is a K-map having three variables, A, B and C. This map contains a pair of 1's that are horizontally adjacent to each other. First 1 represents $\overline{A}\,B\,\overline{C}$ and another 1 represents $AB\overline{C}$. In these two terms only variable A appears in both normal and complemented form while variables B and $\overline{C}$ remain unchanged. These two terms can be looped together to give the resultant expression which eliminates variable A since it appears in both complemented and uncomplemented forms. This is also proved by Boolean law

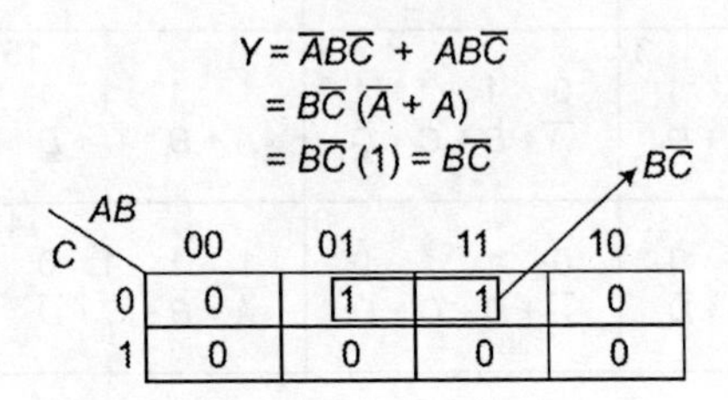

Fig. 6.8 (a) Example of horizontal loop of two cells.

The same principle is true for any vertically adjacent 1's i.e., a pair of vertical group of two cells as shown in Fig. 6.8 (b).

C \ AB	00	01	11	10
0	0	1	0	0
1	0	1	0	0

$\overline{A}B$

Common

$y = \overline{A}\,B\,\overline{C} + \overline{A}\,B\,C$

$= \overline{A}\,B$

Odd

Fig. 6.8 (b) Example of vertical loop of two cells.

In the above example of vertical looping of two cells, the term $\overline{A}\,B$ is common in the below given expression and C is changing.

$$Y = \overline{A}\,B\,\overline{C} + \overline{A}\,B\,C$$
$$= \overline{A}\,B\,(C + \overline{C}) = \overline{A}\,B$$

Therefore, these two variables A and B can be looped to give the output.

Another case of a three-variable K-map is shown in Fig. 6.8(c) in which the left row and right row of squares are considered to be adjacent. Thus, the two 1's in this map can be looped to give the resultant expression $\overline{A}\,\overline{B}\,\overline{C} + A\,\overline{B}\,\overline{C} = \overline{B}\,\overline{C}$.

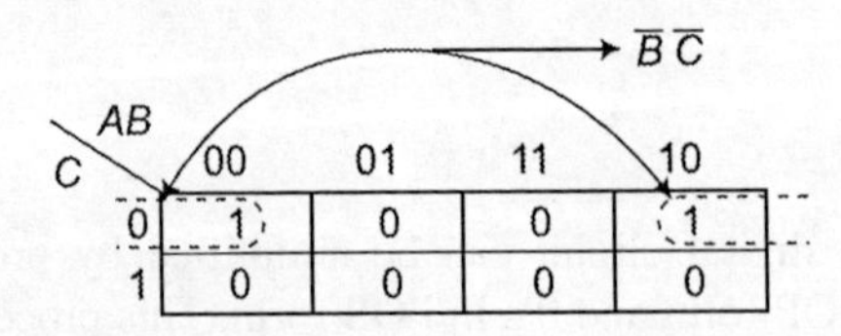

Fig. 6.8 (c) Example of adjacent left and right cells.

$$Y = \overline{A}\,\overline{B}\,\overline{C} + A\,\overline{B}\,\overline{C}$$
$$= \overline{B}\,\overline{C}\,(\overline{A} + A)$$
$$= \overline{B}\,\overline{C}$$

Figure 6.8(d) shows a K-map of four variables, that has two pairs of 1's. The two 1's in top row are horizontally adjacent and two 1's in the bottom are also adjacent as in the previous example of Fig. 6.8 (c). The adjacent condition can be considered in left, right or top, bottom cells similar to a folded page.

When the top pair is looped, B variable is eliminated since it is changing between ($A\,B\,\overline{C}\,\overline{D}$ and $A\,\overline{B}\,\overline{C}\,\overline{D}$) to give $A\,\overline{C}\,\overline{D}$. Looping of bottom pair eliminates the variable A to give term $\overline{B}\,C\,\overline{D}$.

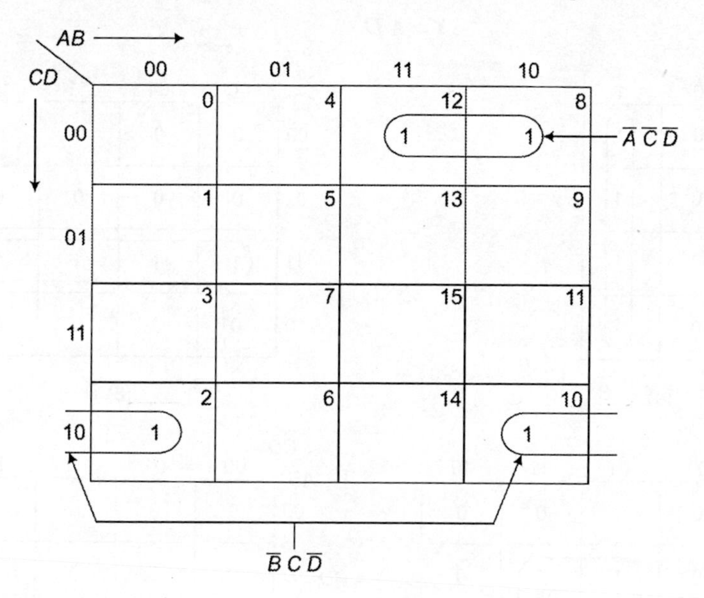

Fig. 6.8 (d) Example of two-cell looping in a four-variable K-map.

(b) Looping of Four: (Quads)

A K-map which contains a group of four 1's that are adjacent to each other is called a quad. Figure 6.9 shows the various possibilities of a quad in a three- or four-variable K-map.

In Fig. 6.9(a), four 1's are horizontally adjacent in a three-variable K-map, whereas in Fig. 6.9(b) four 1's are horizontally adjacent in a four-variable K-map.

In Fig. 6.9(c), four 1's are combined in a square, and they are considered adjacent to each other.

In Fig. 6.9(d), four 1's are also adjacent, as are in Fig. 6.9(e) because, as pointed out in the previous section, the left, right columns or rows and top, bottom rows or columns are considered to be adjacent to each other.

When a quad (a group of four cells) is made, the resultant term will only contain the variables that do not change in all the squares in a quad.

For example, in Fig. 6.9(a), the four squares that contain 1 are $\overline{A}\,\overline{B}\,C$, $\overline{A}\,B\,C$, $A\,B\,C$ and $A\,\overline{B}\,C$. It has been seen that variable C remains unchanged. The resultant expression will be

$$Y = \overline{A}\,\overline{B}\,C + \overline{A}\,B\,C + A\,B\,C + A\,\overline{B}\,C$$
$$= \overline{A}\,C\,(B + \overline{B}) + AC\,(B + \overline{B})$$
$$= \overline{A}\,C + AC$$
$$= C\,(\overline{A} + A)$$
$$= C.$$

which is proved by Boolean laws also.

Another example of Fig. 6.9(d), in which four squares consists are $A\,B\,\overline{C}\,\overline{D}$, $A\,\overline{B}\,\overline{C}\,\overline{D}$, $A\,B\,C\,\overline{D}$ and $A\,\overline{B}\,C\,\overline{D}$. It has been observed that variables A and $\overline{D}$ remain unchanged, so the expression will be

$$Y = A\,\overline{D}.$$

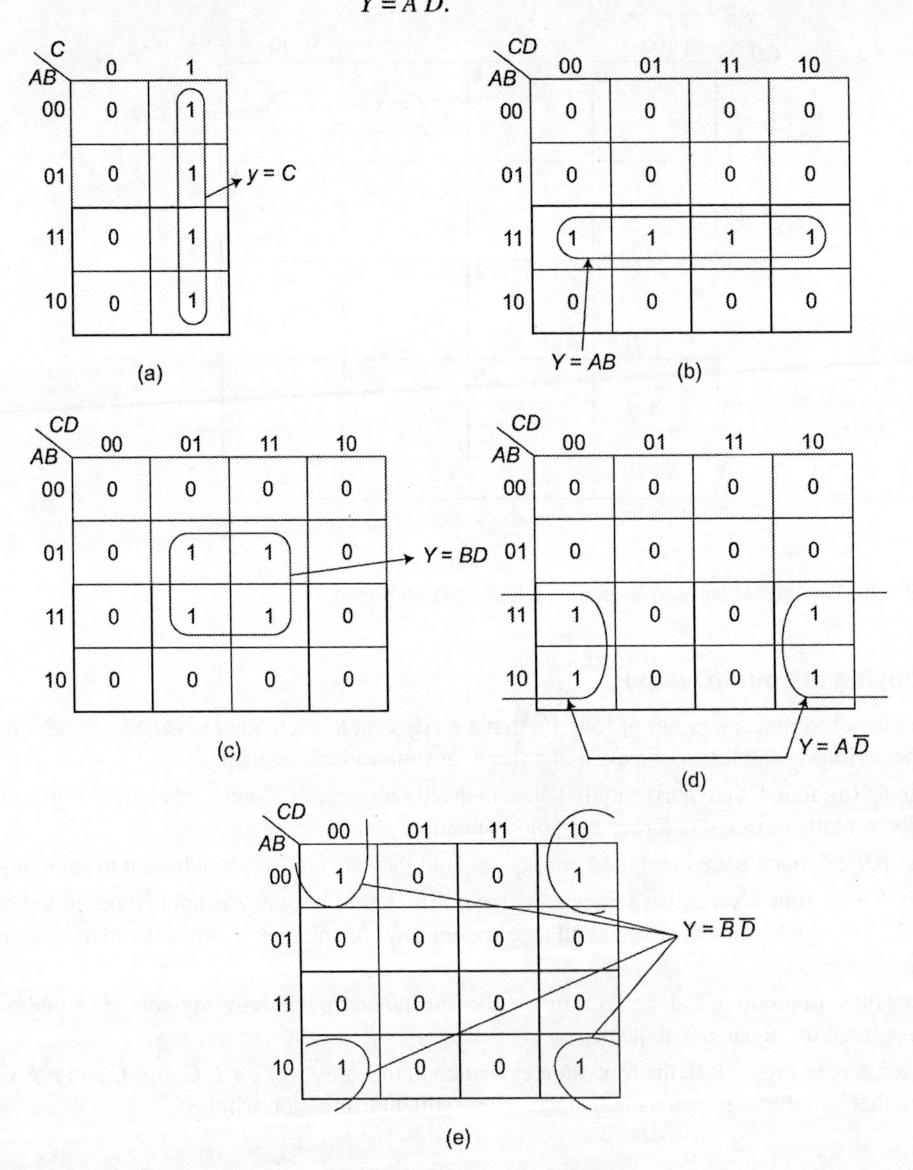

Fig. 6.9 Example of looping of four cells (quad).

(c) Looping of Eight (Octets)

A group of eight 1's that are adjacent to each other is called an octet. Various examples of octets are shown in Fig. 6.10. When an octet is looped in a four-variable map, three or four variables are eliminated because only one variable remains unchanged.

In Fig. 6.10 (a), variable B remains unchanged and all other variables change in complement or uncomplemened form. Therefore, the output expression will be $Y = B$.

In Fig. 6.10 (b), variable C remains unchanged whereas, all other variables change. Therefore, the output expression will be $Y = C$.

Similarly, as shown in Fig. 6.10(c) and 6.10(d), variables $\overline{B}$ and $\overline{D}$ remain unchanged and all other variables are eliminated as these variables are changing in complemented or uncomplemented form. The output expression for K-map of Fig. 6.10(c) is $Y = \overline{B}$ and for Fig. 6.10(d) is $Y = \overline{D}$.

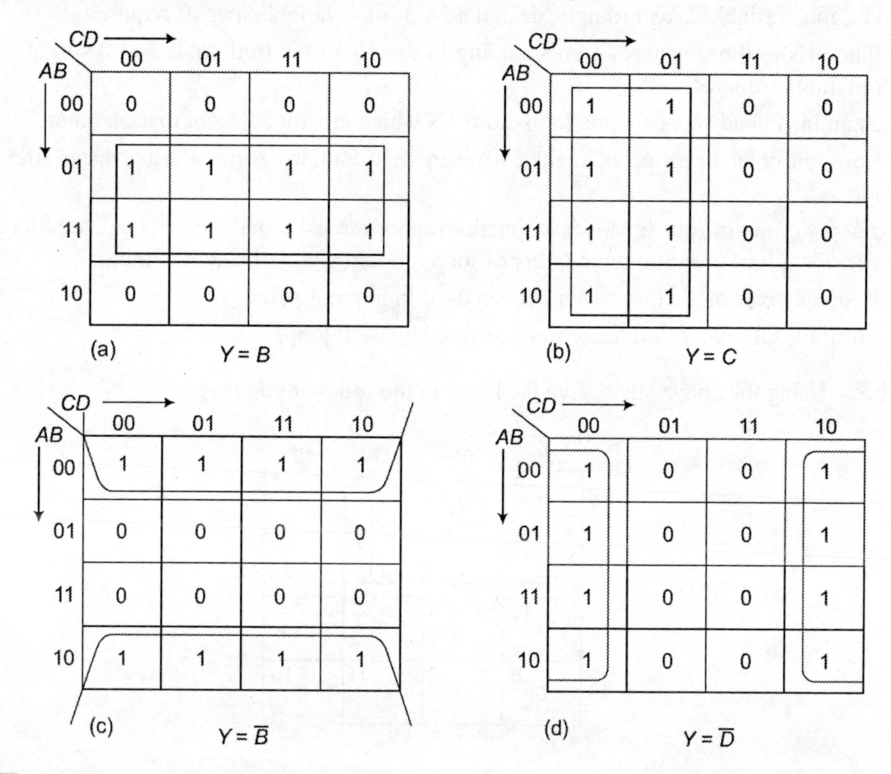

Fig. 6.10 Examples of looping of eight (octets).

To summarize looping in Karnaugh maps the following points should be remembered.

1. Looping a pair of two 1's which are adjacent to each other eliminates *one variable* that appears in complemented and uncomplemented form.
2. Looping a pair of four 1's eliminates *two variables* that appear in complemented and uncomplemented form.
3. Looping a pair of eight 1's eliminates the *three variables* that appear in complemented and uncomplemented form.

6.9 K-MAP SIMPLIFICATION PROCESS

The K-map simplification process depends upon the looping of variables. We have studied the looping of two (pair), four (quads) and eight (octets) 1's on a K-map. The simple concept of looping is the ***elimination of the variable that appears in both complemented and uncomplemented form within a loop and the variables that are same for all cells must appear in the final expression.***

This is to remember that a loop of two 1's eliminates one variable, a loop of four 1's eliminates two variables and a loop of eight 1's eliminates three variables. The following steps are followed for simplifying a Boolean expression using K-Map.

Step I Examine the number of variables (literals) in the given equation. For example, equation $Y = AB + \overline{A}B + A\overline{B}$ contains two variables and equation $Y = ABC + A\overline{B}C + AB\overline{C}$ contains three variables, whereas equation $Y = \overline{A}BD + AB\overline{C} + ABD$ contains variables A, B, C and D i.e., four variables. Accordingly, design a 2-, 3- or 4-variable map as required.

Step II Place 1's in those squares corresponding to the 1's in the truth table and 0's in all the other remaining squares.

Step III Examine the adjacent 1's and loop those 1's which are not adjacent to each other.

Step IV Loop an octet (a group of eight 1's) even if it contains some 1's that have already been looped.

Step V Loop a group of four (quad) that contains one or more 1's that have not yet been looped. But consideration for minimum number of loops should always be kept in mind.

Step VI Loop any pair of 1's that have not been used in looping yet.

Step VII Form the OR sum of all the terms generated by each loop.

Example 6.2 Using the above steps, give the loops in the following K-map.

AB \ CD →	00	01	11	10
00	0 (0)	0 (1)	1 (3)	0 (2)
01	1 (4)	1 (5)	1 (7)	1 (6)
11	1 (12)	1 (13)	0 (15)	0 (14)
10	0 (8)	0 (9)	0 (11)	0 (10)

Solution

$$Y = \overline{A}B + B\overline{C} + \overline{A}CD$$

$\overline{A}B$	$B\overline{C}$	$\overline{A}CD$
↓	↓	↓
Loop of 4, 5, 6, 7	Loop of 4, 5, 12, 13	Loop of 3, 7

1. There is no single 1 in the K-map.
2. The 1 of cell 3 is adjacent to 1 of cell 7. Looping this pair results in $\overline{A}CD$.
3. There are no octets i. e., group of eight 1's.

4. Two quads are there, one of 4, 5, 6, 7 another is made up by 4, 5, 12, 13 which result in $\overline{A}\,B$ and $B\,\overline{C}$ respectively.
5. All 1's have looped.
6. All the terms $\overline{A}\,B$, $B\,\overline{C}$ and $\overline{A}\,CD$ are ORed together to form $Y = \overline{A}\,B + B\,\overline{C} + \overline{A}\,CD$.

6.10 MAPPING OF K-MAPS

The Boolean expression which is to be minimised using K-map is mapped in the K-map by entering 1 in the particular cell. Boolean expression may be expressed in standard SOP form, standard POS form, non-standard SOP or POS form or in the form of truth table. As discussed in the previous section the method to standardize the non-standard Boolean expression in SOP or POS form. This section explains with some examples to map the given Boolean expression in standard, non-standard or truth table form in a K-map.

(a) Mapping of Standard SOP in K-map

Consider the equation

$$Y = \overline{A}\,\overline{B}\,\overline{C} + A\,\overline{B}\,C + \overline{A}\,B\,\overline{C}$$

which requires a three-variable K-map. Select the first product term $\overline{A}\,\overline{B}\,\overline{C}$ and put 1 where this term exists in K-map as shown in Fig. 6.11.

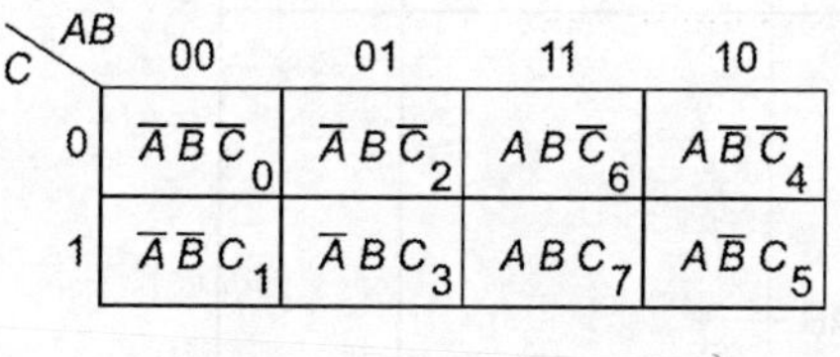

Three-variable K-map format

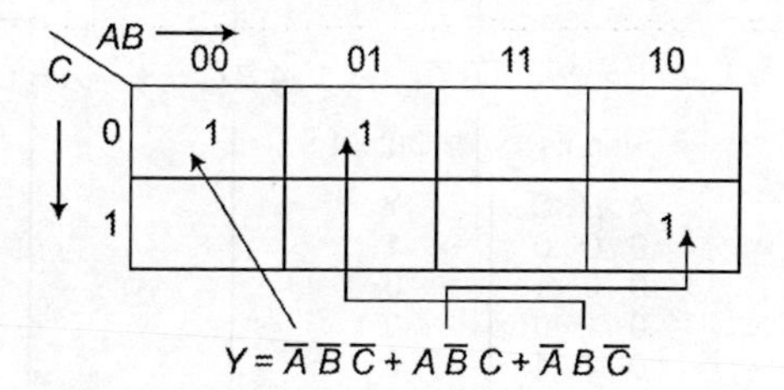

Fig. 6.11 (a) Standard three-variable K-map format and (b) mapping of expression.

(b) Mapping of non-Standard SOP in K-map

The method of converting non-standard SOP in standardized form has been explained in the previous section. Let us consider the following Boolean expression

$$Y = \overline{A} + \overline{A}\,\overline{B} + A\,B\,\overline{C}$$

which is not in standard form. Convert the expression in standard form by multiplying the missing term i.e.,

$$Y = \overline{A}\,(B + \overline{B})\,(C + \overline{C}) + A\overline{B}\,(C + \overline{C}) + AB\overline{C}$$

$$= \overline{A}\,B\,C + \overline{A}\,\overline{B}\,C + \overline{A}\,B\,\overline{C} + \overline{A}\,\overline{B}\,\overline{C} + A\,\overline{B}\,C + \overline{A}\,\overline{B}\,\overline{C} + A\,\overline{B}\,C$$

$$= \overline{A}\,\overline{B}\,\overline{C} + \overline{A}\,\overline{B}\,C + \overline{A}\,B\,\overline{C} + \overline{A}\,B\,C + A\,B\,\overline{C} + A\,\overline{B}\,\overline{C} + A\,\overline{B}\,C.$$

Map the standard SOP form in a K-map.

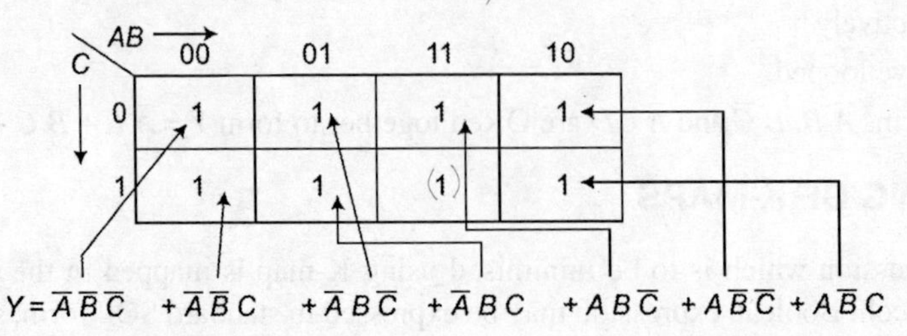

(c) Mapping a Truth Table in K-map

In digital circuits, it is very important to map a truth table in a K-map. Outputs of most of the digital circuits are available in the form of truth tables. For a Boolean expression $Y = \overline{A}\,\overline{B}\,\overline{C} + A\,B\,\overline{C} + A\,\overline{B}\,C$, the truth table can be constructed by putting 1 in the output for the present inputs. A, B and C are three variables present in the equation with its $2^3 = 8$ possible combinations. The output will be 1 for $\overline{A}\,\overline{B}\,\overline{C}$, $A\,B\,\overline{C}$ and $A\,\overline{B}\,C$ and 0 for all other inputs. Enter 1 in the cells of 3-variable K-map for the present inputs and 0 for all other inputs as shown Fig. 6.12.

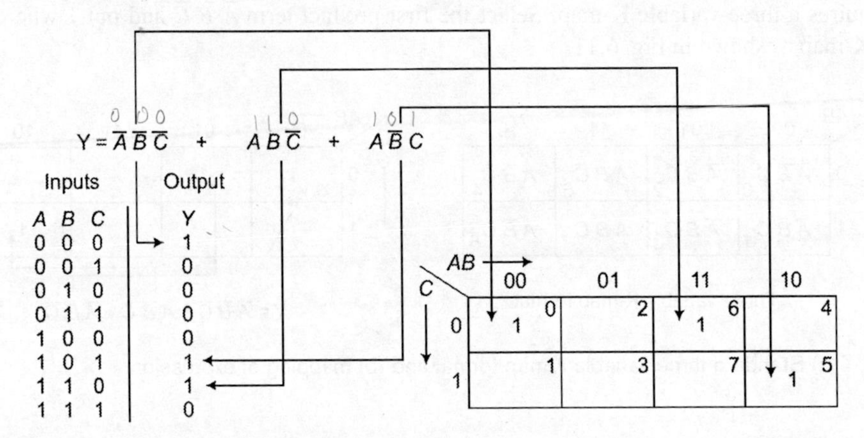

Fig. 6.12 Mapping of a truth table in K-map.

6.11 MINIMIZATION OF PRODUCT OF SUM FORMS

In the previous sections, we have considered the Boolean expression in the sum-of-product form only and minterms to implement a K-map. We can use the same techniques and principle of duality to minimize product of sum (POS) terms also. In this case it is the maxterms for which $f = 0$ that have to be combined into sum form that are as large as possible. A sum term is considered larger if it covers more maxterms. Larger the term, less costly it is to implement.

Consider the POS equation

$$Y = (\overline{A} + B)\,(\overline{A} + C)$$

which contains three variables A, B and C and the circuit of this expression has two OR gates and one AND gate with two inputs of each gate.

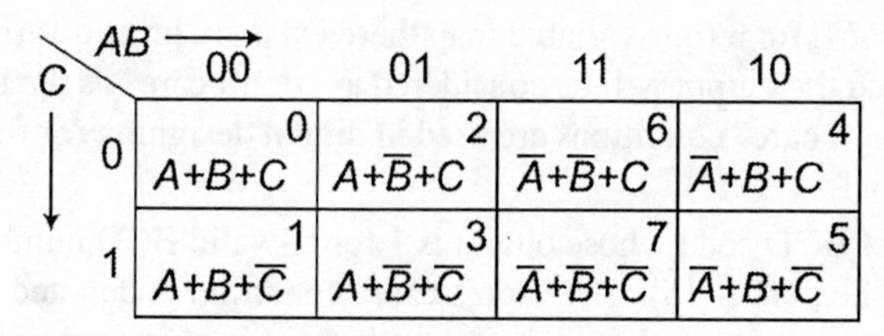

Standard POS form 3-variable K-map

We have seen in the above K-map that the binary value 0 in SOP form discussed earlier is considered 1 in POS form. For example, the input value '000' is considered to be $\overline{A}\ \overline{B}\ \overline{C}$ in SOP form and ABC in POS form, which is shown above in standard POS form 3 variable K-map representation. Usually, when working with POS expressions, the 1's are left off. The following steps are used to minimize the POS equations.

Step I Determine the binary value of each sum term in standard POS form. This is the binary value that makes the term equal to 0.

Step II As the sum term is evaluated, place a 0 on K-map in corresponded cell.

Now, take the expression, we have considered

Problem

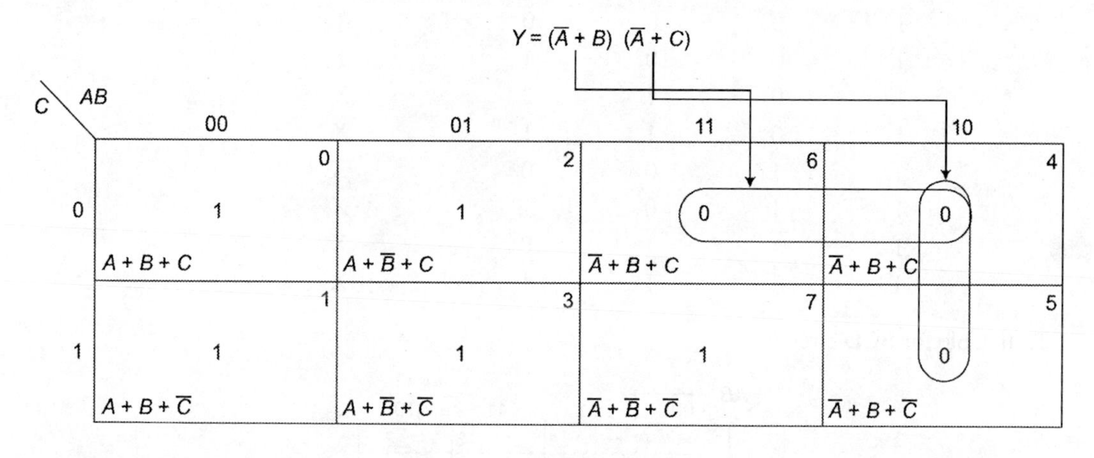

Solution For the Boolean expression

$$Y = (\overline{A} + B)\ (\overline{A} + C)$$

- $\overline{A} + B$, exists in cell no. 4 and 5, hence put 0 in cell no. 4 and 5.
- $\overline{A} + C$, exists in cell no. 4 and 6, hence put 0 in cell no. 6 as cell no. 4 is already filled.
- Put 1 in all other input.

6.12 DON'T CARE CONDITIONS

In the digital hardware, some logic circuits have certain input conditions for which there are no specified levels because these input conditions will never occur. In simple words, there are some digital circuits where we "don't care" whether the output is high or low. For example, a Binary Coded Decimal (BCD code) is valid for 0000, 0001, 0010, 0011, 0100, 0101, 0110, 0111, 1000, 1001 as it is valid for 0 – 9. Along with nine valid combinations, for a four-variable map there exist six invalid variables 1010, 1011, 1100, 1101, 1110, 1111 for which the output will be considered as "don't care" as the invalid combinations are not valid BCD codes. "Don't care" conditions are used in digital designing for fluctuating outputs whose output is uncertain i.e., 00r1.

Consider an example of BCD code whose output is 1 for all valid BCD numbers and "don't care" for all invalid BCD numbers i.e., (10 – 15). The "don't care" condition is denoted by XO Red. The "don't care's" can be looped in K-map in a pair, quad or octet if there is a presence of a single 1 or more in a K-map. "Don't care" conditions are considered to be 1 in SOP K-maps and 0 is POS K-maps.

A	*B*	*C*	*D*	*BCD output*
0	0	0	0	1
0	0	0	1	1
0	0	1	0	1
0	0	1	1	1
0	1	0	0	1
0	1	0	1	1
0	1	1	0	1
0	1	1	1	1
1	0	0	0	1
1	0	0	1	1
1	0	1	0	X
1	0	1	1	X
1	1	0	0	X
1	1	0	1	X
1	1	1	0	X
1	1	1	1	X

Truth table for BCD circuit

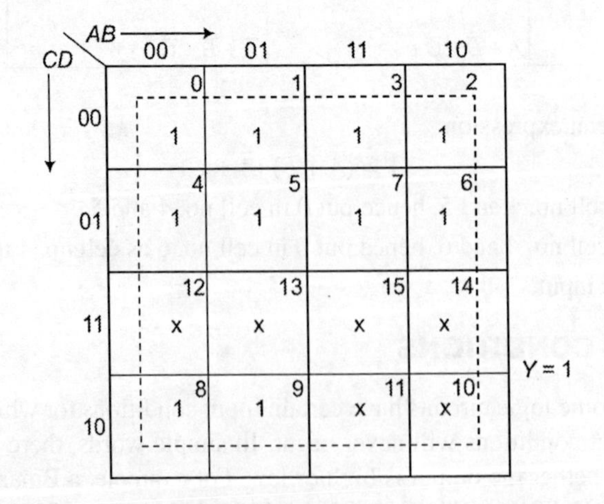

Fig. 6.13 Truth table and K-map for BCD circuit.

Example 6.3 Using K-map simplify the Boolean expression

$$Y = \overline{C}\,(\overline{A}\,\overline{B}\,\overline{D} + D) + A\,\overline{B}\,C + \overline{D}$$

Solution The Boolean expression

$$Y = \overline{C}\,(\overline{A}\,\overline{B}\,\overline{D} + D) + A\,\overline{B}\,C + \overline{D}$$

contains four variables A, B, C and D. Simplify the expression by logical multiplication.

$$Y = \overline{A}\,\overline{B}\,\overline{C}\,\overline{D} + \overline{C}\,D + A\,\overline{B}\,C + \overline{D}$$

The output is 1 for the above input conditions and 0 for all remaining inputs.

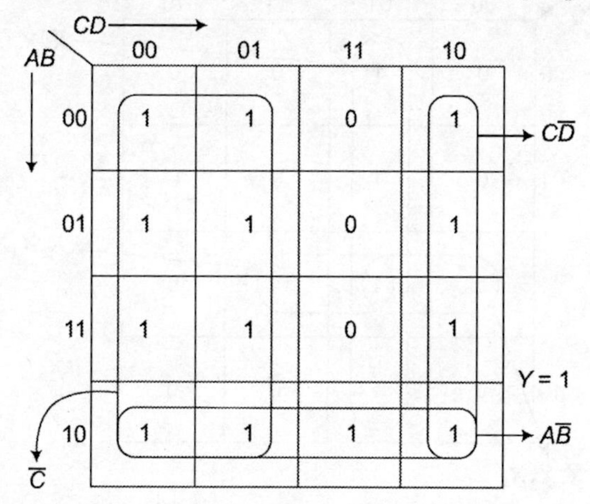

The output expression consists of

an octet $\overline{C}$

a quad $A\overline{B}$

and another quad $\overline{C}\overline{D}$

$\therefore$ SOP minimisation of the above expression is $Y = \overline{C} + A\overline{B} + \overline{C}\overline{D}$.

Example 6.4 Simplify the following SOP expression using K-map.

$$Y = B\,\overline{C}\,\overline{D} + \overline{A}\,B\,\overline{C}\,D + A\,B\,\overline{C}\,D + \overline{A}\,B\,C\,D + A\,B\,C\,D$$

Solution The expression is of four variables A, B, C and D.

CD \ AB →	00	01	11	10
00	0 (0)	1 (4)	1 (12)	0 (8)
01	0 (1)	1 (5)	1 (13)	0 (9)
11	0 (3)	1 (7)	1 (15)	0 (11)
10	0 (2)	0 (6)	0 (14)	0 (10)

→ $B\overline{C}$

→ BD

The minimised expression is

$$Y = B\,\overline{C} + B\,D.$$

Example 6.5 Minimise using K-map

$$Y = \overline{X}_1\,X_2\,\overline{X}_3\,\overline{X}_4 + X_1\,\overline{X}_2\,\overline{X}_3\,\overline{X}_4 + \overline{X}_1\,\overline{X}_2\,\overline{X}_3\,X_4 + X_1\,X_2\,\overline{X}_3\,X_4 + X_1\,X_2\,\overline{X}_3\,\overline{X}_4 + X_1\,X_2\,X_3\,X_4$$

Solution The expression contains four variables X_1, X_2, X_3 and X_4.

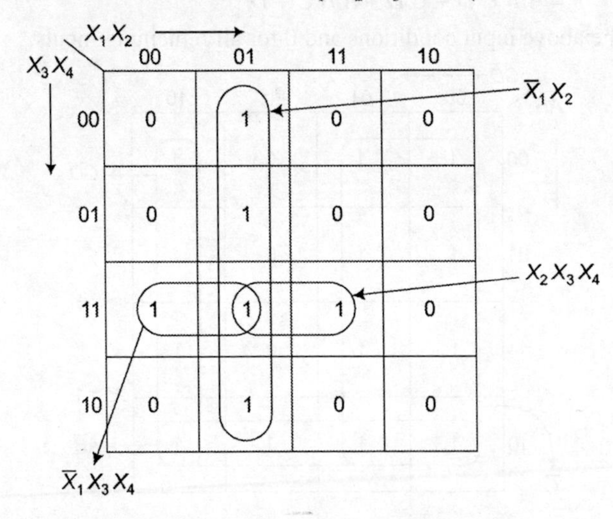

The output expression will be

$$Y = \overline{X}_1\,X_2 + X_2\,X_3\,X_4 + \overline{X}_1\,X_3\,X_4.$$

6.13 REPRESENTATION OF A DIGITAL HARDWARE

Boolean expressions can be represented by the following methods in SOP or POS forms.

In SOP form

$$Y = AB + \overline{A}\,C + ABC$$

$$\Sigma m = (1, 3, 5, ...)$$

or $\Sigma\,(A, B, C) = (1, 3, 5....)$ or a truth table

In POS form

$$Y = (A + B).\,(\overline{A} + C).\,(A + B + C)$$

$$\Pi\,m = (0, 1, 3, 5\ .....)$$

or $\Pi\,(A, B, C) = (1, 3, 5....)$

or a truth table.

Example 6.6 Reduce using K-map and design it with gates $F = X\,Y\,\overline{Z} + X\,Y\,Z + \overline{X}\,Y\,Z.$

Solution The expression $F = X\,Y\,\overline{Z} + XYZ + \overline{X}\,Y\,Z$ contains three variables X, Y, Z.

F X(YZ̄ + YZ)   YZ(X+X̄)

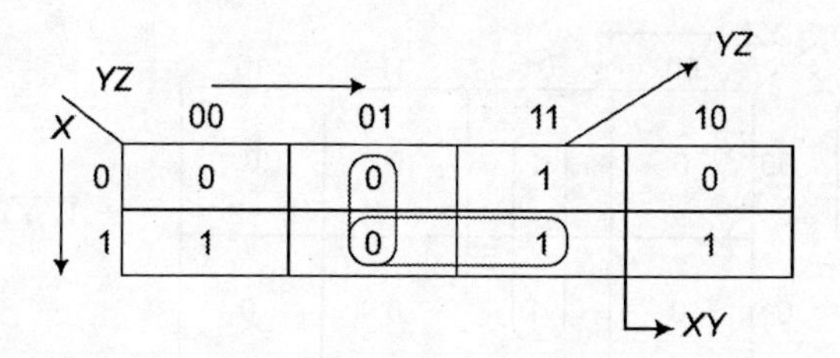

$F = XY + YZ$ is the reduced form.

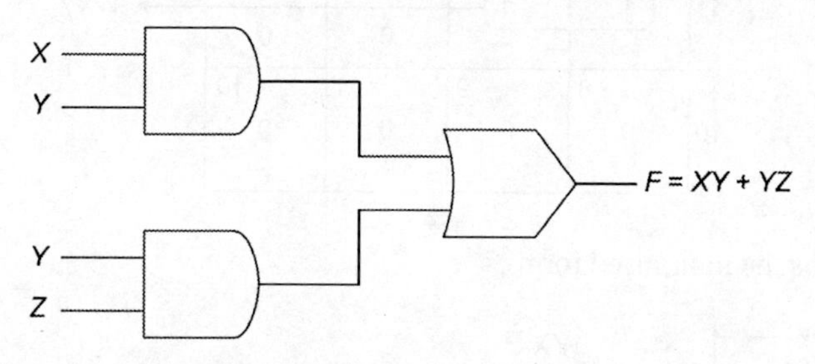

Representation of $Y = XY + YZ$ using gates.

Example 6.7 Reduce the following using K-map $f = \Sigma m$ (1, 3, 4, 5, 6, 7)

Solution The expression

$$f = \Sigma m\ (1, 3, 4, 5, 6, 7)$$

ranges from 0 to 7. So, 8 cells are required to implement it. Three-variable K-map is used to solve this expression.

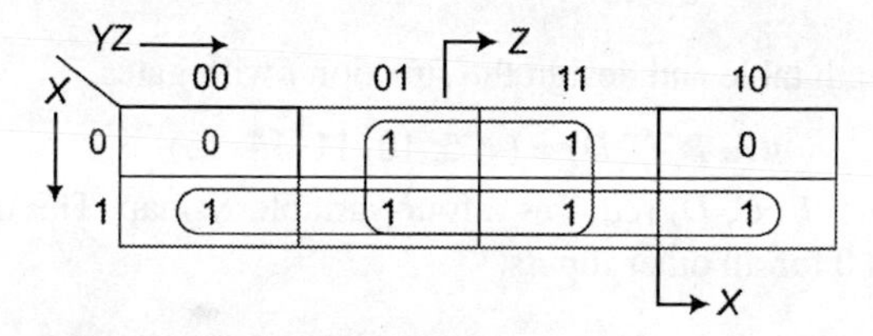

$f = X + Z$ is the minimised form of the above expression.

Example 6.8 Reduce using K-map and design using gates

$$F(W, X, Y, Z) = (1, 5, 12, 13)$$

Solution As the expression indicates that W, X, Y, Z are four variables of the above expression. Also the function ranges upto 13. $2^3 = 8$ and $2^4 = 16$, Therefore, 3-variable K-map consists of 8 cells but, here we require 13 cells. So, 4-variable K-map is used to implement

$$F(W, X, Y, Z) = (1, 5, 12, 13)$$

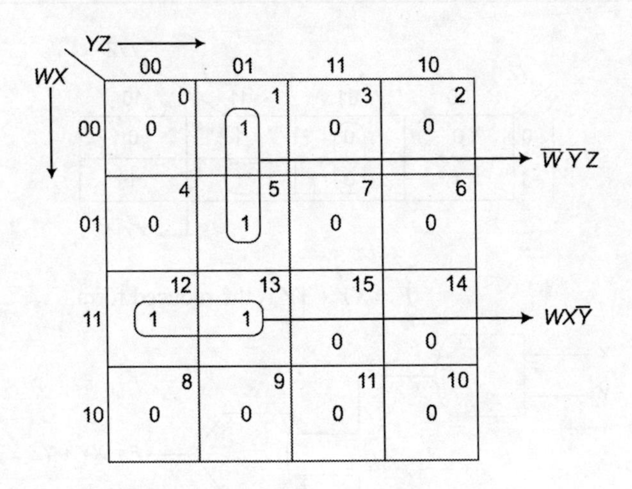

$f = W X \overline{Y} + \overline{W}\,\overline{Y} Z$ is the minimised form

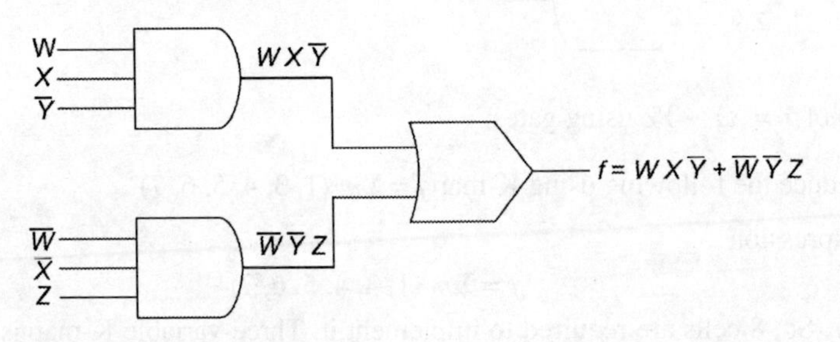

Design of $f = W X \overline{Y} + \overline{W}\,\overline{Y} Z$ using gates.

Example 6.9 Draw the truth table and design the function f with gates.

$$f(A, B, C, D) = (4, 5, 10, 11, 14, 15)$$

Solution The function $f(A, B, C, D)$ requires a four-variable K-map. The truth table's output is 1 for (4, 5, 10, 11, 14 and 15) and 0 for all other inputs.

A	B	C	D	f
0	0	0	0	0
0	0	0	1	0
0	0	1	0	0
0	0	1	1	0
0	1	0	0	1
0	1	0	1	1
0	1	1	0	0
0	1	1	1	0
1	0	0	0	0
1	0	0	1	0

Truth table

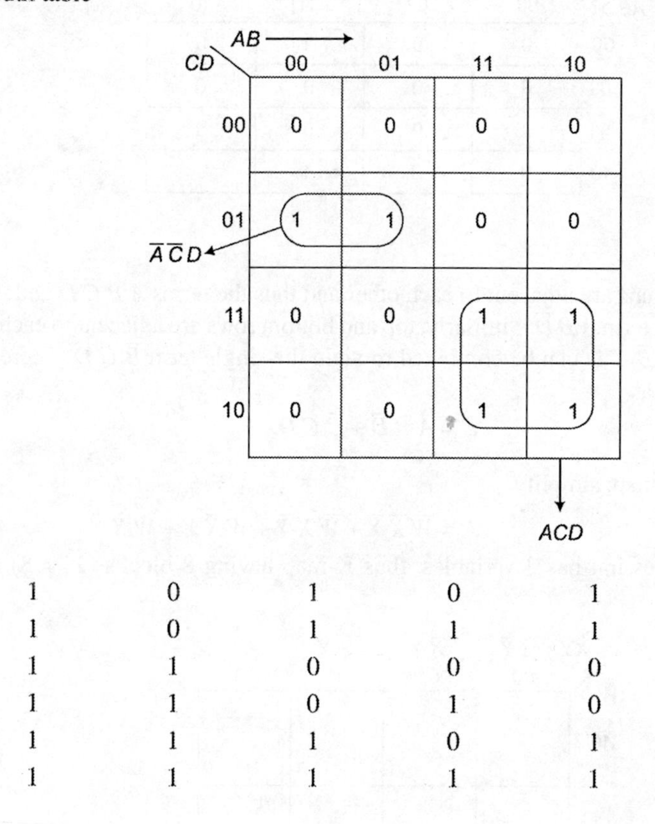

$f = AC + \overline{A}\,\overline{C}\,D$ is minimised function

A
C
AC
$F = AC + \overline{A}\,\overline{C}\,D$
$\overline{A}$
$\overline{C}$
$\overline{D}$
$\overline{A}\,\overline{C}\,D$

Design with gates.

Example 6.10 Using K-map simplify the Boolean expression

$$E = \overline{A}\,\overline{B}\,C\,D + A\,B\,\overline{C}\,\overline{D} + A\,B\,C\,\overline{D} + A\,\overline{B}\,C\,D$$

Solution The expression contains four variables and thus can be mapped into a K-map of $2^4 = 16$ squares as shown below:

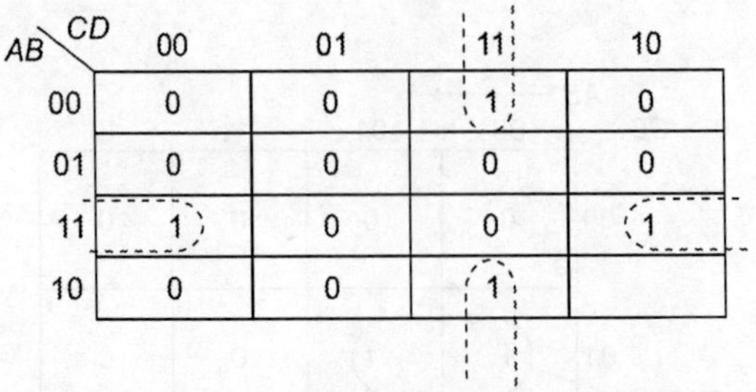

AB \ CD	00	01	11	10
00	0	0	1	0
01	0	0	0	0
11	1	0	0	1
10	0	0	1	

The left and right columns are adjacent to each other and thus the terms $A\,B\,\overline{C}\,\overline{D}$ and $A\,B\,C\,\overline{D}$ can be combined to yield a single term $A\,B\,\overline{D}$. Similarly, top and bottom rows are adjacent to each other and thus the terms $\overline{A}\,\overline{B}\,C\,D$ and $A\,\overline{B}\,C\,D$ can be combined to yield the single term $\overline{B}\,C\,D$. Therefore, the given expression can be written as

$$E = A\,B\,\overline{D} + \overline{B}\,C\,D.$$

Example 6.11 Using K-map, simplify

$$A = \overline{W}\,\overline{X}\,\overline{Y} + \overline{W}\,X\,\overline{Y} + W\,\overline{X}\,\overline{Y} + W\,X\,\overline{Y}$$

Solution

Step I The above expression has 3 variables, thus K-map having 8 blocks ($2^3 = 8$) can be used to represent it.

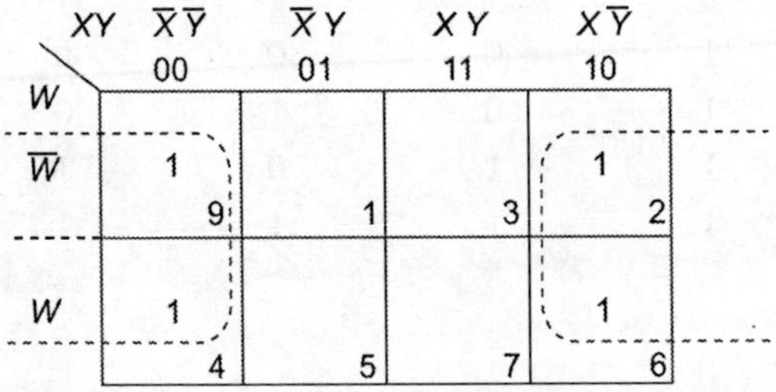

W \ XY	$\overline{X}\,\overline{Y}$ 00	$\overline{X}\,Y$ 01	$X\,Y$ 11	$X\,\overline{Y}$ 10
$\overline{W}$	1 (9)	(1)	(3)	1 (2)
W	1 (4)	(5)	(7)	1 (6)

Step II Writing down the 1's (SOP form, minterms) in the blocks satisfying the above expression i.e. $\overline{W}\,\overline{X}\,\overline{Y}$, block numbered 0, will be assigned (1) and so on.

Step III Grouping the 1's: In the above map, a single quad can be formed, in a way represented by dotted lines.

Step IV Eliminating the fluctuating terms within the quad, we get simplified expression as $A = \overline{Y}$ (simplified expression for the given expression).

EXERCISES

1. Draw a truth table (three-variable) that represents the Boolean expression $A\overline{B}\overline{C} + \overline{B}\overline{C} + \overline{A}\overline{B} = Y$.
2. Describe the various steps for simplifying a Boolean expression using K-map.
3. Use a K-map to simplify the Boolean expression $ABC + \overline{A}\overline{B}\overline{C} + \overline{A}\overline{B}C + A\overline{B}\overline{C} = Y$. Write the simplified Boolean expression in the minterm form.

4. Use K-map to simplify the Boolean expression

$$P\overline{Q}\,\overline{R}\,\overline{S} + P\overline{Q}RS + \overline{P}\,\overline{Q}RS + PQRS = Z$$

5. Minimize the Boolean function

$$f(A, B, C) = \Sigma\, 0, 1, 5 + \Sigma\, 2, 6$$

using the mapping method in both minimized SOP and POS forms.

6. Use a K-map to simplify the Boolean expression

$$Y = \overline{A}\,\overline{B}\,C + AB\overline{C} + \overline{A}\,\overline{B}\,\overline{C} + A\overline{B}C + ABC$$

7. Write the simplified Boolean expression given by K-map

CD \ AB				
	1	1		1
	1	1		1
			1	1

8. $\overline{A} \cdot B + C \cdot D$ is a simplified Boolean expression of the expression $ABCD + \overline{A}\,\overline{B}CD + \overline{A}B$. Determine if there are any "don't care" entries.
9. Write a simplified maxterm Boolean expression for π 0, 4, 5, 6, 7, 10, 14 using a K-Map.
10. Simplify the Boolean functions using K-map method

$$f(A, B, C, D, E, F) = \Sigma\,(6, 9, 13, 18, 19, 25, 26, 27, 29, 41, 45, 57)$$

11. Use K-map to simplify the given function

$$f(A, B, C, D, E, F, G) = \Sigma\,(20, 21, 22, 23, 28, 29, 52, 53, 60, 61)$$

12. Simplify the Boolean function $f(X, Y, Z) = \overline{X}\,\overline{Z} + YZ$ for don't care conditions expressed as $\overline{X}\,\overline{Y}Z + XY\overline{Z} + X\overline{Y}$.
13. Simplify the function given by $f(P, Q, R) = (P + Q + R)\,(\overline{P} + Q + \overline{R})\,(P + \overline{Q} + R)$ for "don't care" condition expressed as $(\overline{P} + \overline{Q}) \cdot (\overline{P} + Q + R)$.
14. Draw the K-maps for 2 input OR and XOR gates.
15. Draw the K-maps for 2 input NAND and NOR gates.
16. Draw the truth table and K-Map for $Y = AB + \overline{AB}$.
17. Draw the K-map for $Y = A\overline{B} + AB$ and then use it to obtain a minimised expression.
18. Draw K-map for the function
    (a) $Y = AB + \overline{A}\,\overline{C}$
    (b) $Y = ABC + \overline{A}B\overline{C}$

19. What expression is represented by a Karnaugh map?
20. Simplify the Boolean function $F(u, v, w, x) = \Sigma\ (0, 1, 2, 5, 7, 10, 11, 15)$
21. Simplify in POS & SOP $F(A, B, C, D) = \Sigma\ (1, 2, 5, 9, 10)$

CD \ AB				
			1	1
	1	1	1	1
			1	1
			1	1

7

Sequential Logic Circuits

INTRODUCTION

Digital circuits include sequential and combinational logic. Sequential circuits are those in which the output(s) of the logic circuit not only depends upon the inputs applied to the circuit but also on the past output(s). Flip-flops, binary counters, shift registers, ring counters and ring oscillators come under sequential logic circuits.

7.1 SEQUENTIAL CIRCUITS

Logic circuits are two types: "combinational" and "sequential". A combinational circuit is one whose output(s) depends only on the function of its inputs. A sequential circuit is one whose output(s) depends not only on the function of its inputs but also on the past sequence of outputs. The concept of combination and sequential logic design can be easily understood by a heater fan speed selection knob of an older car and a newer car. In older cars the output of fan speed depends only on the position of the knob either fan speed or heater whereas in the new cars the output depends on an arbitrarily long sequence of up/down pushes, beginning when the heater is turned on. The older car is a combinational design based whereas the new car is based on sequential logic circuit.

As discussed in the above example, the output of the newer car fan control depends upon the long sequences of up/down pushes. Therefore, it is necessary to make a state table which lists the output for the various input states to design a sequential logic circuit.

We can determine the output of fan speed control circuit if the current state is known. The state of a sequential circuit is defined as the collection of state variables whose values at any one time contain all the information about the past.

In the example of a newer car fan speed control, fan speed is the current state. A three-speed fan might be stored, two-binary state variables representing a decimal number between 0 and 3, with 0 corresponding to "OFF" and 3 to the highest speed. If we know the current state we can predict the next state as the function of inputs.

In the previous chapter a combinational logic circuit was defined whose output(s) is strictly a function of its inputs. A *sequential circuit* is a logic circuit whose output(s) is a function of its input(s) and also its internal state(s). The (internal) state of a sequential logic circuit is either a logic 0 or a logic 1, and because sequential circuits are able to maintain a state, it is also called a *memory circuit.* Generally, sequential circuits have two outputs, one of which is the complement of the other. Sequential circuits may or may not include logic gates.

Flip-flops, also known as *bistable multivibrators*, are electronic circuits with two stable outputs one of which is the complement of the other. The outputs will change only when directed by an input command.

There are 4 types of flip-flops and these are listed below:

1. Set-Reset (SR) or Latch
2. D Type (Data or Delay)
3. JK
4. T (Toggle)

7.2 SET-RESET (SR) FLIP-FLOP

A basic *Set-Reset (SR) flip-flop* constructed with two NAND gates is shown in Fig. 7.1. The figure shows the symbol for the SR flip-flop where *S* stands for Set and *R* stands for reset.

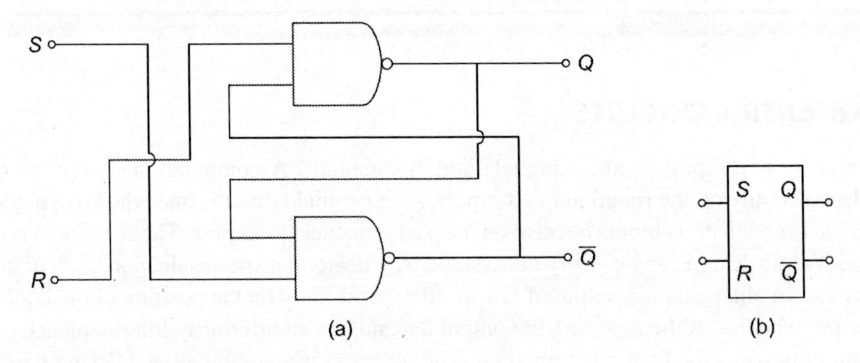

Fig. 7.1 Design and symbol for the SR flip-flop.

For a 2-input NAND gate the output is logic 0 when both inputs are high (1) and the output is logic 1 otherwise. *Present state* of the flip-flop is represented by Q_n and the next state as Q_{n+1}. Figure 7.1(a) shows the behaviour of SR flip-flop. The state table of SR flip-flop is shown in Table 7.1.

TABLE 7.1 State table for the SR flip-flop with NAND gates

Inputs		*Outputs*			
Set (S)	*Reset (R)*	*Present State (Q_n)*	*Next State (Q_{n+1})*	*Behaviour of Flip-Flop*	
0	0	0	1	But $\overline{Q_{n+1}} = 1$ also	The condition where
0	0	1	1	But $\overline{Q_{n+1}} = 1$ also	$S = R = 0$ must be avoided
0	1	0	0	No change	
0	1	1	0	Reset (or clear)	
1	0	0	1	Set	
1	0	1	1	No change	
1	1	0	0	No change	
1	1	1	1	No change	

Operation of SR flip-flop

1. ***Invalid State:*** As shown in the state table when both inputs S and R are at logic 0 simultaneously, both outputs Q and $\overline{Q}$ are at logic 1, which is an invalid condition, since Q and $\overline{Q}$ must be complements of each other. Therefore, the condition when $S = R = 0$ must be avoided during flip-flop operation with NAND gates.
2. ***Reset:*** When $R = 1$ and $S = 0$, the next state output Q_{n+1} becomes logic 0 irrespective of the previous state Q_n which is known as the reset state or clear conditions of flip-flop.
3. ***Set:*** When $R = 0$ and $S = 1$, the next state output Q_{n+1} becomes logic 1 irrespective of the previous state Q_n which is known as the preset state or set condition of the flip-flop.
4. ***No Change:*** When $R = 1$ and $S = 1$, the next state output Q_{n+1} remains the same as the present state, i.e., there is no state change.

7.3 CLOCKED SET-RESET (SR) FLIP-FLOP

Digital circuitry uses synchronous circuits rather than asynchronous circuits. Synchronous circuits are those circuits which are synchronized with the same clock pulse. Figure 7.1 shows an asynchronous flip-flop in which the inputs and the present state determine the next state. It is practical to synchronize the inputs of flip-flop with an external lock pulse and two AND gates before the asynchronous flip-flop circuit. With this modification, the circuit behaves as synchronous SR flip-flop, in which CLK refers to external clock pulse.

SR clocked flip-flop can be constructed by using NAND gates or NOR gates. Implementation of SR clocked flip-flop using NAND gates is shown in Fig. 7.2.

*SET and CLEAR

Most flip-flops are provided with a *set direct* (SD) and a *reset direct* (RD) commands which are asynchronous inputs, that is, they function independently of the clock pulse. Figure 7.3 shows the symbol of an SR flip-flop with set direct, reset direct, and clock pulse (CP) indicated by the small triangle. These inputs are known as SET and CLEAR.

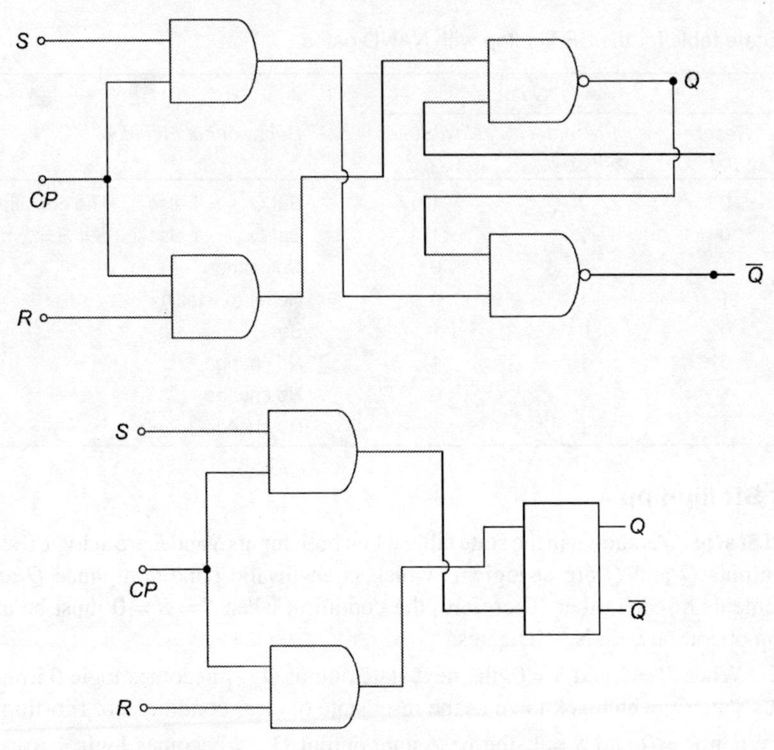

Fig. 7.2 Synchronous SR flip-flop.

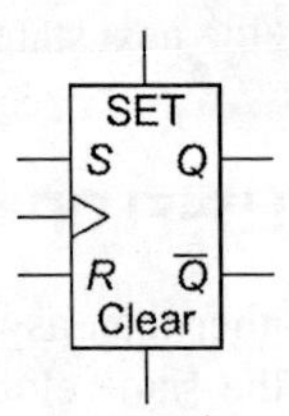

Fig. 7.3 Symbol for a typical flip-flop with clock pulse, set direct, and reset direct inputs.

We should remember that the flip-flop inputs are used to tell the flip-flop what to do, whereas the clock pulse (CLK) tells the flip-flop when to do it, and the set and clear commands bypass the CLK command.

7.4 DELAY (D) FLIP-FLOP

The D flip-flop is a clocked flip-flop which is a modification of SR flip-flop. This flip-flop can be constructed either with NAND or NOR gates. This is basically known as *D* (delay) flip-flop. The state table of D flip-flop shown in Fig. 7.4 shows that the output i.e., next state Q_{n+1} is same as the input *D*. The state of this flip-flop after the clock pulse is same as before the clock pulse. If $D = 1$, before the clock pulse and the present state $Q_{n=1}$. D flip-flop considers the same (previous state) after applying clock pulse.

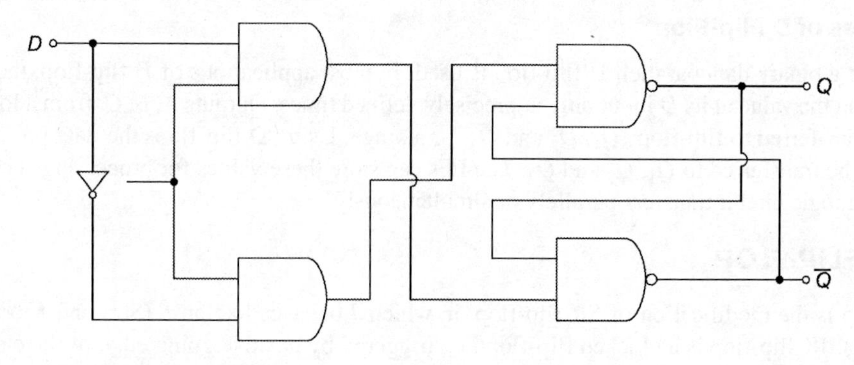

Fig. 7.4 Construction of a *D* flip-flop.

Symbol of D flip-flop is shown in Fig. 7.5, in which a small triangle inside the rectangle shows that this type of flip-flop is triggered during the leading edge of the clock pulse, which is shown in Table 7.2.

TABLE 7.2 Characteristic table for the D flip-flop with NAND gates

CLK	*Input*	*Output*
	D	*Q*
↑	0	0
↑	1	1

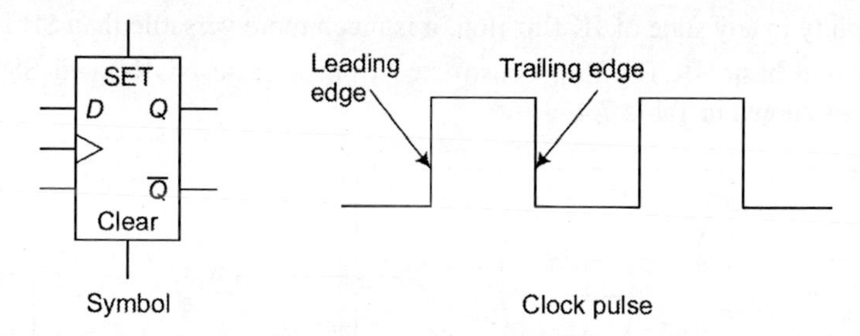

Fig. 7.5 D flip-flop symbol activated during the leading edge of the clock pulse.

The operation of D flip-flop is as below.

1. Thus, if *D* is 1 and clock pulse is high, the output goes to 1 irrespective of what it is at the present state.

   The purpose of DFF is to provide delay in the input and output, but the output is same as the input applied.

2. Similarly, if *D* is 0 and clock pulse is high the output becomes 0.

Device SN7474 is a D flip-flop with asynchronous set and clear commands.

Usefulness of D Flip-Flop

To transmit a binary data parallel, D flip-flop is used. In most applications of D flip-flop, the output Q must take on the value at its D input only at precisely defined times. Outputs A, B, C, from a logic circuit are to be transferred to flip-flops Q_1, Q_2 and Q_3 for storage. Using D flip-flops the data present at A, B and C will be transferred to Q_1, Q_2 and Q_3. The FFs can store these values for processing. This way the data from a logic circuit transfers parallely or simultaneously.

7.5 JK FLIP-FLOP

JK flip-flop is the modification of SR flip-flop, in which J behaves like an S (Set) and K behaves like an R (reset). JK flip-flop is a clocked flip-flop i.e., triggered by positive going edge of the clock signal. For any combination of the input J and K produces a valid output, which is important characteristic of JK flip-flop, which is not possible in SR flip-flop. We recall the $S = R = 0$ state of SR flip-flop (NAND gated) and $S = R = 1$ state of SR flip-flop (NOR gated), which provides invalid outputs and must be avoided. In JK flip-flop, it is perfectly valid to have $J = K = 0$ or $J = K = 1$ simultaneously. Table 7.4 shows the state table of JK flip-flop and Fig. 7.6 shows the construction and symbol of JK flip-flop. The working of JK flip-flop is given below:

1. When $J = K = 0$, the output is assumed to be 1
2. When the positive going edge of the first clock pulse exhibits, i.e. $J = 0$ and $K = 1$, flip-flop will be cleared or reset.
3. On the occurrence of the 2nd clock pulse when $J = 1$ and $K = 0$, flip flop will observe set state.
4. When $J = K = 1$ exists, this flip-flop exhibits the valid state i.e., it complements the present state output.

Due to unambiguity in any state of JK flip-flop, it is much more versatile than SR flip-flop.

Figure 7.6 shows a basic JK flip-flop constructed from a basic NOR-gated SR flip-flop and its characteristic table is shown in Table 7.4.

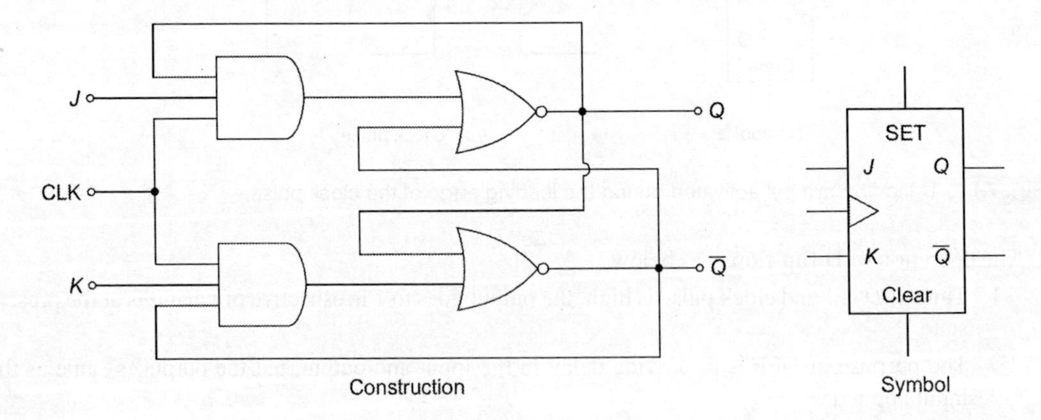

Fig. 7.6 Construction of a JK flip-flop and its symbol.

TABLE 7.3 State table for the JK flip-flop with NOR gates

Inputs		*Present state*	*Next state*	
J	*K*	Q_n	Q_{n+1}	
0	0	0	0	No change
0	0	1	1	No change
0	1	0	0	No change
0	1	1	0	Reset (or clear)
1	0	0	1	Set
1	0	1	1	No change
1	**1**	**0**	**0**	**Toggle (State change)**
1	**1**	**1**	**1**	**Toggle (State change)**

In JK flip-flop, malfunctioning of output will occur, if the clock pulse has a long time duration. This is because the feedback connection will cause output to change continuously whenever the clock pulse remains high while also at the same time, we have $J = K = 1$. To avoid this undesirable operation, the clock pulses must have a time duration, shorter than the propagation delay through flip flop. This effect of malfunctioning is eliminated by using master slave or edge triggered JK flip-flop.

7.6 TOGGLE (T) FLIP-FLOP

T flip-flop derives from the toggle function of the flip-flop. The toggle means a state change when the T input is ASSERTED. The T flip-flop is a single input JK flip-flop, which is obtained from basic JK flip-flop by connecting the *J* and *K* inputs together, which is designated as T. The state table of T flip-flop is shown in Table 7.5. Figure 7.7 shows the construction and symbol of T flip-flop.

T flip-flops are most widely used in counters, shift registers and frequency drivers.

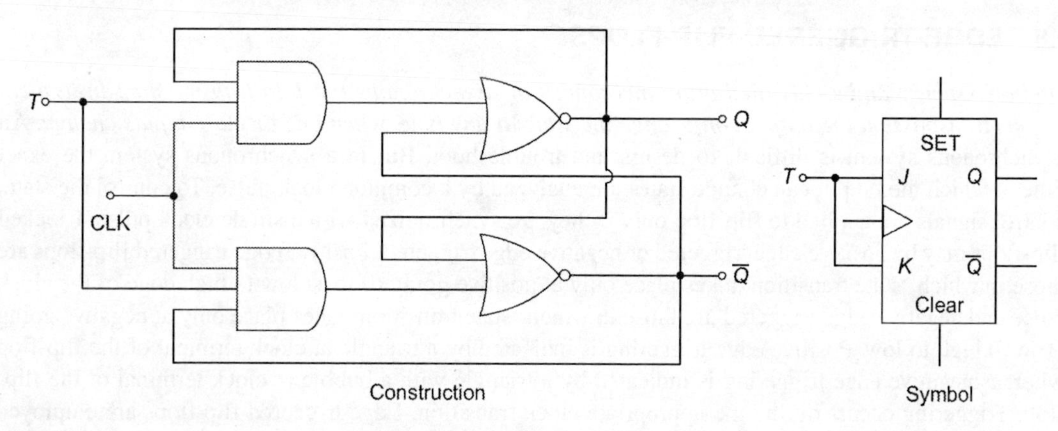

Fig. 7.7 Construction of a T flip-flop and its symbol.

TABLE 7.4 Characteristic table for the T flip-flop with NOR gates

Input T	*Present state* Q_n	*Next state* Q_{n+1}
0	0	0
0	1	1
1	0	1
1	1	0

Table 7.4 indicates that the output state changes whenever the input T is logic 1 and the clock pulse is high.

7.7 FLIP-FLOP TRIGGERING

In the previous section, we mentioned that a timing problem exists with the basic JK flip-flop. This problem can be eliminated by making the flip-flop sensitive to pulse transition. Figure 7.8 shows a *positive pulse* and a *negative pulse* transition points of a clock pulse.

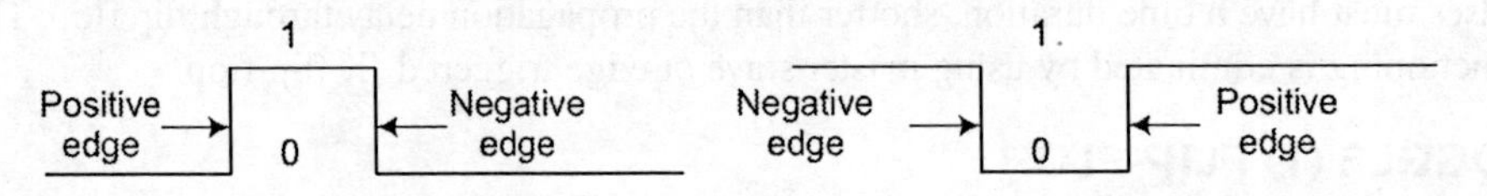

Fig. 7.10 Positive and negative pulses.

A positive pulse is a waveform in which the normal state is logic 0 and changes to logic 1. Whereas, a negative pulse is a waveform in which the normal state is 1. In either case, the positive edge is the transition from 0 to 1 and the negative edge is the transition from 1 to 0.

7.8 EDGE-TRIGGERED FLIP-FLOPS

Digital systems can be classified into synchronous or asynchronous for data transmission. In asynchronous transmission the output can change the state at any time, when one or more inputs change. An asynchronous system is difficult to design and troubleshoot. But in a synchronous system the exact time at which the output can change states are analyzed by a common clock pulse. To control the state, control signals are applied to flip-flop only if they are synchronized with a single clock pulse. Clocked flip-flops may be positive edge triggered or negative edge triggered. Positive edge triggered flip-flops are those in which 'state transition' takes place only at positive going (0 to 1) low to high edge of the clock pulse and negative edge triggered are those in which 'state transition' takes place only at negative going (1 to 0) high to low. Positive edge triggering is indicated by a triangle at clock terminal of the flip-flop whereas negative edge triggering is indicated by a triangle with a bubble at clock terminal of the flip-flop. Triggering occurs during the appropriate clock transition. Edge triggered flip-flops are employed in applications where incoming data may be random. The SN74LS74 IC device shown in Fig. 7.9 is a positive edge triggered D type flip flop and the timing diagram of Fig. 7.10 shows that the output Q goes from low to high or from high to low at the positive edge of the clock pulse. Three basic FFs are available in edge-triggering SR, JK & D. JK and D FF are readily available in IC form whereas SR FF is not readily available in IC form.

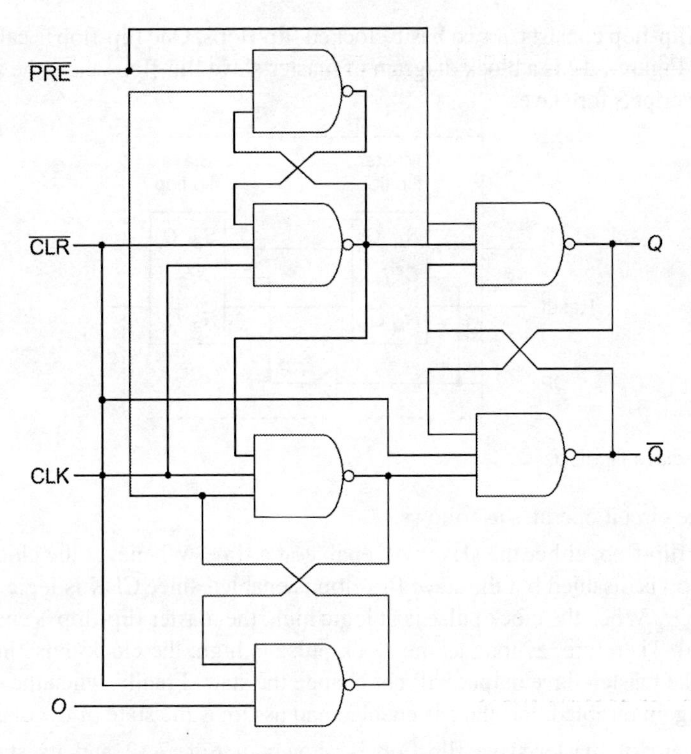

Fig. 7.9 The SN74LS74 positive edge triggered D-type flip-flop with preset and clear.

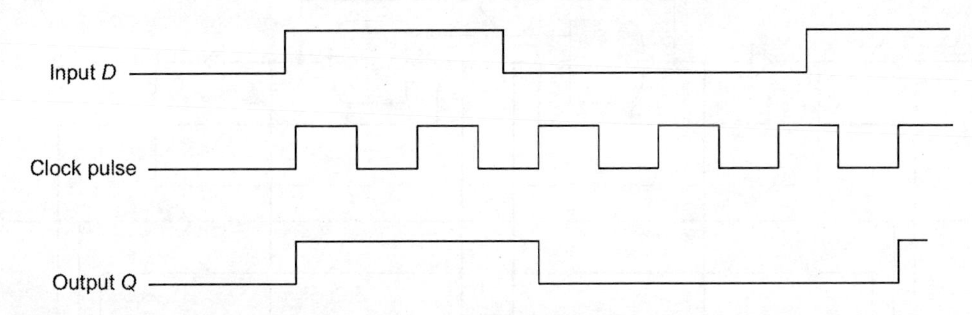

Fig. 7.10 Timing diagram for positive edge triggered D flip-flop.

7.9 MASTER/SLAVE FLIP-FLOPS

The problem in JK flip-flop, i.e. malfunctioning of output will occur, if the clock pulse has a long time duration. This is because the feedback connection will cause output to change continuously whenever the clock pulse remains high while also at the same time, we have $J = K = 1$ which is described in the previous section. To avoid this undesirable operation, the clock pulses must have a time duration, shorter than the propagation delay through flip-flop. A master-slave is the remedy to avoid the malfunctioning of output.

A master-slave flip-flop consists of two basic clocked flip-flops. One flip-flop is called the *master* and the other the *slave*. Figure 7.12 is a block diagram of master-slave flip-flop where the subscript M stands for master and subscript S for slave.

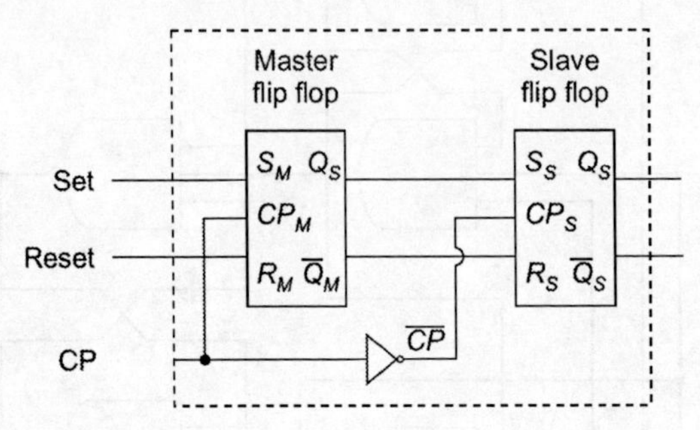

Fig. 7.11 Master-slave flip-flop.

The master-slave circuit operates as follows:

In master-slave flip-flop, either the slave will enable at a time. Whenever the clock pulse is at logic 0 the master flip-flop is disabled but the slave flip-flop is enabled since $\overline{\text{CLK}}$ is logic 1. Therefore, $Q_S = Q_M$, and also $\overline{Q}_S = \overline{Q}_M$. When the clock pulse is at logic high, the master flip-flop is enabled and the slave flip-flop is disabled. Therefore, as long as the clock pulse is high, the clock pulse input to the slave is disabled and thus the master-slave output will not change the state. Finally, when the clock pulse returns to 0 the master is again disabled, the slave is enabled and assumes the state of the master.

The logic diagram of master-slave flip-flop is shown in Fig. 7.12 and its symbol is shown in Fig. 7.13.

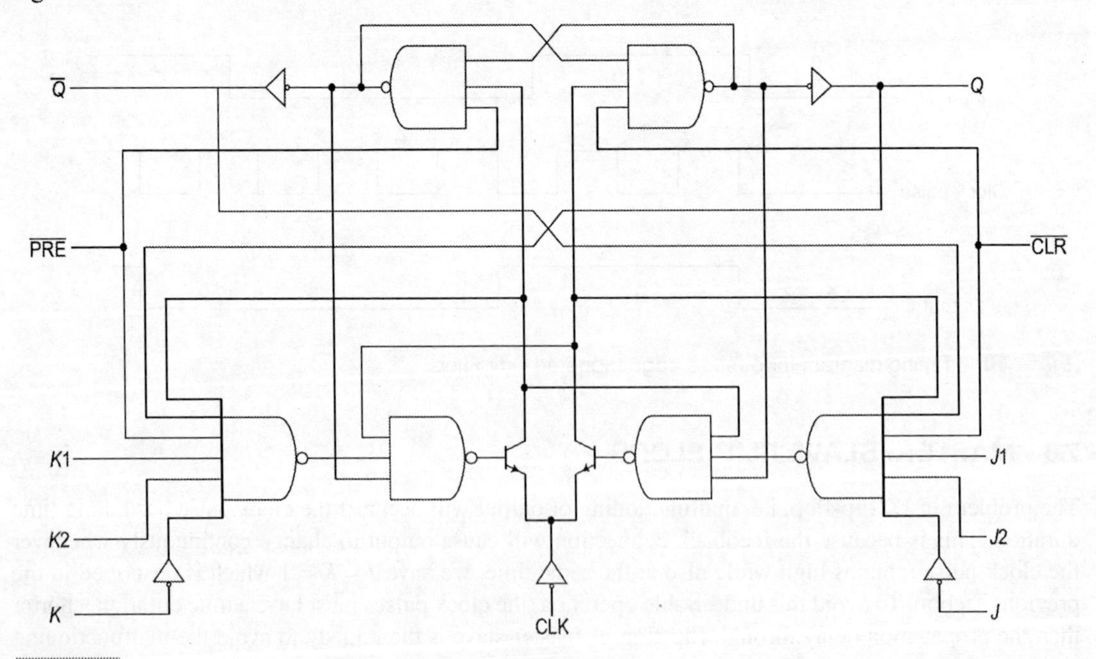

Fig. 7.12 Logic diagram for the JK master-slave flip-flop.

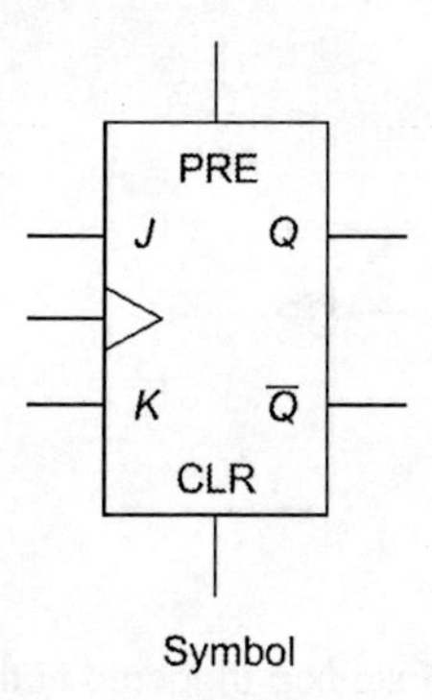

Fig. 7.13 Symbol for the JK master-slave flip-flop triggered at the positive edge of the clock pulse.

Logic diagram and the symbol of the master-slave JK flip-flop triggered at the negative edge are shown in Figs. 7.14 and 7.15 respectively.

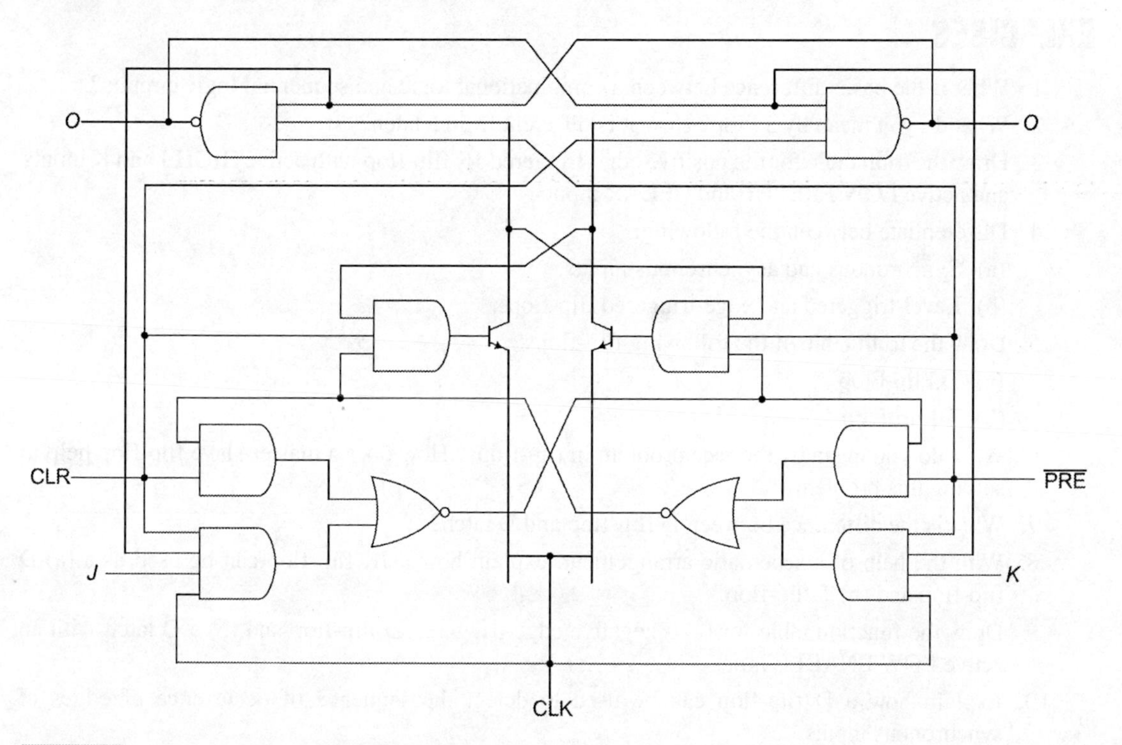

Fig. 7.14 Logic diagram for the master-slave JK flip-flop triggered at the negative edge.

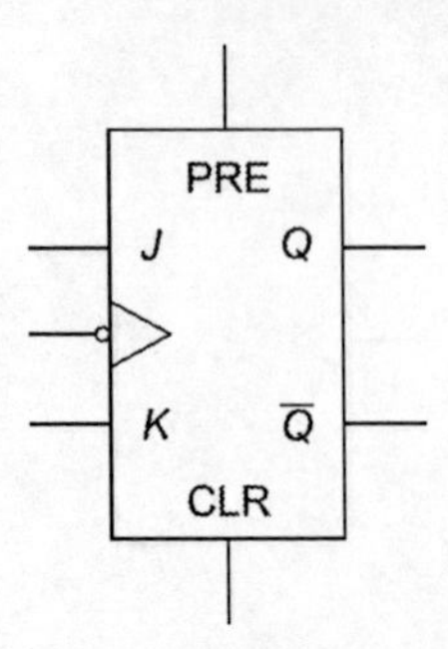

Fig. 7.15 Symbol for the master-slave JK flip-flop triggered at the negative edge of the clock pulse.

The positive edge triggered JK flip-flop is preferable in applications where the incoming data is not synchronized with the clock whereas the *negative edge triggered master-slave* JK flip-flop is more suitable for synchronous operations like synchronous counters.

EXERCISES

1. What is the basic difference between a combinational logic and sequential logic circuits?
2. What do you mean by a Flop? How it is different from a latch?
3. Draw the truth table for the positive edge-triggered JK flip-flop with active HIGH J and K inputs and active LOW PRESET and CLEAR inputs
4. Differentiate between the following:
   (a) Synchronous and asynchronous inputs.
   (b) Level-triggered and edge-triggered flip-flops.
5. Draw the truth table of the following flip-flops:
   (a) D Flip-Flop
   (b) T Flip-Flop
6. What do you mean by the race problem in flip-flops? How does a master-slave flip-flop help in solving this problem?
7. What is the difference between D flip-flop and D latch?
8. With the help of a schematic arrangement, explain how a JK flip-flop can be used as a (a) D flip-flop and (b) T flip-flop.
9. Draw the function table for (a) a negative edge-triggered D flip-flop and (b) a D latch with an active LOW ENABLE input.
10. Explain how a D flip-flop can be used to detect the sequence of occurrence of edges of synchronous inputs.
11. What is a clocked JK flip-flop? What improvement does it have over a clocked *R-S* flip-flop?
12. Describe the logic implementation of RS flip-flop having active HIGH R and S inputs. Also draw its truth table.
13. Present the general form of a sequential logic circuit.
14. Why a sequential circuit is also known as memory circuit?

15. Give the design and symbol for SR flip-flop. Also show the state table for the SR flip-flop with NAND gates.
16. Define edge-triggered flip-flops in detail. Present the logic diagram and timing for positive edge triggered D flip flop.
17. Explain the operation of master-slave flip-flop along with its truth table, logic diagram and timing diagram.
18. Give the conversions for the following:
    (a) SR flip-flop to JK flip-flop
    (b) D flip-flop to JK flip-flop
    (c) T flip-flop to SR flip-flop
    (d) JK flip flop to SR flip-flop
19. How is a JK flip-flop made to toggle?
20. In case of master slave flip-flop, when is the master enabled?
21. Four positive edge-triggered D flip-flops are used to store a 4-bit binary number as shown below. Determine if the circuit is functioning properly, and if not, what might be wrong.

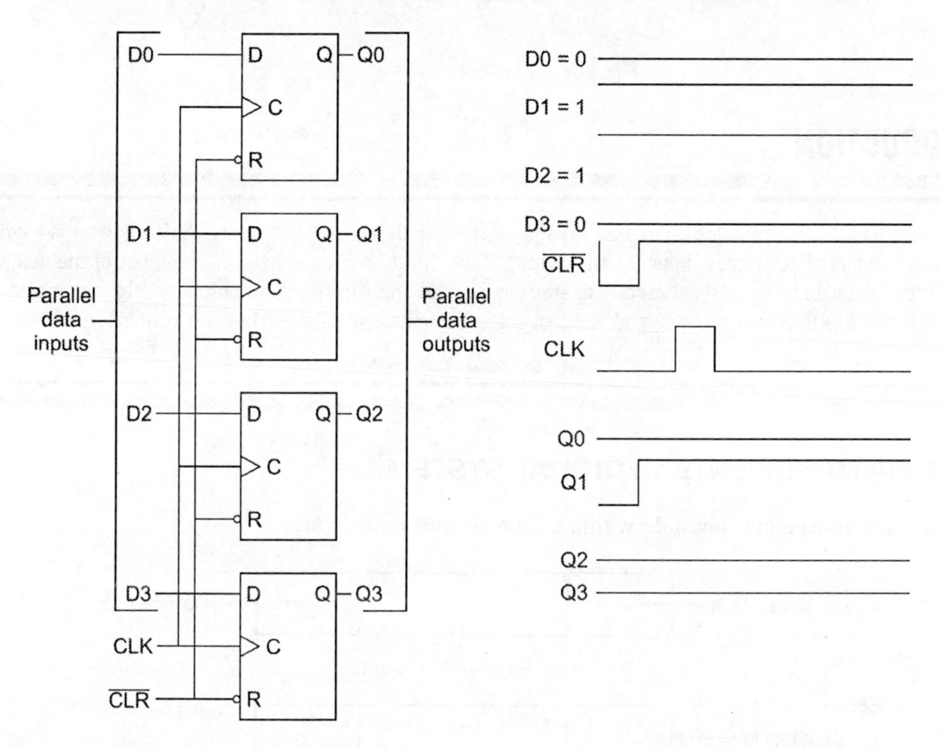

22. What are the advantages/drawbacks of JK flip-flop when compared to D flip-flop?
23. Why D flip-flop is preferred for circuits implementation?

8

Shift Registers

INTRODUCTION

Shift registers are sequential logic circuits, capable of shifting data one bit at a time. Shift register is a chain of flip-flops connected in cascade, in which the output of one flip-flop is connected to the input of the neighbour. The shift register behave as synchronous operations i.e., all the flip-flops are connected to a common clock pulse. All the flip-flops perform set and reset operations simultaneously as they are connected with a single clock pulse.

8.1 DATA MOVEMENTS IN DIGITAL SYSTEMS

The basic data movements possible within a four-bit shift register are:

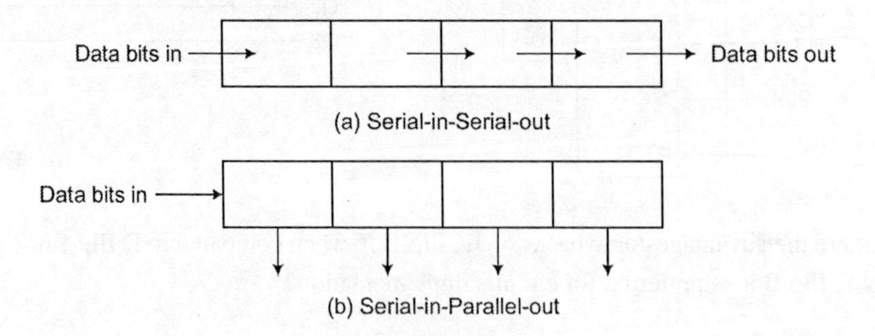

(a) Serial-in-Serial-out

(b) Serial-in-Parallel-out

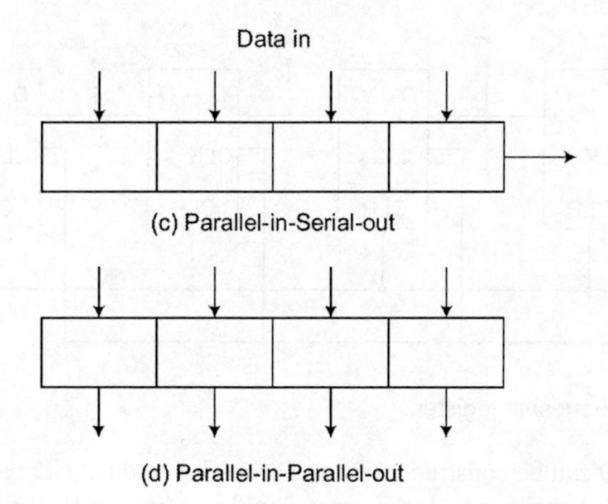

Fig. 8.1 Types of shift registers.

8.2 SERIAL-IN-SERIAL-OUT SHIFT REGISTER

A 4-bit serial-in-serial-out shift register constructed using D flip-flop is shown in Fig. 8.2. Assume a data word 1001 is to be transmitted through SISO. The LSB of the data has to be shifted through the register from FF0 to FF3. During each clock pulse, one bit is transmitted from left to right. The clocked behaviour of SISO shift register using D flip-flop is shown in Fig. 8.2.

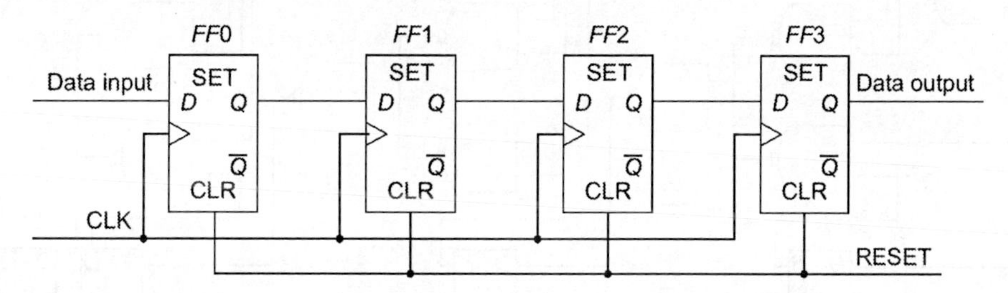

Fig. 8.2

8.3 SERIAL-IN-PARALLEL-OUT SHIFT REGISTER

A serial in parallel out shift register (SIPO) stores one bit at a time into a register. After the data is stored in the register, it can be read from the parallel output lines $Q_1 - Q_4$.

Reset (active low) line is connected with all flip-flops to clear the data before the incoming data. Once the data is stored, each bit appears on its respective output line, and all bits are available simultaneously. A construction of 4-bit SIPO shift register using D flip-flop is shown in Fig. 8.3.

8.4 PARALLEL-IN-SERIAL-OUT (PISO) SHIFT REGISTER

A 4-bit parallel in serial-out shift register using D flip-flop is shown in Fig. 8.4.

The data can be entered in parallel and can be taken out serially.

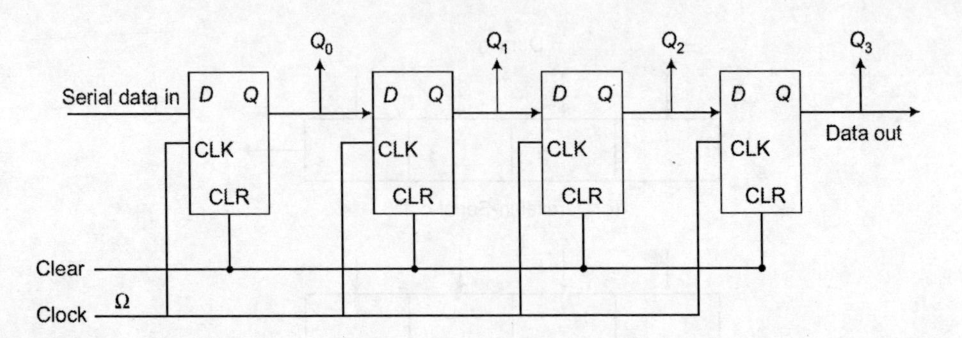

Fig. 8.3 Serial-in-parallel-out shift register.

In Fig. 8.4 below, PISO can be constructed using D flip-flop and NAND gates. In which D_0, D_1, D_2, and D_3 are the parallel inputs, where D_0 is the most significant bit and D_3 is the least significant bit. To input the data, and to shift the data the following procedure is used:

1. To input the data parallel, the control bit write/shift is taken to LOW and the data is clocked in.
2. To shift the data in serial fashion, the control bit is taken to HIGH, and the data can be taken out from the output data terminal of FF_4.

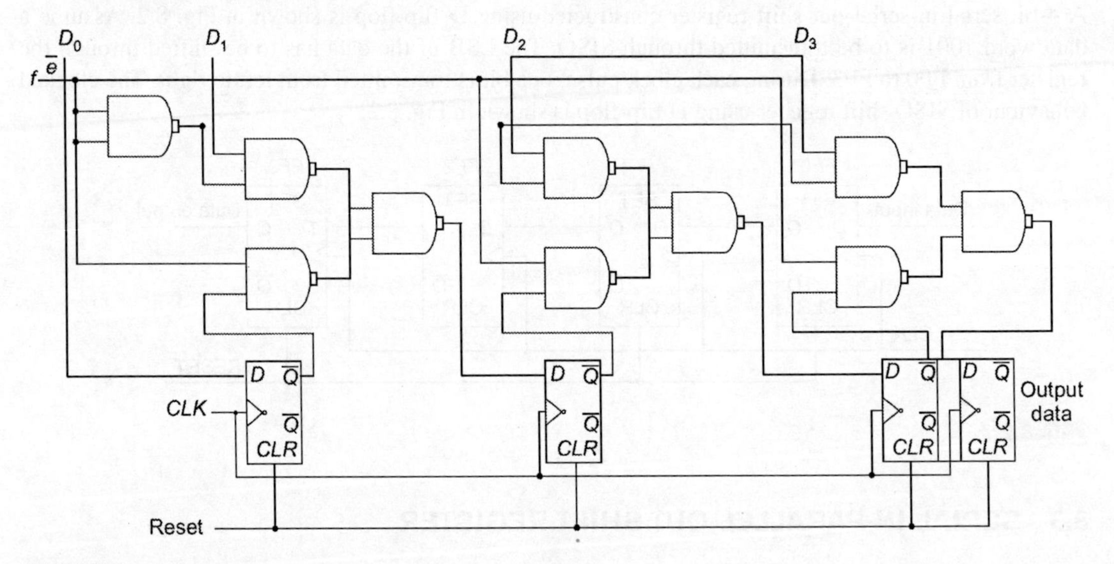

Fig. 8.4 Parallel-in serial-out shift register.

8.5 PARALLEL-IN-PARALLEL-OUT SHIFT REGISTER

In parallel-in-parallel-out shift registers, all data bits appear on parallel outputs simultaneously on the entry of data bits. D_1, D_2, D_3 and D_4 are the parallel inputs and Q_1, Q_2, Q_3 and Q_4 are the parallel outputs. Fig. 8.5 shows the P_1P_0 shift register using D flip flop.

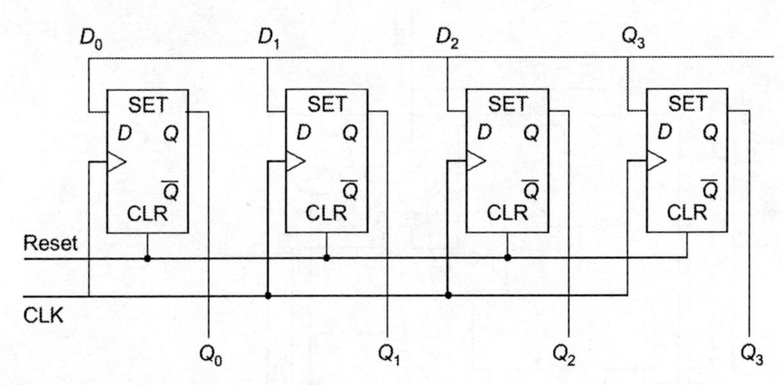

Fig. 8.5 Parallel-in parallel-out shift register.

8.6 BIDIRECTIONAL SHIFT REGISTER

A bidirectional, or reversible shift register is one in which the data can be shifted either to left or right. A bidirectional shift register is a universal shift register having capability of shift-left and shift-right, synchronous parallel and serial data transfer and is easily expanded for both serial and parallel operation.

In nutshell, the bidirectional shift register is designed to incorporate virtually all the features one wants in a shift register. The register has four distinct modes of operation, namely:

1. Parallel load
2. Right shift (in direction Q_A towards Q_D)
3. Left shift (in direction Q_D towards Q_A)
4. Inhibit clock (idle mode)

Synchronous i.e., parallel loading is accomplished by applying the four bits of data and both control inputs S_0 and S_1 high. The data is loaded into the associated flip-flops and appear at the output after the positive edge of clock pulse. Serial flow of data is inhibited during loading.

Right shift is accomplished synchronously with the rising edge of clock pulse when S_0 is high and S_1 is low. Serial data for this mode is entered at the shift-right data input. When S_0 is low and S_1 is high data shifts left synchronously and the new data is entered at the shift left serial input.

Flip-flop clocking is inhibited when both mode control inputs are low. A four-bit bidirectional shift register is shown in Fig. 8.6.

8.7 APPLICATIONS OF SHIFT REGISTERS

Shift registers are sequential logic circuits. These are designed using flip-flops. They can be found in many digital applications to produce time delay, to simplify combinational logic and to convert serial data to parallel data.

Time delay: The SISO shift register can be used as a time delay device. The amount of delay can be controlled by:

1. Number of stages in the register
2. Clock frequency.

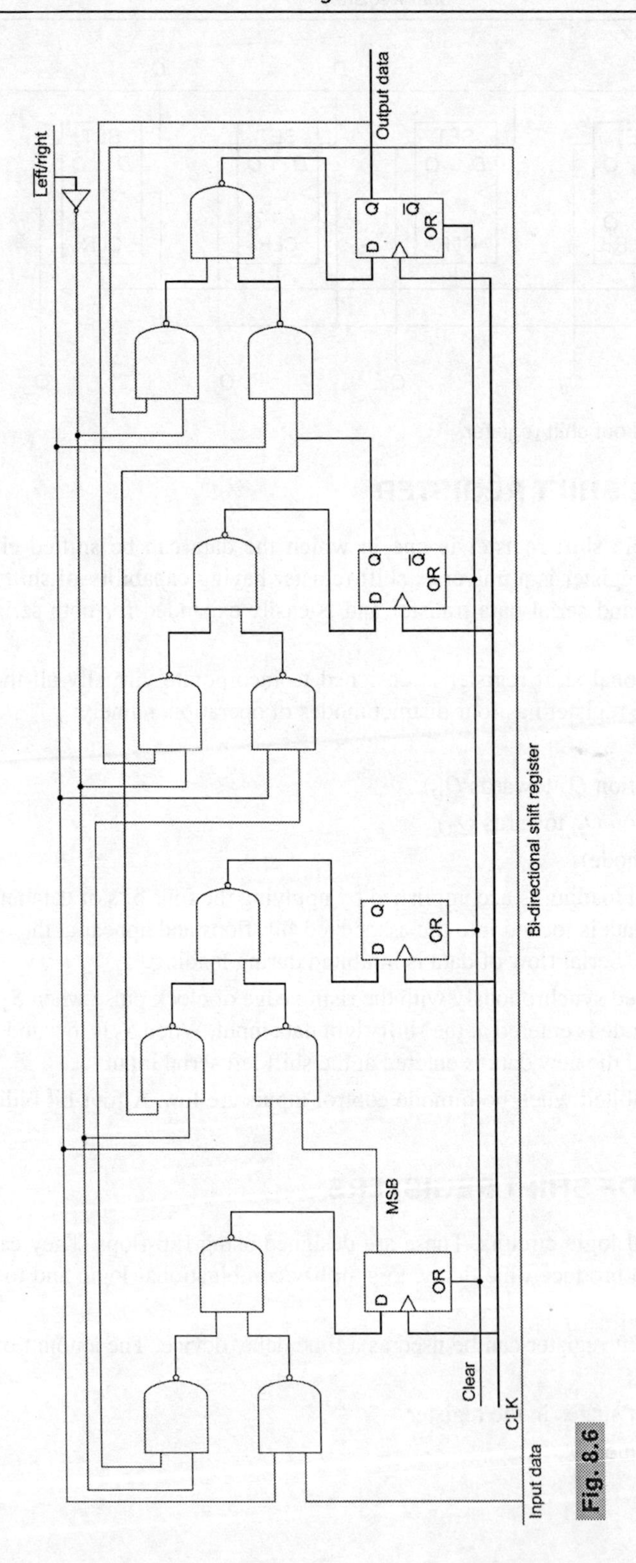
Left/right
Output data
D
Q
$\bar{Q}$
OR
MSB
Clear
CLK
Input data
Bi-directional shift register

Fig. 8.6

Simplification of Combinational Logic

Shift registers can be used to simplify the combinational logic circuit. A combinational circuit, where it is required to shift data parallel or serial, digital circuitry can be simplified using shift registers.

Conversion of Serial Data to Parallel Data

Modern computerised or microprocessor based system requires incoming data in parallel form. But frequently, these systems must communicate with external devices that send or receive serial data. So serial to parallel conversion is required which can be achieved by the shift registers (Refer to Fig. 8.3).

8.8 SHIFT REGISTER COUNTERS

(Ring and Johnson Counters)

Ring counters and Johnson counters are basically shift registers with serial outputs connected back to the serial inputs in order to produce particular sequences. A four-bit ring counter is shown in Fig. 8.8 and its state table is shown in Table 8.1.

For a ring counter shown in Fig. 8.7, it is assumed that flip-flops D_1 through D_3 are initially reset to 0 and flip-flop D_4 is preset to logic 1. It is compulsory that for any ring counter, atleast one flip-flop of the ring counter must be preset to 1 and at least one flip-flop must be reset to 0.

The ring counter shown Fig. 8.7 has 4 distinct states; the counter can be considered a mod-4 counter. The clocked behaviour of the ring counter can be easily understood from Table 8.1 and the timing diagram shown in Fig. 8.8.

A Johnson counter is a modified form of ring counter in which the complemented output of the last flip-flop is connected to the input of the first flip-flop.

A ring counter and Johnson counter are two shift register counters. They differ from each other in two respects. First, a Johnson counter may start with all flip-flops in the reset condition, whereas a ring counter requires at least one flip-flop to logic 1 and at least one flip-flop at logic 0. Second, a Johnson counter generates $2n$ timing signals with n flip-flops whereas a ring counter generates n timing signals with n flip-flop. If 8 timing signals are to be generated, one would require a ring counter with 8 flip-flops whereas a Johnson counter require 4 flip-flop. Johnson counter be realized using D flip-flops as shown in Figure 8.9 and the clocked behaviour of Johnson counter is shown in Fig. 8.10, Table 8.2.

Counters: As the name indicates, a counter is an important tool in digital circuits, which has the ability to count the pulses. Nowadays, we see the application of counters even in our home appliances such as microwave oven, washing machines, etc. We have studied flip-flops in the previous section, which is used to store one bit or one binary information. To store many binary informations a group of flip-flops is required which is known as register.

8.9 COUNTERS

A counter is a register, capable to count number of clock/pulses arrived at its clock input. The counters are most widely used to count the pulse in large number of control system design, computers, scientific equipment, healthcare equipment, etc. Counters can be classified based on its construction:

1. Clock: Asynchronous or synchronous
2. Clock Trigger: Positive edged or negative edged
3. Counts: Binary, decade
4. Count Direction: UP, Down, UP/Down
5. Flip-Flops: JK, T, D

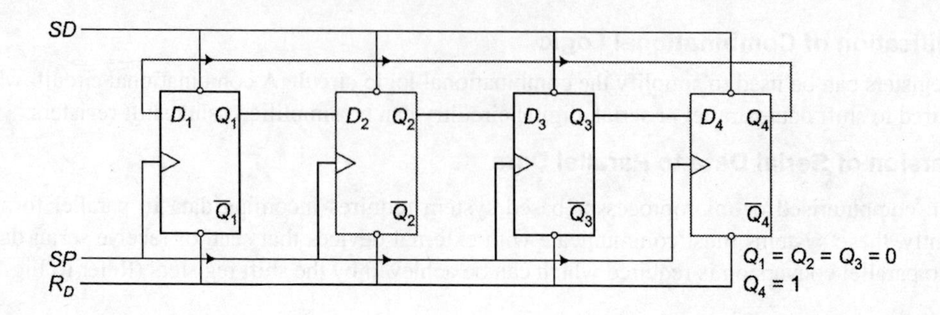

Fig. 8.7 Ring counter.

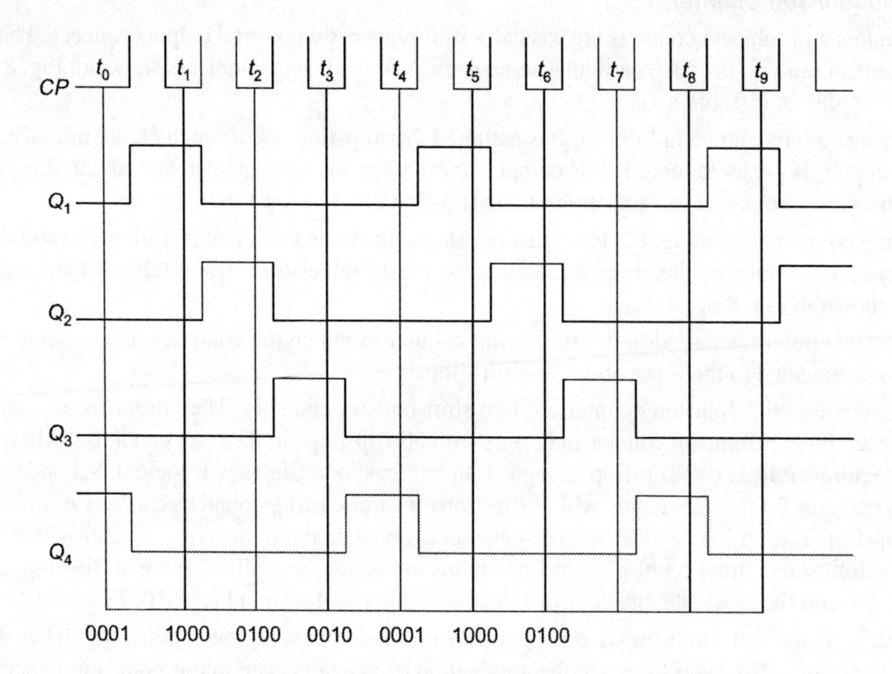

Fig. 8.8 Timing diagram of ring counter.

TABLE 8.1 Clocked behaviour of ring counter.

Time count		State			
		Q_1	Q_2	Q_3	Q_4
t_0	0	0	0	0	1
t_1	1	1	0	0	0
t_2	2	0	1	0	0
t_3	3	0	0	1	0
t_4	4	0	0	0	1
t_5	5	1	0	0	0
and so on					

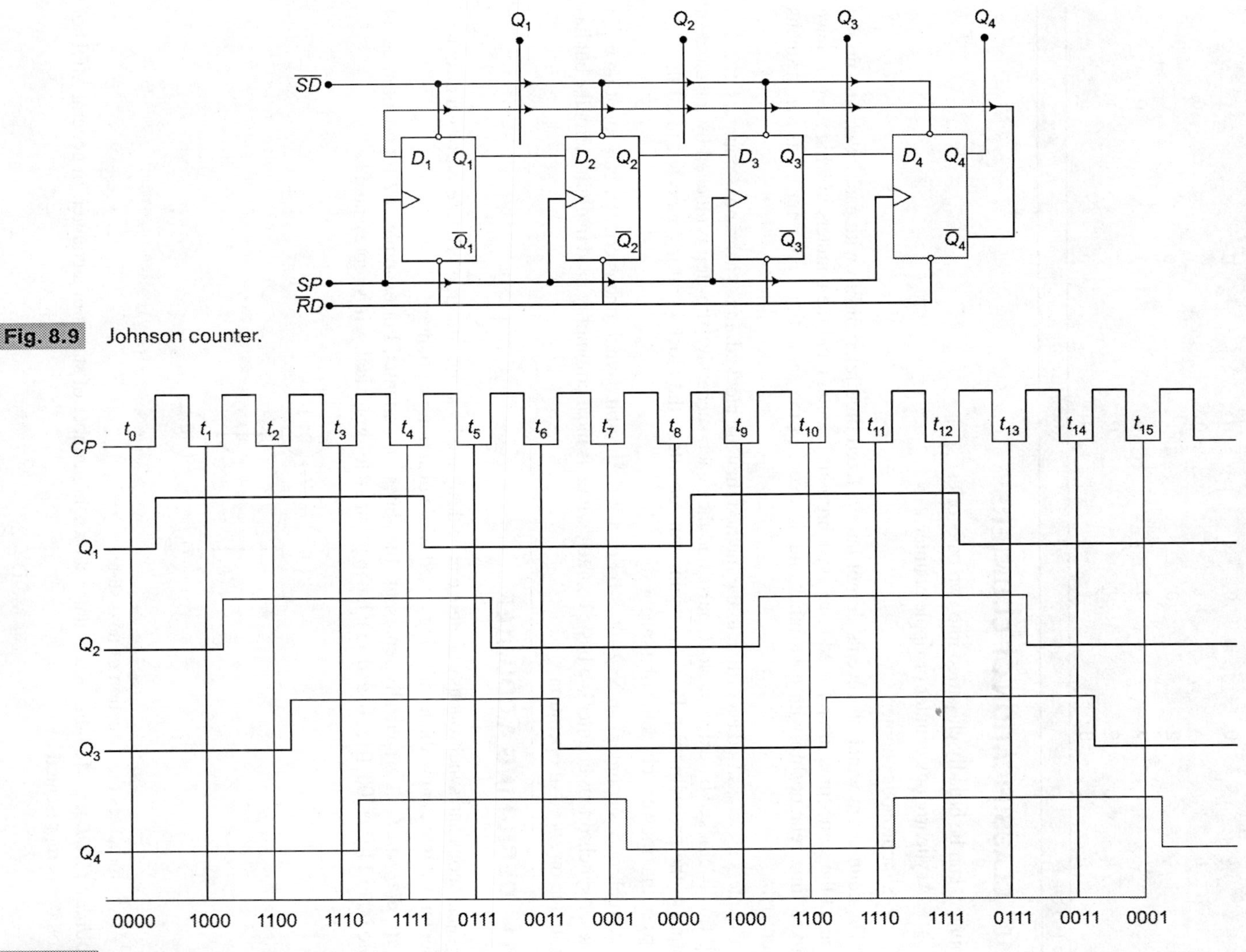

Fig. 8.9 Johnson counter.

Fig. 8.10 Clocked behaviour of Johnson counter.

Time count		State			
		Q_1	Q_2	Q_3	Q_4
t_0	0	0	0	0	1
t_1	1	1	0	0	0
t_2	2	0	1	0	0
t_3	3	0	0	1	0
t_4	4	0	0	0	1
t_5	5	1	0	0	0
and so on					

8.10 CLASSIFICATION OF COUNTERS

Counters can be broadly classified into two categories:

- Asynchronous counter or ripple counter
- Synchronous counter

The counters in which the events do not have a fixed time relationship with each other and do not occur at the same time, are referred to as asynchronous counters and the counters in which events have a fixed time relationship with each other and do occur at the same time, are referred as synchronous counters.

Counters are classified according to the method they are clocked. In asynchronous counters, the first flip-flop is clocked by external clock pulse and then each successive flip-flop is clocked by the clocked output of the preceding flip-flop. In synchronous counters, the clock input is connected to all the flip-flops so that they are clocked simultaneously.

Asynchronous counter can be 2-bit, 3-bit or 4-bit. An n-bit asynchronous counter can count up to $2n$ states. Asynchronous counter is a basic counter, which is most commonly used in digital circuits, but has limitation on speed of operation.

8.11 DESIGNING A COUNTER

Asynchronous and synchronous counters can be designed by a state diagram. Before starting a design of counter, we must know the terminology used while designing a counter.

State diagram: It graphically represents the states of a counter. For example, a 2-bit counter, count upto 00 to 11 (i.e., 00, 01, 10, and 11), having four states graphically, it is represented as

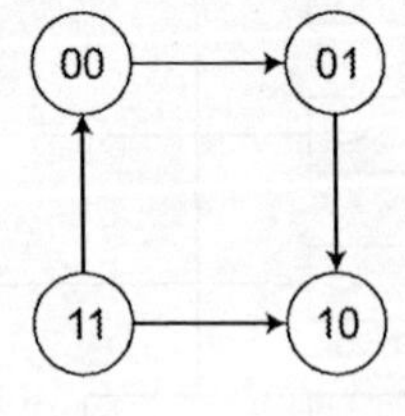

Arrow indicates the direction of next state.

Modulus (MOD): Modulus of a counter shows the number of stages of a counter. In general MOD of a counter is represented by

$$\text{MOD} = 2^n$$

where n represents the number of flip-flops required to design a counter. For example, if we want to count up to 16 i.e., (0000 to 1111) then we require 2^4, as (4-bit count) 4 flip-flops to design a counter.

8.12 DESIGN OF A ASYNCHRONOUS COUNTER

While designing a asynchronous counter, the following steps are undertaken. Remember that the clock pulse to the first flop is clocked externally and to the successive flip-flops is provided by the clocked output of the preceding flip-flop.

Step I Determine the Mod of a counter to know, how many flip-flops are used to design a particular counter using Mod = 2^N; where N is no. of flip-flops.

Step II Draw a state diagram to know the stages of a counter.

Step III Design a truth table for reset logic using valid states.

Step IV Simplify the truth table using K-map and design a Boolean expression.

Step V Draw a logic diagram using Boolean expression.

Example 8.1 Design a Mod-3 asynchronous counter.

Solution Mod-3 counter means a counter used to count 3 states.

1. State diagram:

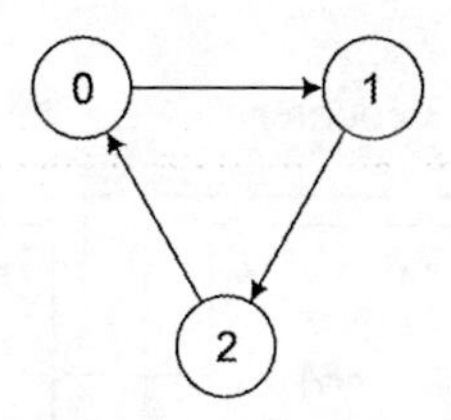

2. The number of flip-flops used is given by $m \leq 2^N$

   where m represents modulus and N = number of flip-flops so for $m = 3$ therefore $\Rightarrow 3 \leq 2^2$

   i.e., $N = 2$

   So, the number of flip-flops used = 3.

3. Two flip-flops can count up to 4 states but a Mod-3 counter has three states, so after the third state the counter is reset to zero.

   This is done by applying a *reset logic's output* to the *clear input* of all the flip-flops used.

4. The block diagram of Mod-3 counter is:

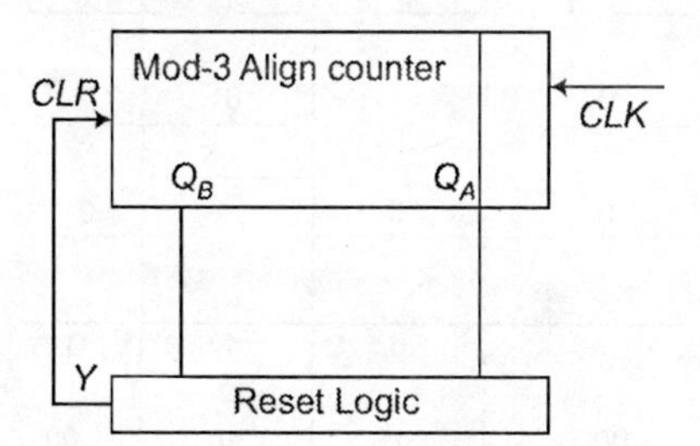

5. Truth table of reset logic

State	Flip-flop outputs Q_B	Q_A	Output of Reset Logic (Y)
0	0	0	1
1	0	1	1
2	1	0	1
3	1	1	0

States 0, 1, 2: Valid states are represented by '1'

State 3: Invalid state because the counter is reset after the state '3'

6. Draw K-maps

For the output of Reset Logic (Y)

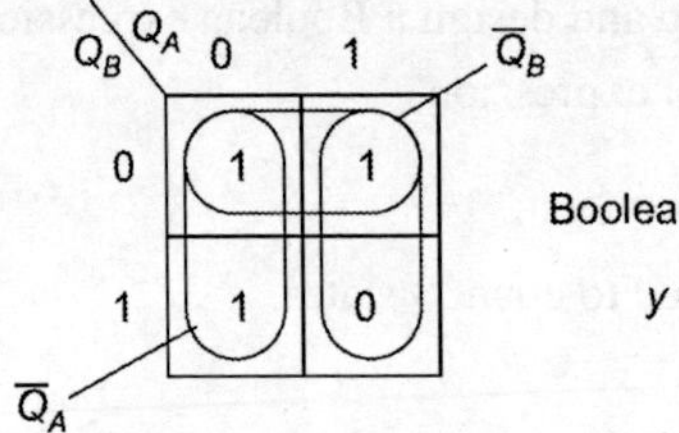

Boolean expression is given as:

$$y = \overline{Q}_B + \overline{Q}_A = \overline{Q_B \cdot Q_A}$$

7. Realize the logic circuit of Mod-3 counter.

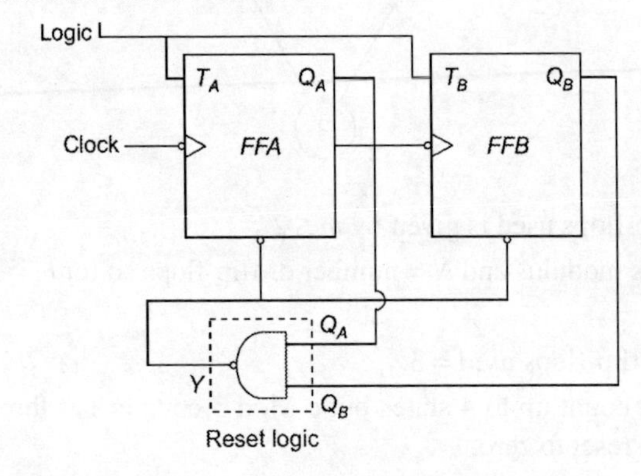

8. Timing diagram of Mod-3 counter:

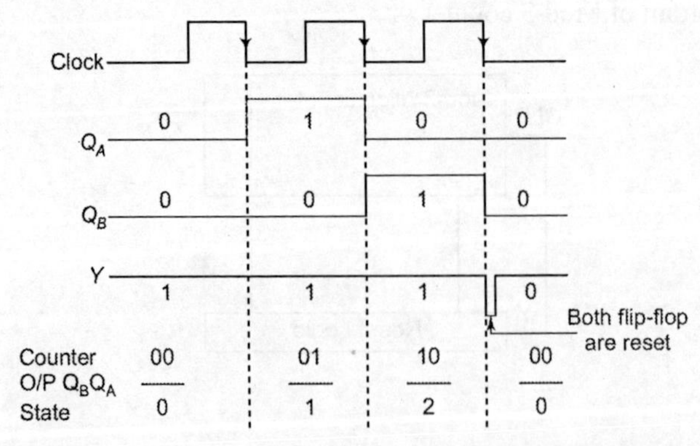

Example 8.2 Design a Mod-6 asynchronous counter.

Solution Mod-6 counter means counter with 6 states.

1. State diagram:

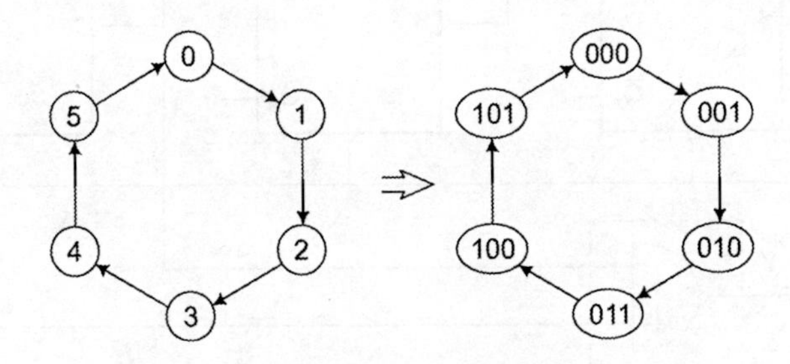

2. To design a reset logic, truth table for the reset logic is:

State	*Flip flop output* Q_C	Q_B	Q_A	*Output of Reset Logic (Y)*
0	0	0	0	1
1	0	0	1	1
2	0	1	0	1
3	0	1	1	1
4	1	0	0	1
5	1	0	1	1
6	1	1	0	0
7	1	1	1	0

3. K-maps for simplification to derive Boolean expression for output (Y) of reset logic

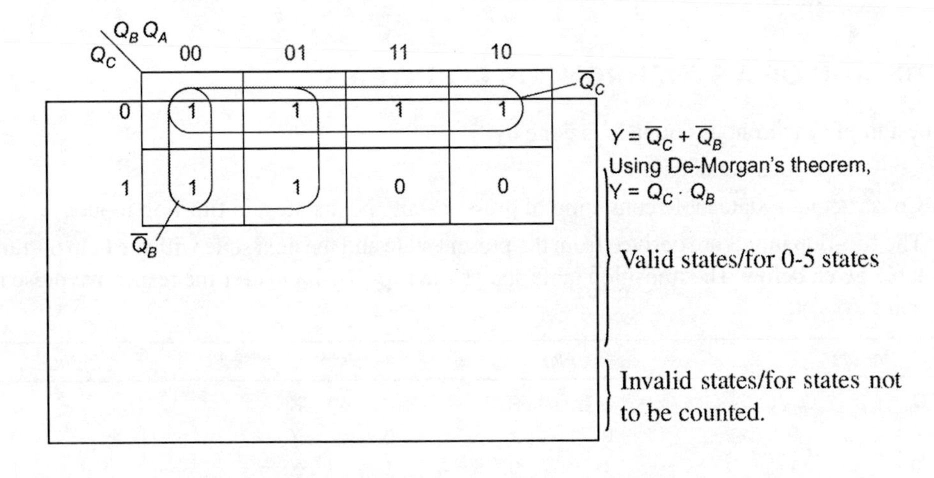

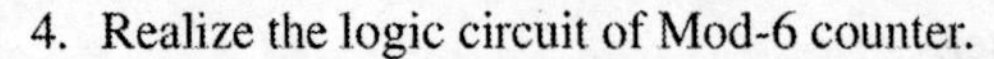

4. Realize the logic circuit of Mod-6 counter.

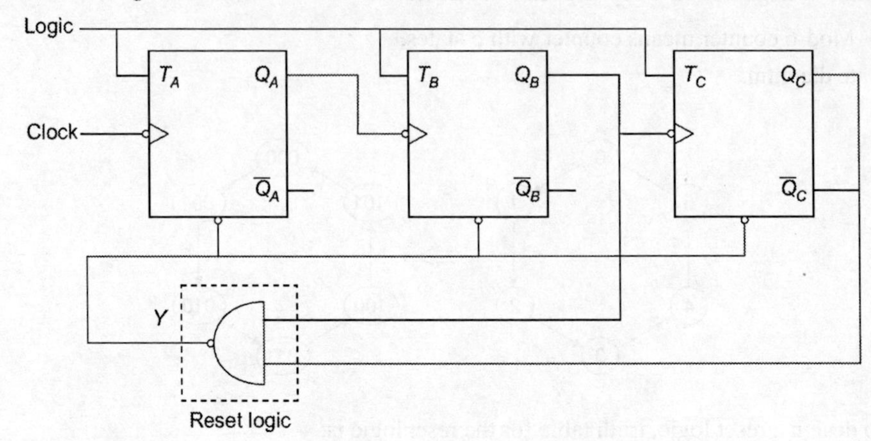

5. Timing diagram:

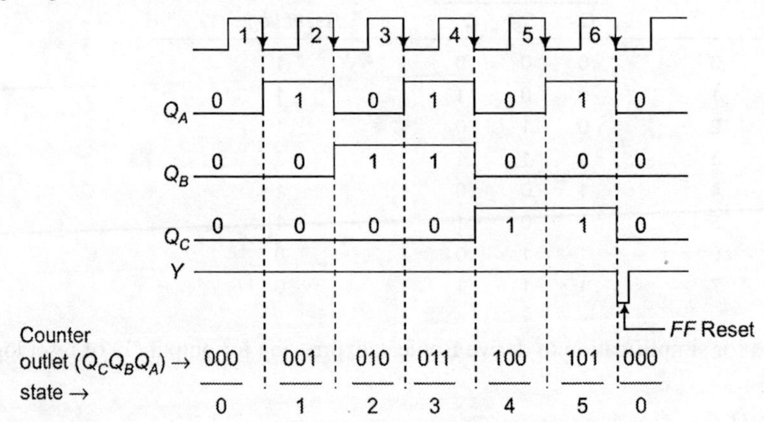

8.13 DESIGN OF A SYNCHRONOUS COUNTER

The design of synchronous counters is done by:

- Constructing a state diagram
- Constructing a state table consisting of present state, next state and flip-flop inputs,
- The flip-flop inputs are derived from the present state and the next state with the help of transition table given below. The transition table shows the flip-flop inputs for the respective present state and next state.

Inputs		*Flip*		*Flop*		*Inputs*	
Q_n	Q_{n+1}	S	R	J	K	T	D
0	0	0	X	0	X	0	0
0	1	1	0	1	X	1	1
1	0	0	1	X	1	1	0
1	1	X	0	X	0	0	1

The design of a synchronous counter is illustrated in the examples given below.

Example 8.3 Mod-5 synchronous counter.

Solution Mod-5 means a counter will have 5 output states and the number of flip-flops required is 3

1. State diagram:

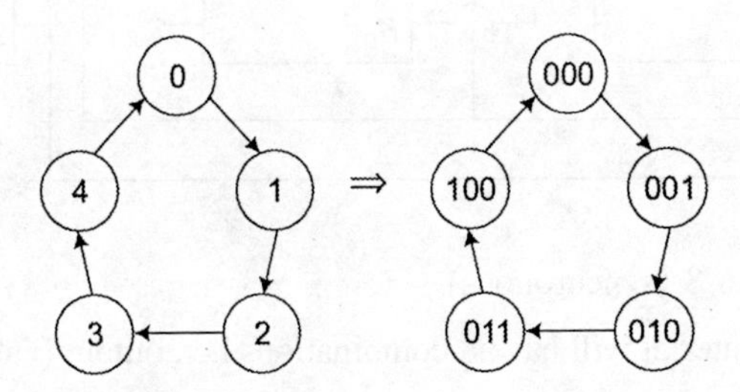

i.e., 0 – 4 states are VALID $ 5 – 6 – 7 are don't cares.

2. Now truth table for Mod-5 counter is:

States	*Present Flip-Flop*			*Desired Flip-Flop outputs (next)*			*Present inputs to Flip-Flops*					
	Q_C	Q_B	Q_A	Q_C	Q_B	Q_A	J_C	K_C	J_B	K_B	J_A	K_A
0	0	0	0	0	0	1	0	X	0	X	1	X
1	0	0	1	0	1	0	0	X	1	X	X	1
2	0	1	0	0	1	1	0	X	X	0	1	X
3	0	1	1	1	0	0	1	X	X	1	X	1
4	1	0	0	0	0	0	X	1	0	X	0	X

3. Now draw K-maps for Boolean expression of $J_A/K_B/J_B/K_B/J_C/K_C$

$Q_C \backslash Q_BQ_A$	00	01	11	10
0	1	x	x	1
1	0	x	x	x

$J_A = \overline{Q}_C$

$Q_C \backslash Q_BQ_A$	00	01	11	10
0	x	1	1	x
1	x	x	x	x

$K_A = 1$

$Q_C \backslash Q_BQ_A$	00	01	11	10
0	0	x	1	0
1	x	x	x	x

$J_B = Q_A$

$Q_C \backslash Q_BQ_A$	00	01	11	10
0	x	x	1	0
1	x	x	x	x

$K_B = Q_A$

$Q_C \backslash Q_BQ_A$	00	01	11	10
0	0	0	1	0
1	x	x	x	x

$J_A = Q_C + Q_A$

$Q_C \backslash Q_BQ_A$	00	01	11	10
0	x	x	x	x
1	1	x	x	x

$K_A = 1$

4. Realization of logic circuit using Boolean expression:

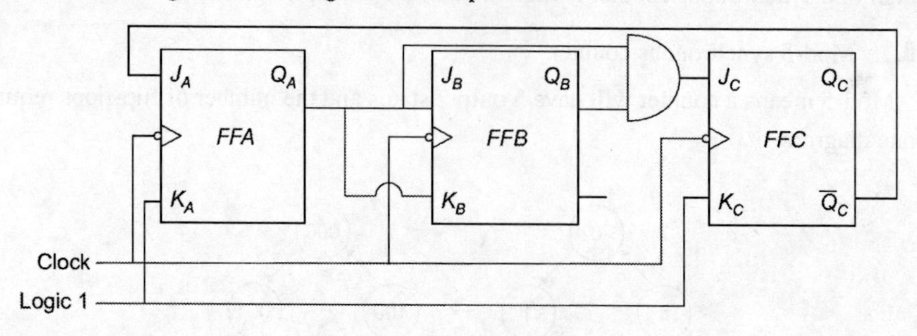

Example 8.4 Mod-8 counter 8-(synchronous)

Solution For a Mod-8 counter, it will have 8 combinations i.e., output (valid) states and it will require 3 flip-flops

1. State diagram:

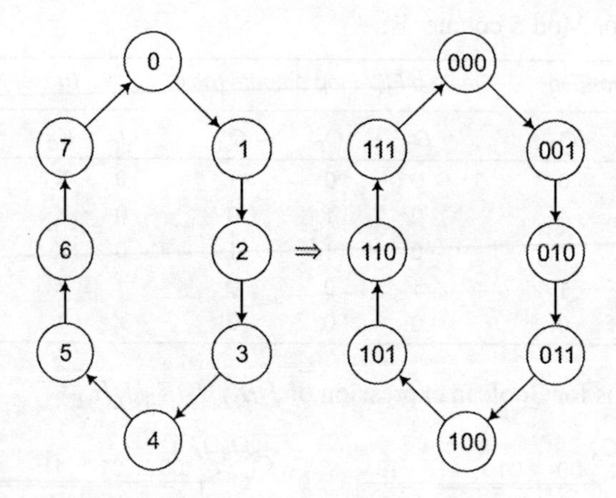

i.e., it will have 8 valid states from 0-7.

2. Truth table for deriving the Boolean expressions:

States	Present State			Desired Outputs			Present Inputs					
	Q_C	Q_B	Q_A	Q_C	Q_B	Q_A	J_C	K_C	J_B	K_B	J_A	K_A
0	0	0	0	0	0	1	0	X	0	X	1	X
1	0	0	1	0	1	0	0	X	1	X	X	1
2	0	1	0	0	1	1	0	X	X	0	1	X
3	0	1	1	1	0	0	1	X	X	1	X	1
4	1	0	0	1	0	1	X	0	0	X	1	X
5	1	0	1	1	1	0	X	0	1	X	X	1
6	1	1	0	1	1	1	X	0	X	0	1	X
7	1	1	1	0	0	0	X	1	X	1	X	1

3. K-maps:

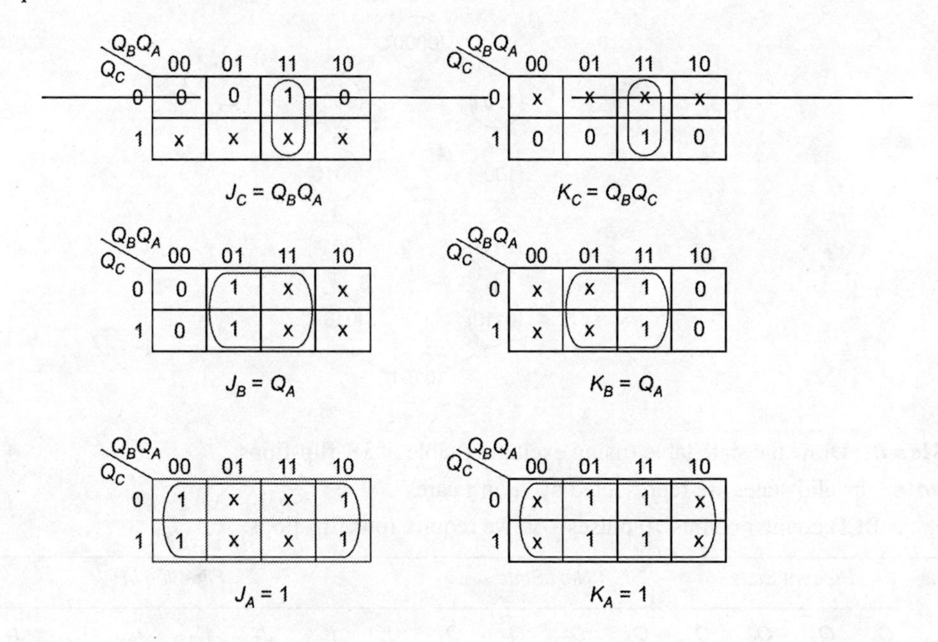

4. Realize the logic circuit of Mod-8 counter:

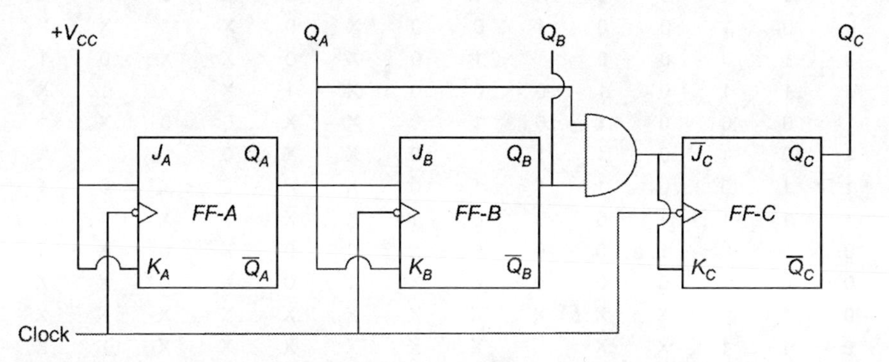

Example 8.5 Design a synchronous BCD upcounter.

Solution A BCD counter consists of 10 states i.e., the states from 0000 to 1001 are valid and 1010 to 1111 are invalid states. (The invalid states are represented by "don't care" conditions.)

Case I Implementation of BCD upcounter using JK FF:

Steps

1. Draw state diagram
2. Construct state table
3. Use K-map for simplification and solving algebraic expressions of flip-flop I/P.
4. Implement using combinational and sequential circuits.

Step I Construction of state diagram for BCD upcounter.

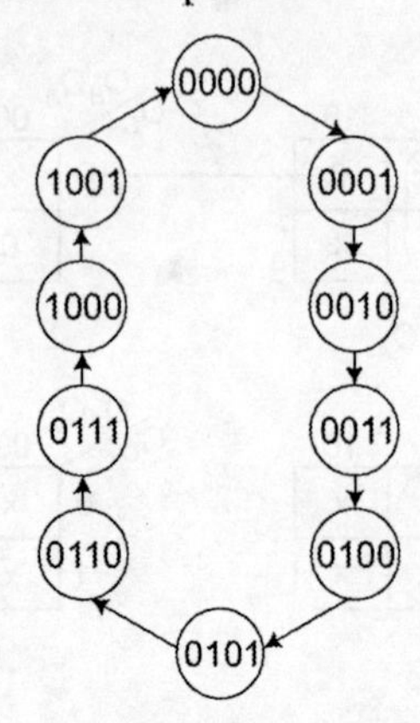

Step II Draw the state table (using excitation table of JK flip-flops.

Note Invalid states are represented by "don't cares".

As BCD counter counts 10 pulses—so we require four flip-flops.

Present State				*Next State*				*Flip-Flop I/P*							
Q_4	Q_3	Q_2	Q_1	Q_4	Q_3	Q_2	Q_1	J_4	K_4	J_3	K_3	J_2	K_2	J_1	K_1
0	0	0	0	0	0	0	1	0	X	0	X	0	X	1	X
0	0	0	1	0	0	1	0	0	X	0	X	1	X	X	1
0	0	1	0	0	0	1	1	0	X	0	X	X	0	1	X
0	0	1	1	0	1	0	0	0	X	1	X	X	1	X	1
0	1	0	0	0	1	0	1	0	X	X	0	0	X	1	X
0	1	0	1	0	1	1	0	0	X	X	0	1	X	X	1
0	1	1	0	0	1	1	1	0	X	X	0	X	0	1	X
0	1	1	1	1	0	0	0	1	X	X	1	X	1	X	1
1	0	0	0	1	0	0	1	X	0	0	X	0	X	1	X
1	0	0	1	0	0	0	0	X	1	0	X	0	X	X	1
1	0	1	0	X	X	X	X	X	X	X	X	X	X	X	X
1	0	1	1	X	X	X	X	X	X	X	X	X	X	X	X
1	1	0	0	X	X	X	X	X	X	X	X	X	X	X	X
1	1	0	1	X	X	X	X	X	X	X	X	X	X	X	X
1	1	1	0	X	X	X	X	X	X	X	X	X	X	X	X
1	1	1	1	X	X	X	X	X	X	X	X	X	X	X	X

Step III After using the state table to derive expressions for the flip-flops inputs as function of the present state inputs Q_1, Q_2, Q_3, Q_4 simplify the expressions using K-maps.

(a) K-map for J_1 and K_1:

No K-map is required as the "don't cares" of the last two columns of the state table can be treated as logic ones and thus $J_1 = K_1 = 1$.

(b) K-map for J_2 and K_2:

Q_4Q_3 \ Q_2Q_1	00	01	11	10
00		1	X	X
01		1	X	X
11	X	X	X	X
10			X	X

$J_2 = Q_1\overline{Q}_4$

Q_4Q_3 \ Q_2Q_1	00	01	11	10
00	X	X	1	
01	X	1	X	
11	X	X	X	X
10	X	X	X	X

$K_2 = Q_1$

(c) K-map for J_3 and K_3:

Q_4Q_3 \ Q_2Q_1	00	01	11	10
00			1	
01	X	X	X	X
11	X	X	X	X
10			X	X

$J_3 = Q_1Q_2$

Q_4Q_3 \ Q_2Q_1	00	01	11	10
00	X	X	X	X
01			1	
11	X	X	X	X
10	X	X	X	X

$K_3 = Q_1Q_2$

(d) K-map for J_4 and K_4

Q_4Q_3 \ Q_2Q_1	00	01	11	10
00				
01			1	
11	X	X	X	X
10	X	X	X	X

$J_4 = Q_1Q_2Q_3$

Q_4Q_3 \ Q_2Q_1	00	01	11	10
00	X	X	X	X
01	X	X	X	X
11	X	X	X	X
10		1	X	X

$K_4 = Q_1$

Step IV Implementation of expressions using JK flip-flops

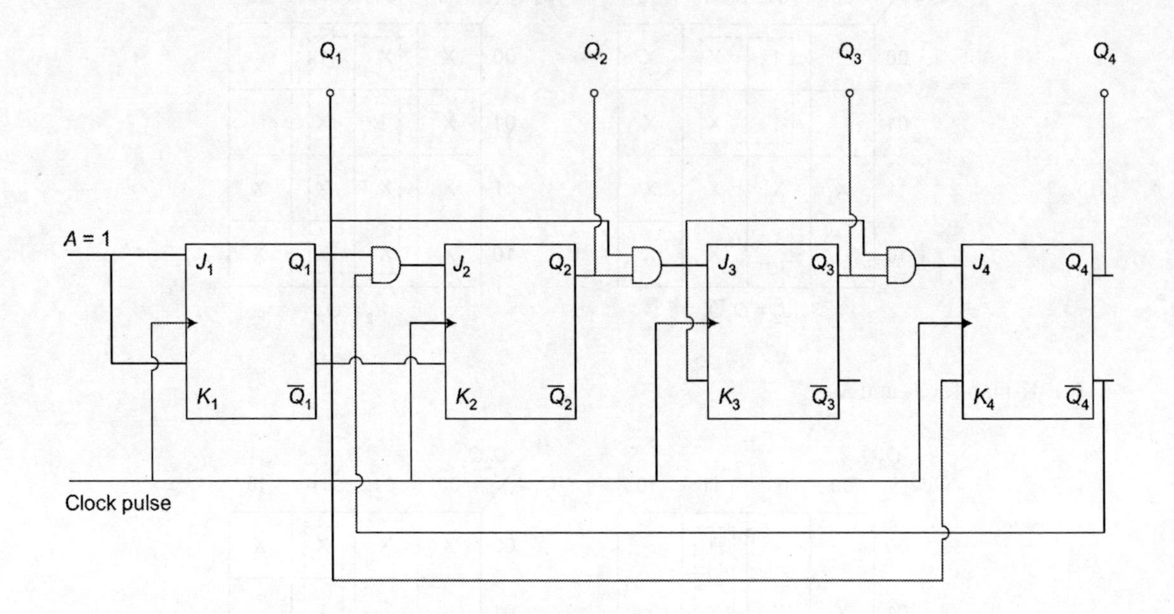

Case II Implementation of BCD counter using T flip-flop

Step I

Present State				*Next State*				*Flip-Flop Input*			
Q_4	Q_3	Q_2	Q_1	Q_4	Q_3	Q_2	Q_1	T_4	T_3	T_2	T_1
0	0	0	0	0	0	0	1	0	0	0	1
0	0	0	1	0	0	1	0	0	0	1	1
0	0	1	0	0	0	1	1	0	0	0	1
0	0	1	1	0	1	0	0	0	1	1	1
0	1	0	0	0	1	0	1	0	0	0	1
0	1	0	1	0	1	1	0	0	0	1	1
0	1	1	0	0	1	1	1	0	0	0	1
0	1	1	1	1	0	0	0	1	1	1	1
1	0	0	0	1	0	0	1	0	0	0	1
1	0	0	1	0	0	0	0	1	0	0	1
1	0	1	0	X	X	X	X	X	X	X	X
1	0	1	1	X	X	X	X	X	X	X	X
1	1	0	0	X	X	X	X	X	X	X	X
1	1	0	1	X	X	X	X	X	X	X	X
1	1	1	0	X	X	X	X	X	X	X	X
1	1	1	1	X	X	X	X	X	X	X	X

Step II Simplify the expressions using K-maps:

(a) K-maps for T_1

Q_4Q_3 \ Q_2Q_1	00	01	11	10
00	1	1	1	1
01	1	1	1	1
11	X	X	X	X
10	1	1	X	X

$T_1 = 1$

(b) K-map for T_2

Q_4Q_3 \ Q_2Q_1	00	01	11	10
00	0	1	1	0
01	0	1	1	0
11	X	X	X	X
10	0	0	X	X

$T_2 = Q_1\overline{Q}_4$

(c) K-map for T_3

Q_4Q_3 \ Q_2Q_1	00	01	11	10
00	0	0	1	0
01	0	0	1	0
11	X	X	X	X
10	0	0	X	X

$T_3 = Q_1Q_2$

(d) K-map for T_4

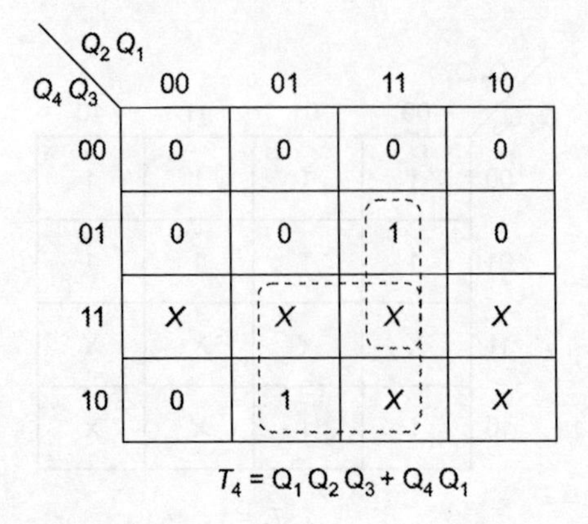

$$T_4 = Q_1 Q_2 Q_3 + Q_4 Q_1$$

Step III Implementation using T flip-flops.

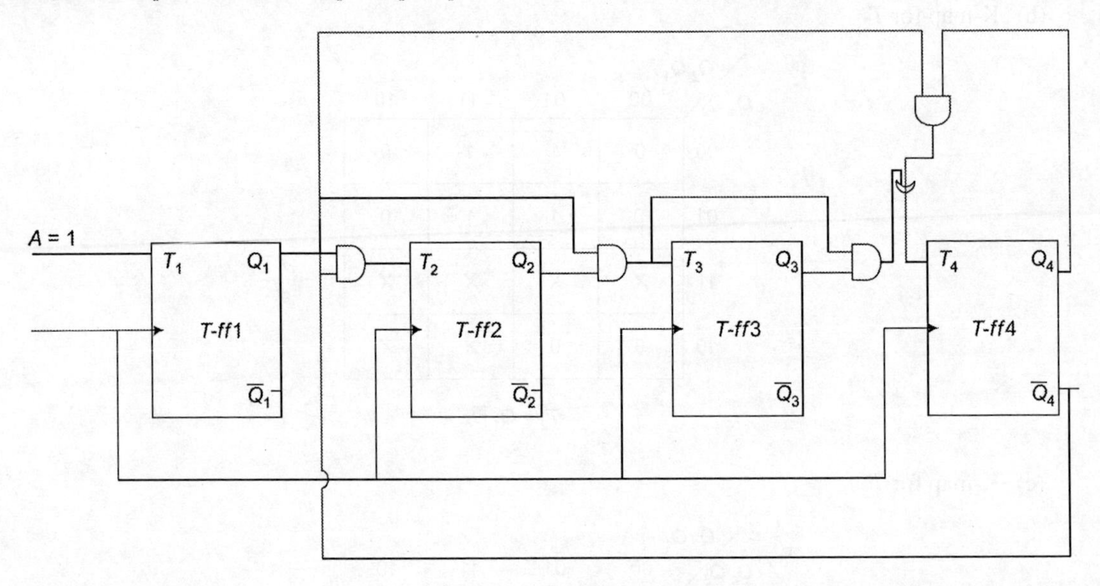

Example 8.6 Design a synchronous modulo-6 gray code counter using T flip-flops.

Solution

1. Determine the number of flip-flops.
2. Draw state diagram
3. Draw state table for T flip-flop
4. Use K-map to simplify the expressions
5. Implementation of the expression using T flip-flops.

Step I Modulo-6 means 6-states i.e., $n = 3$ flip-flops

Step II

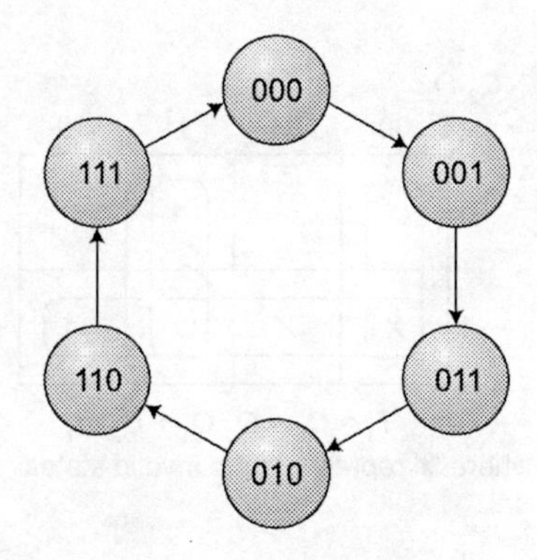

Step III

Present State			*Next State*			*Flip-Flop Input*		
Q_3	Q_2	Q_1	Q_3	Q_2	Q_1	T_3	T_2	T_1
0	0	0	0	0	1	0	0	1
0	0	1	0	1	1	0	1	0
0	1	1	0	1	0	0	0	1
0	1	0	1	1	0	1	0	0
1	1	0	1	1	1	0	0	1
1	1	1	0	0	0	1	1	1

Step IV

(a) K-map for T_3

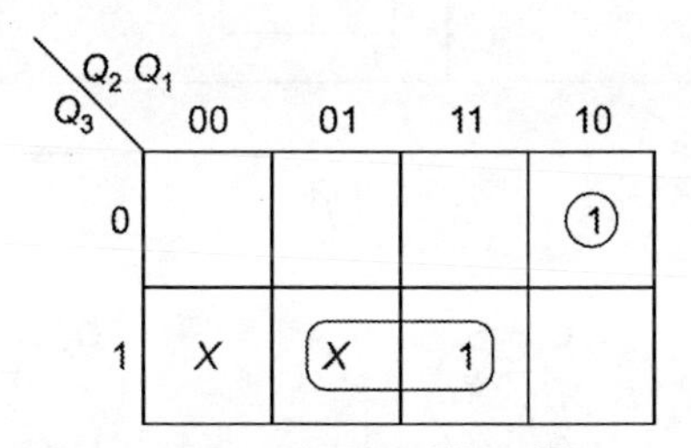

$T_3 = Q_3 Q_1 + \overline{Q}_3 Q_2 \overline{Q}_1$

where 'x' represents the invalid states

(b) K-map for T_2

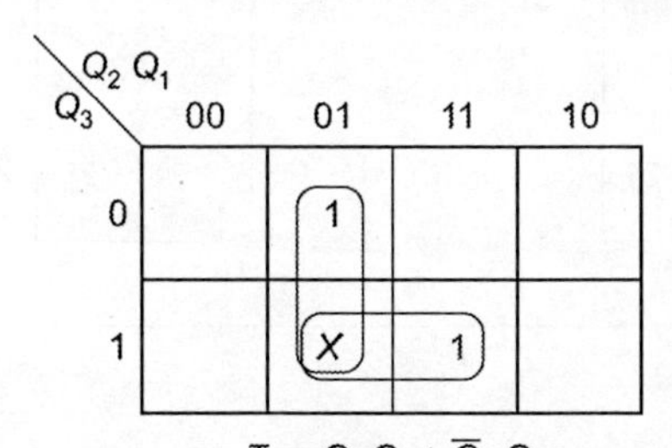

$T_2 = Q_3 Q_1 + \overline{Q}_2 Q_1$

where 'x' represents the invalid states

(c) K-map for T_1

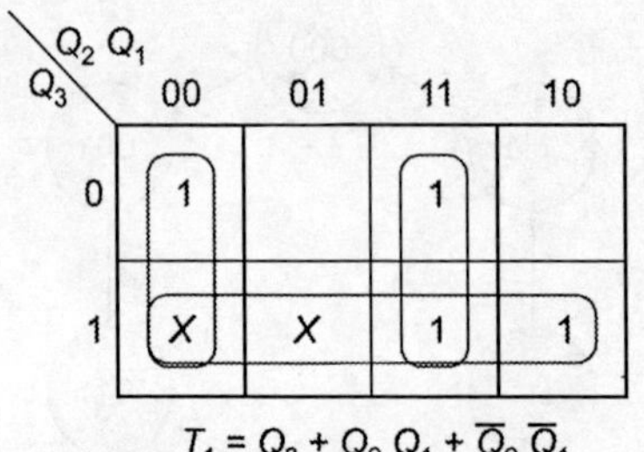

$T_1 = Q_3 + Q_2 Q_1 + \overline{Q}_2 \overline{Q}_1$
where 'x' represents the invalid states

Step V

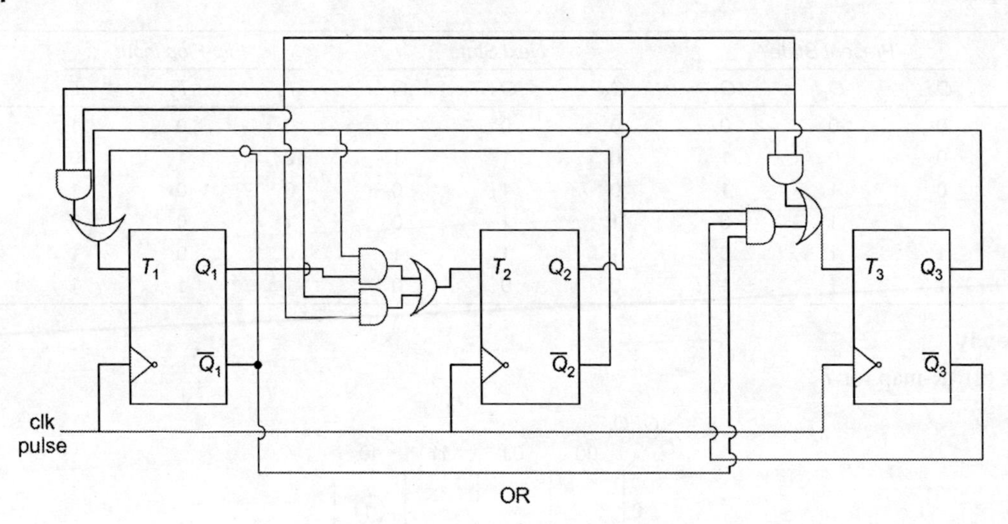

OR

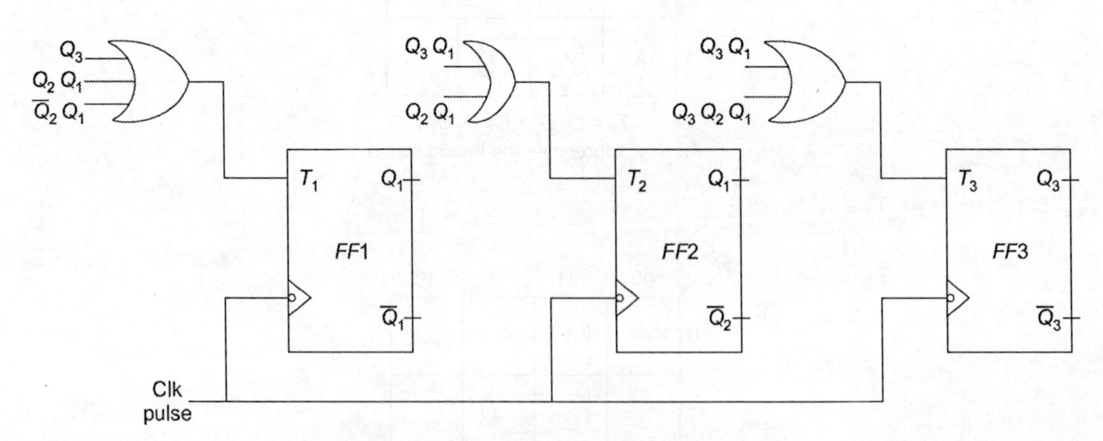

Example 8.7 Design a 3-bit binary up/down counter using JK flip-flops.

Solution For selecting an up and down mode, a mode signal is required.

For M = 0: The counter acts is an UP counter.

M = 1: The counter acts as a DOWN counter.

Step I The number of flip-flops required is 3.

Step II Draw state diagram

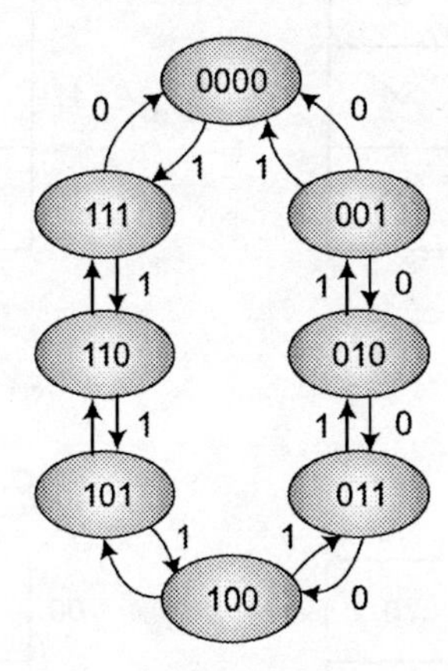

Step III Draw state table for JK flip-flops

Mode Control (M)	Present State			Next State			Flip-Flop Inputs					
	Q_3	Q_2	Qj_1	Q_3	Q_2	Q_1	J_1	K_1	J_2	K_2	J_3	K_3
0	0	0	0	0	0	1	1	X	0	X	0	X
0	0	0	1	0	1	0	X	1	1	X	0	X
0	0	1	0	0	1	1	1	X	X	0	0	X
0	0	1	1	1	0	0	X	1	X	1	1	X
0	1	0	0	1	0	1	1	X	0	X	X	0
0	1	0	1	1	1	0	X	1	1	X	X	0
0	1	1	0	1	1	1	1	X	X	0	X	0
0	1	1	1	0	0	0	X	1	X	1	X	0
1	0	0	0	1	1	1	1	X	1	X	1	X
1	1	1	1	1	1	0	X	1	X	0	X	0
1	1	1	0	1	0	1	1	X	X	1	X	0
1	1	0	1	1	0	0	X	1	0	X	X	0
1	1	0	0	0	1	1	1	X	1	X	X	1
1	0	1	1	0	1	0	X	1	X	0	0	X
1	0	1	0	0	0	1	1	X	X	1	0	X
1	0	0	1	0	0	0	X	1	0	X	0	X

Step IV Draw K-maps.

1. For $J_1 = K_1 = 1$

2. For J_2 and K_2

MQ_3 \ Q_2Q_1	00	01	11	10
00	0	1	X	X
01	0	1	X	X
11	1	0	X	X
10	1	0	X	X

$$J_2 = Q_1\overline{M} + \overline{Q}_1 M$$

MQ_3 \ Q_2Q_1	00	01	11	10
00	X	X	1	0
01	X	X	1	0
11	X	X	0	1
10	X	X	0	1

$$K_2 = Q_1\overline{M} + \overline{Q}_1 M$$

3. For J_3 and K_3

MQ_3 \ Q_2Q_1	00	01	11	10
00	0	0	1	X
01	X	X	X	X
11	X	X	X	X
10	1	0	0	0

$$J_3 = \overline{M}Q_2Q_1 + M + \overline{Q}_2\overline{Q}_1$$

MQ_3 \ Q_2Q_1	00	01	11	10
00	X	X	X	X
01	0	0	1	0
11	1	0	0	0
10	X	X	X	X

$$K_3 = \overline{M}\,Q_2\,Q_1 + M\overline{Q}_2\,\overline{Q}_1$$

Step V Implementation using JK flip-flops.

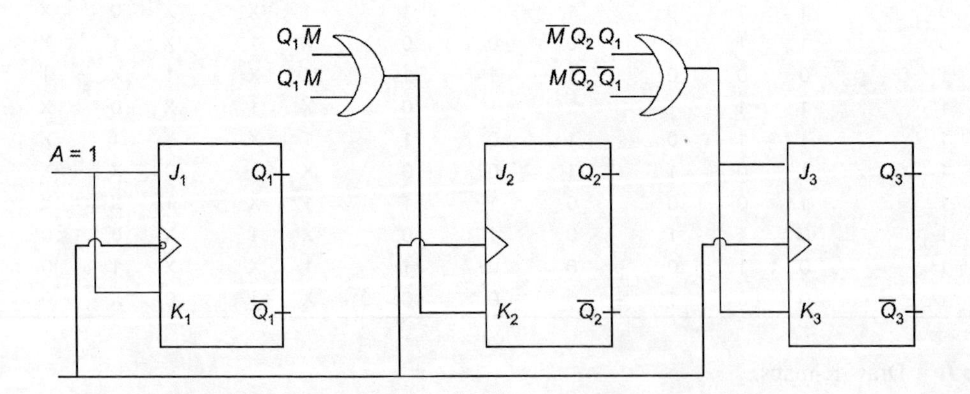

EXERCISES

1. What do you mean by shift register? State its types and applications.
2. Draw circuit diagram of 3-bit SIPO shift register using D flip-flops and explain its working.
3. What type of register is needed to convert serial data to parallel and parallel data to serial.
4. What is the basic difference between a serial and a parallel transfer?
5. Indicate the type of register:
   (a) into which a complete binary number can be loaded in one operation and then shifted out one bit at a time?
   (b) into which data can be entered only one bit at a time but have all data bits available as outputs?
   (c) in which we have access to only the leftmost or rightmost flip-flop?
6. What do you understand by a bidirectional shift register? Explain its working in detail.
7. Discuss the operation of 4-bit serial-in-parallel out (SIPO) register. Give the truth table and timing diagram and truth table.
8. Differentiate between PIPO and PISO with the help of neat diagram.
9. Define counter and give the steps for designing asynchronous counters.
10. Design a Mod-3 asynchronous counter with the help of a state diagram.
11. Design a Mod-6 asynchronous counter with the help of a state diagram.
12. Differentiate between synchronous and asynchronous counters.
13. Explain the advantages and disadvantages of the synchronous counter over the asynchronous counter.
14. Describe the operation of a 4-bit ripple counter with the help of T flip-flop.
15. How does a counter differ from a register?
16. How does a counter work as frequency divider?
17. How many flip-flops are required to construct the following modulus counter:
    (a) 5 (b) 38 (c) 92 (d) 10
18. Design steps for 3-bit up/down synchronous counter.
19. For Mod-10 asynchronous upcounter
    (a) Write the truth table.
    (b) Draw its timing diagram.
    (c) Draw circuit diagram using T flip-flops
    (d) If the output frequency is 21 kHz. What is the clock input?
20. Draw the divide by 7 asynchronous counter using T flip-flop. Writer truth table, draw the timing diagram.

9

Combinational Logic Circuits

INTRODUCTION

Logic circuits are broadly categorized into combinational circuits and sequential circuits. In a combinational circuit the output is always a function of inputs only at that instant. As it does not require any past values, it does not have any memory. On the other hand, a sequential logic circuit has a memory because its output at any instant is a function of the present inputs as well as the past values of output. So the elements are required to store the past values of output (past information) and hence, require memory.

The requirements for the design of any combinational circuit may be represented in the following ways:

1. A set of statements
2. Boolean expressions
3. Truth table

9.1 COMBINATIONAL LOGIC CIRCUITS

There are two different approaches that a designer can adopt for designing combinational circuits.

1. Traditional Method

In traditional method, the Boolean expression or the truth table needs to be simplified by using standard methods and the simplified expression is then implemented by using gates.

2. Use of MSI

In this method, there is need for simplification of Boolean expressions or truth table. Instead of this, the complex logic methods available in medium scale integration (MSI) can be directly used in this approach.

Computer Aided Designing (CAD) tools are used for the design using PLDs and FPGAs. We will focus on the design of combinational circuits using the Traditional Design Approach, but surely the concepts learnt here will help in understanding and developing of design using MSs, PLDs and FPGAs etc.

The following are the various methods that we can use to simplify the Boolean functions:

1. Algebraic method
2. Karnaugh-map Technique
3. Crime-Mecluskey method
4. Variable entered (YEM) mapping technique.

But before we start dealing with the design procedure of combinational logic circuits. Let us have a glimpse over the basic features of combinational and sequential circuits.

TABLE 9.1 Comparison of combinational and sequential logic circuit

Combinational Logic Circuit	*Sequential Logic Circuit*
1. Only the present status of inputs determines the value of output.	1. Present status of inputs as well as the past values of output decide the present output.
2. It does not have a memory.	2. It has a memory.
3. It requires no feedback path from output to input.	3. It has a feedback path from output to input.
4. No clock signal is required.	4. It may or may not have a clock signal.
5. A truth table is only required to describe the operation.	5. Its operation can be described by a truth table and a timing diagram.
6. The clock transition has no effect on its action.	6. Its action is governed by clock transition.
7. It has a simple circuit.	7. It has a circuit comparatively complex than that of a combinational logic circuit.
8. It is built using basic gates, i.e., NOT, NAND, NOR, AND OR and XOR gates.	8. It is built using basic gates and combinational logic circuits.
9. Examples: adders, subtractors, multiplexers, demultiplexers, decoders, comparators.	9. Examples: flip-flops, counters, shift registers, etc.

9.2 DESIGN PROCEDURE

A combinational logic circuit is designed to give the desired output for specified input variable. So either a statement for the design problem is given or a truth table is provided. If the problem is given in the statement form, first of the corresponding truth table needs to be prepared by analyzing and investigating thoroughly the problem statement. Now from the prepared truth table, the Boolean expression is extracted. If the problem is given directly in truth-table form, find out the corresponding Boolean expression that supports it. Now implement the logic design for the circuit using gates.

Example 9.1 The operating condition of three lights (red, green, yellow, blue) are to be monitored. $X = 1$ means red light is ON, $B = 1$ means green light is ON and $C = 1$ means yellow light is ON. OFF = 0 for any lights.

The overall indicator LED blue should glow only when majority of the lights are OFF. Design the logic circuit.

Solution 1. Obtain the truth table.

The output will be 1 only when either two or all the lights are OFF. The required truth table is:

A *Blue*	*B* *Green*	*C* *Yellow*	*Y* *Output*
0	0	0	1
0	0	1	1
0	0	0	1
0	1	0	1
0	1	1	0
0	0	0	1
1	0	0	0
1	0	1	0
1	1	0	0
1	1	1	0

2. Simplify to get the Boolean expression using K-map,

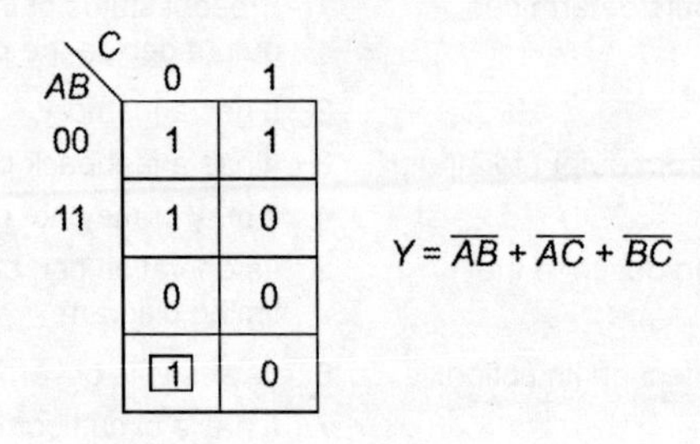

$Y = \overline{AB} + \overline{AC} + \overline{BC}$

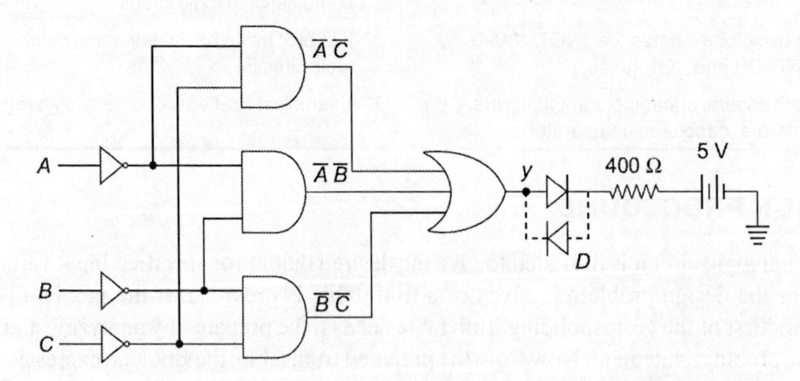

Logic circuit shown in the figure has a voltage drop of 1V and current of 10 mA, its ON-resistance

is $\dfrac{1}{(10 \times 10^{-3})} = 100\ \Omega$. To limit the currents from 5V source to 10 mA, it is necessary to connect a 400 Ω resistance as shown. When the majority of lights fail, $y = 1$ and LED will glow otherwise $Y = 0$ and LED will be off. The diode D is connected in antiparallel to LED against high reverse voltage.

Example 9.2 An electric power generating station supplies current to three loads. A single generator is required when only one load is switched ON. If more than one load is switched ON an auxiliary generator must be started. Design a logic circuit to provide the output signal to start the auxiliary generator.

Solution Let the loads be designated as A, B, C. The logic circuit should give a low output (i.e., u) if $A = B = C = 0$ and if any two of A, B, C are low or if only a single input is high. However, if any of the two or all the inputs are at high logic (1), the output is 1. The truth table is as given below.

A	*B*	*C*	*Y*	
0	0	0	0	
0	0	1	0	
0	1	0	0	
0	1	1	1	$\overline{A}BC$
1	0	0	0	
1	0	1	1	$A\overline{B}C$
1	1	0	1	$AB\overline{C}$
1	1	1	1	ABC

$$
\begin{aligned}
Y &= \overline{A}\,BC + A\overline{B}\,C + AB\overline{C} + ABC \\
&= \overline{A}\,BC + ABC + A\overline{B}\,C + ABC + AB\overline{C} + ABC \\
&= BC\,(A + \overline{A}) + AC\,(B + \overline{B}) + AB\,(C + \overline{C}) \\
&= BC + AC + AB
\end{aligned}
$$

The logic circuit is as shown:

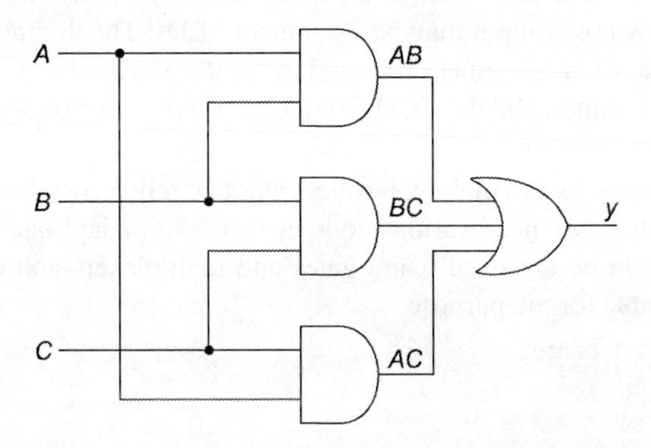

Example 9.3 In a fertilizer plant, a liquid chemical is used in the manufacturing process. The chemical is stored in three different tanks. A level sensor in each tank produces a high voltage when the level of chemical in the tank drops below a specified point.

Design a circuit that monitors the chemical level in each tank and indicates when the level in any two drops below the specified point.

Solution Let P, Q and R be the inputs from the sensors in tanks A, B and C. The AND gate G_1 checks the levers in tanks A & B, gate G_2 checks tanks A & C and gate G_3 checks tanks B & C. When the chemical level in any two of the tanks gets too low, one of the AND gates will have HIGH on both of its inputs causing its output to be HIGH, and so the final output X from the OR gate is HIGH, which is used to activate an indicator (lamp or an audible alarm).

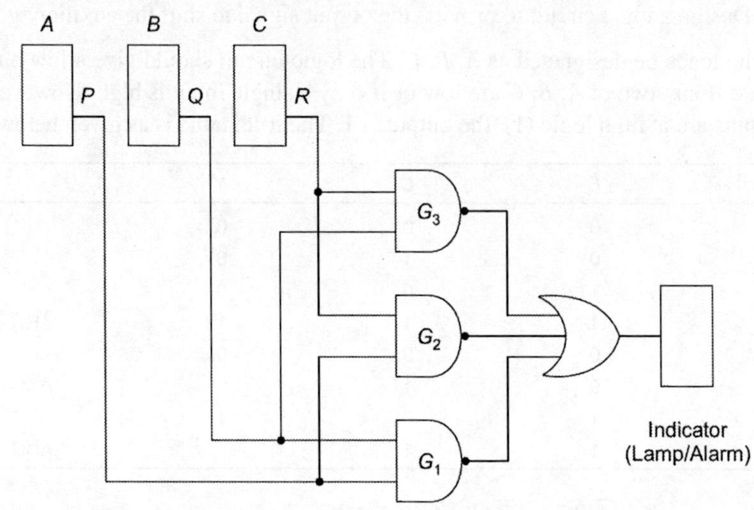

9.3 CODE CONVERTERS

A digital system deals with many varieties of binary codes. Some examples of these binary codes are: Binary Coded Decimal (BCD), Excess-3 (× 5.3) Gray, octal, hexadecimal, etc. But certain systems require their conversion from one form to another. Take for example, there might be a system hexing its input in natural BCD and whose output may be 7-segment LEDs. The digital system must therefore be capable of processing data in straight binary format. Hence, it is required that the data must be converted from BCD to binary at the output, and the BCD O/P has to be converted to 7-segment code before it can drive the LEDs.

Similarly the microprocessors and microcontrollers work largely on octal and hexadecimal platforms. In our real world of practice, we need various code converters to adapt one system with another. The various code converters can be designed using gates and multiplexers and demultiplexers. There are some MSI ICS also available for the purpose.

The various code converters are:

1. Binary to Excess-3
2. Binary to Gray
3. Gray to Binary
4. Binary to BCD
5. BCD to Binary

9.3.1 Binary to Excess-3 Converter

A number in excess-3 is obtained by adding $(3)_{10}$ or $(0011)_2$ in the binary number. Let us first make the truth table for conversion of binary code to XS-3.

1. Truth table

Binary			XS-3			
B_2	B_1	B_0	E_3	E_2	E_1	E_0
0	0	0	0	0	1	1
0	0	1	0	1	0	0
0	1	0	0	1	0	1
0	1	1	0	1	1	0
1	0	0	0	1	1	1
1	0	1	1	0	0	0
1	1	0	1	0	0	1
1	1	1	1	0	1	0

2. Find the Boolean expression for E_3, E_2, E_1, and E_0 in terms of B_2, B_1 and B_0 using K-map.

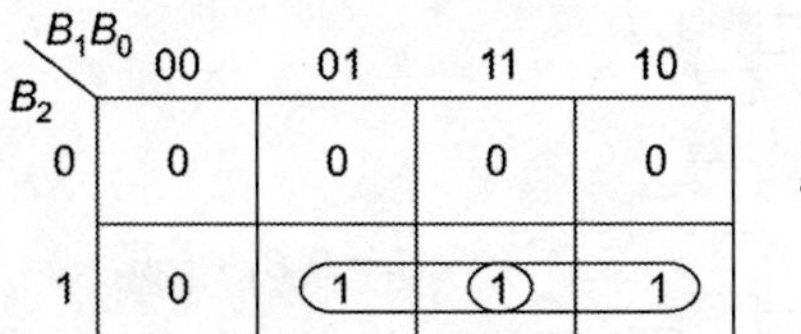

$E_3 = B_0B_2 + B_1B_2$

B_2 \ B_1B_0	00	01	11	10
0	0	1	1	1
1	1	0	0	0

$E_2 = B_0\overline{B}_2 + B_1\overline{B}_2 + \overline{B}_0\overline{B}_1B_2$

B_2 \ B_1B_0	00	01	11	10
0	1	0	1	0
1	1	0	1	0

$E_1 = B_0B_1 + \overline{B}_0\overline{B}_1$

B_2 \ B_1B_0	00	01	11	10
0	1	0	0	1
1	1	0	0	1

$E_0 = \overline{B}_0B_2$

3. Implement using gates:

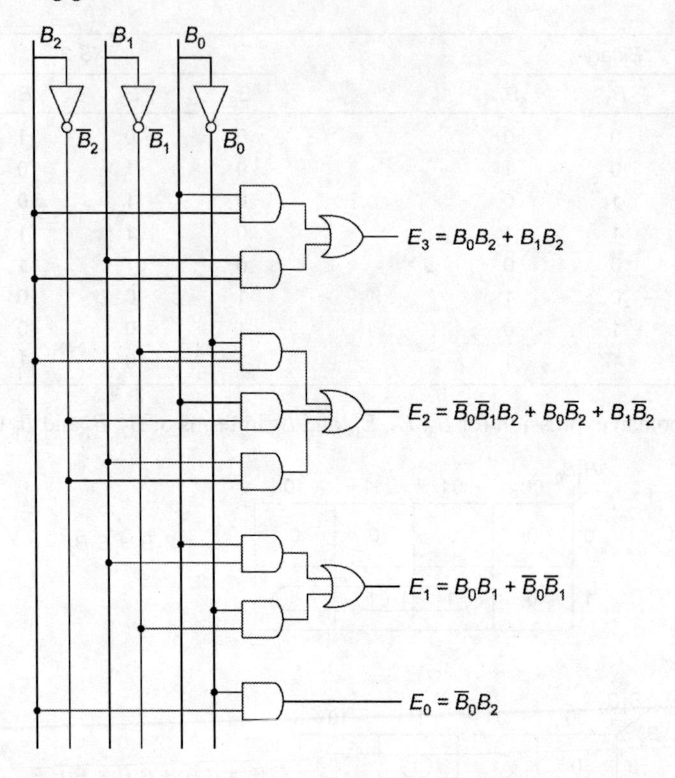

9.3.2 Binary to Gray

1. Make the truth table.

Binary				Gray			
B_3	B_2	B_1	B_0	G_3	G_2	G_1	G_0
0	0	0	0	0	0	0	0
0	0	0	1	0	0	0	1
0	0	1	0	0	0	1	1
0	0	1	1	0	0	1	0
0	1	0	0	0	1	1	0
0	1	0	1	0	1	1	1
0	1	1	0	0	1	0	1
0	1	1	1	0	1	0	0
1	0	0	0	1	1	0	0
1	0	0	1	1	1	0	1
1	0	1	0	1	1	1	1
1	0	1	1	1	1	1	0
1	1	0	0	1	0	1	0
1	1	0	1	1	0	1	1
1	1	1	0	1	0	0	1
1	1	1	1	1	0	0	0

2. Extract the Boolean expression for G_3, G_2, G_1 and G_0 using K-map.

B_3B_2 \ B_1O	00	01	11	10
00	0	0	0	0
01	0	0	0	0
11	1	1	1	1
10	1	1	1	1

$G_3 = B_3$

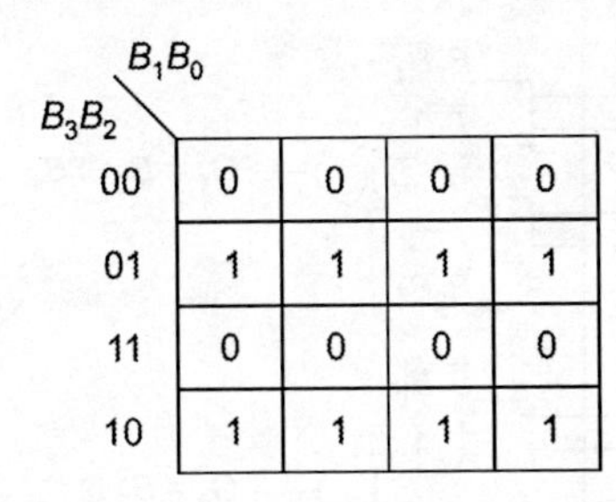

B_3B_2 \ B_1B_0				
00	0	0	0	0
01	1	1	1	1
11	0	0	0	0
10	1	1	1	1

$G_2 = \overline{B}_3B_2 + B_3\overline{B}_2$

$G_2 = B_3 \oplus B_2$

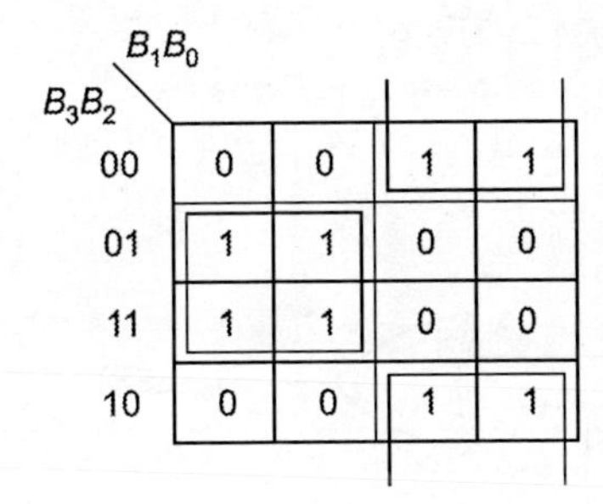

B_3B_2 \ B_1B_0				
00	0	0	1	1
01	1	1	0	0
11	1	1	0	0
10	0	0	1	1

$G_1 = \overline{B}_2B_1 + \overline{B}_2B_1$

$= B_2 \oplus B_1$

B_3B_2 \ B_1B_0				
00	0	1	0	1
01	0	1	0	1
11	0	1	0	1
10	0	1	0	1

$G_0 = \overline{B}_1B_0 + B_1\overline{B}_0$

$= B_1 \oplus B_0$

3. Implementation using gates:

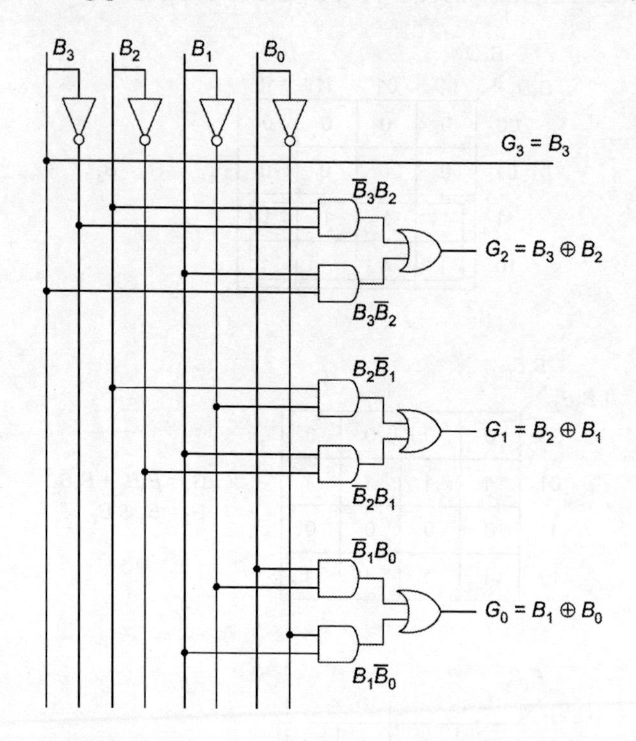

Implementation using XOR gates

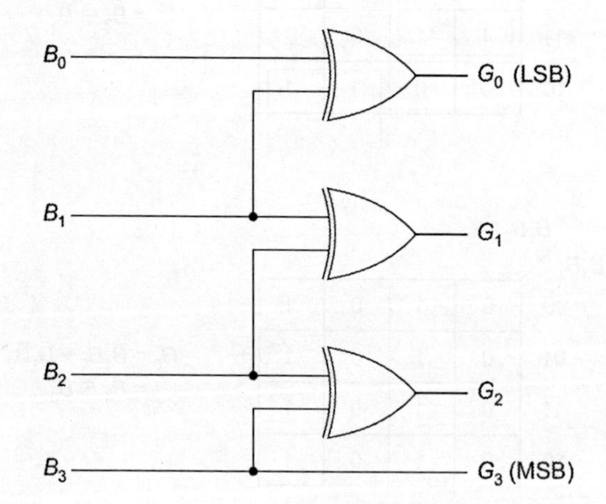

9.3.3 Gray to Binary Conversion

Gray				Binary			
G_3	G_2	G_1	G_0	B_3	B_2	B_1	B_0
0	0	0	0	0	0	0	0
0	0	0	1	0	0	0	1
0	0	1	0	0	0	1	1
0	0	1	1	0	0	1	0
0	1	0	0	0	1	1	1
0	1	0	1	0	1	1	0
0	1	1	0	0	1	0	0
0	1	1	1	0	1	0	1
1	0	0	0	1	1	1	1
1	0	0	1	1	1	1	0
1	0	1	0	1	1	0	0
1	0	1	1	1	1	0	1
1	1	0	0	1	0	0	0
1	1	0	1	1	0	0	1
1	1	1	0	1	0	1	1
1	1	1	1	1	0	1	0

G_3G_2 \ G_1G_0	00	01	11	10
00	0	0	0	0
01	0	0	0	0
11	1	1	1	1
10	1	1	1	1

$B_3 = G_3$

G_3G_2 \ G_1G_0	00	01	11	10
00	0	0	0	0
01	1	1	1	1
11	0	0	0	0
10	1	1	1	1

$$B_2 = G_3\overline{G_2} + \overline{G_3}G_2$$
$$= B_3\,\overline{G_2} + \overline{B_3}G_2$$
$$= B_3 \oplus G_2$$

G_3G_2 \ G_1G_0	00	01	11	10
00	0	0	1	1
01	1	1	0	0
11	0	0	1	1
10	1	1	0	0

$$B_1 = \overline{G_3}\,\overline{G_2} + G_1 + \overline{G_3}G_2\overline{G_1} + G_3G_2G_1$$
$$+ G_3\overline{G_2}\,\overline{G_1}$$
$$B_1 = B_2 \oplus G_1$$

G_3G_2 \ G_1G_0	00	01	11	10
00	0	(1)	0	(1)
01	1	0	(1)	0
11	0	(1)	0	(1)
10	(1)	0	(1)	0

$$B_0 = \overline{G}_3\overline{G}_2\overline{G}_1G_0 + \overline{G}_3\overline{G}_2G_1\overline{G}_0 + \overline{G}_3G_1\overline{G}_1\overline{G}_0$$
$$+ \overline{G}_3G_2G_1G_0 + G_3G_2\overline{G}_1G_0 + G_3G_2G_1\overline{G}_0$$
$$+ G_3\overline{G}_2\overline{G}_1\overline{G}_0 + G_3\overline{G}_2G_1G_0$$
$$\Rightarrow B_0 = B_2\,B_1 \oplus G_0$$

3. Implementation using XOR gates:

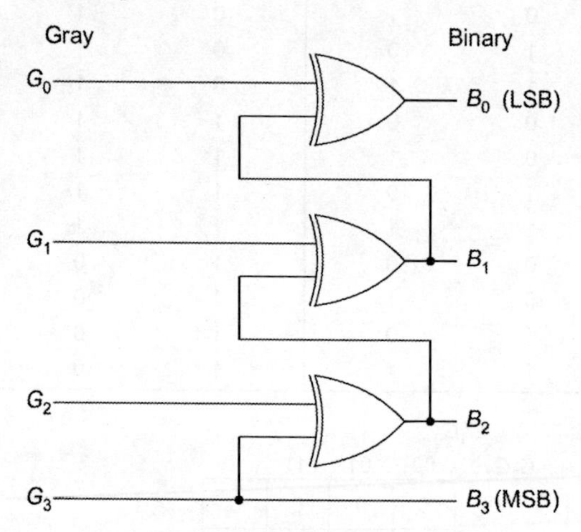

9.4 MULTIPLEXERS

One of the most important and widely used standard circuits in digital design is a multiplexer, in short written as MUX. A multiplexer is a combinational logic circuit that selects one out of many inputs and produces a single output.

To be more precise, a multiplexer (MUX) is a device that allows digital formation from several sources to be routed onto a single line for transmission over that line to a common destination. The basic MUX has several data input lines and a single output line. It also has data-select inputs, which allow digital data on any one of the inputs to be switched to the output line. MUXs are also known as data selectors.

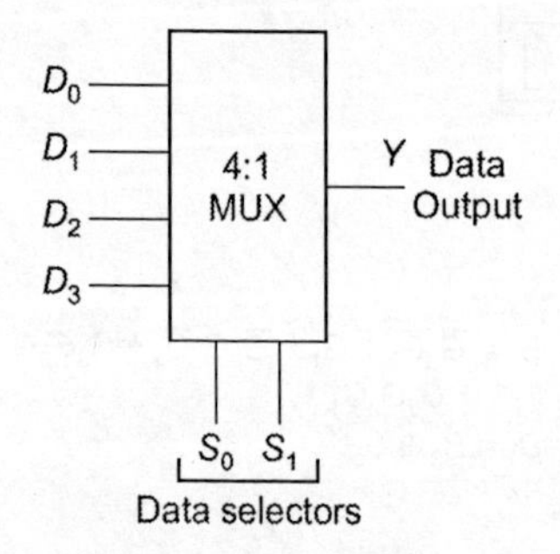

Fig. 9.1 4:1 Multiplexer

Table 9.2 Truth table for 4:1 MUX

Data-select Inputs		Input selected
S_1	S_0	Y
0	0	D_0
0	1	D_1
1	0	D_2
1	1	D_3

A 2-bit code and data. select(s) inputs controls the data on the input to pass through to the data output line. Binary $S_1 = 0$, $S_0 = 0$ would route D_0 on output line. Binary $S_1 = 0$, $S_0 = 1$ would route D_1 on output line and $S_1 = 1$ $S_0 = 0$ and $S_1 = 1$, $S_0 = 1$ would route D_2 and D_3 respectively on the output line.

Now let us synthesize the logic circuitry for the multiplexer. For this, we require, first of all, the Boolean expression governing the function of the MUX.

The data output is equal to D_0 only if $S_1 = 0$, $S_0 = 0$; $Y = D_0\,\overline{S}_1\,\overline{S}_0$

The data output is equal to D_1 only if $S_1 = 0$, $S_0 = 1$; $Y = D_1\,\overline{S}_1\,\overline{S}_0$

The data output is equal to D_2 only if $S_1 = 1$, $S_0 = 0$; $Y = D_2\,S_1\,\overline{S}_0$

The data output is equal to D_3 only if $S_1 = 1$, $S_0 = 1$; $Y = D_3\,S_1\,S_0$

We OR these teams together to get the total expression for the data output.

$$Y = D_0\,\overline{S}_1\,\overline{S}_0 + D_1\,\overline{S}_1\,S_0 + D_2\,S_1\,\overline{S}_0 + D_3\,S_1\,S_2$$

The expression is now implemented using 3-input AND gates, a 4-input OR gate and two inverters to generate the complements of S_1 and S_0, as shown in the figure below. As it is clear, that data can be selected from any of the given input lines, this circuit is also referred to as data selector.

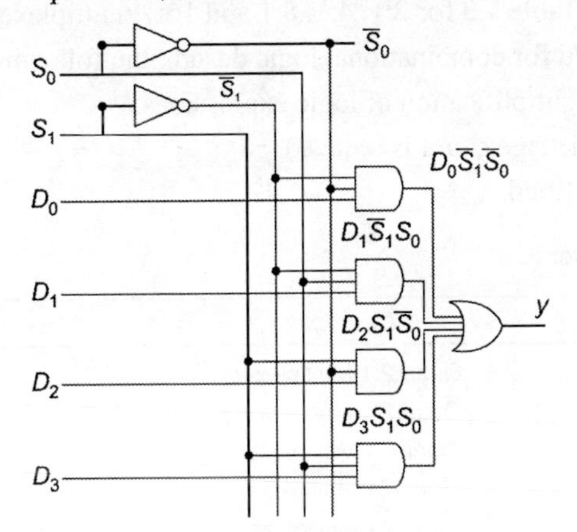

A multiplexer can be 4:1, 8:1, 16:1 and soon basically, a multiplexer is always of the form 2^m: 1, where it has m-select lines. A 4:1 multiplexer has 2 (2^2:5) select lines; 8:1 and 16:1 have respectively 3 and 4 select lines. The input which is to be output or routed to the output line is decided by the select input. A 4:1 MUX has four inputs, 8:1 has 8 inputs and a 16:1 has 16-inputs. More complex multiplexers are generally avoided and they are realized using lower order MUX. Same is the case with DEMUX. A DEMUX can be 1:4, 1:8, 1:16 and so on. It has always 1 input line and 2^m output lines with m select lines, which decides on which output line the input data is to be routed.

To generalize the concept of MUX, we can say that a MUX has n input lines and one output line. For selecting one out of n input lines for routing them to output line, a set of m select inputs is required, where $2^m = n$.

Now depending on the digital inputs applied on select lines, one of the n input lines is selected and transmitted over the output line. Normally a strobe (or ENABLE) input G is embedded which helps in cascading and it is generally an Active-Low signal, which means it performs its intended operation when it's low.

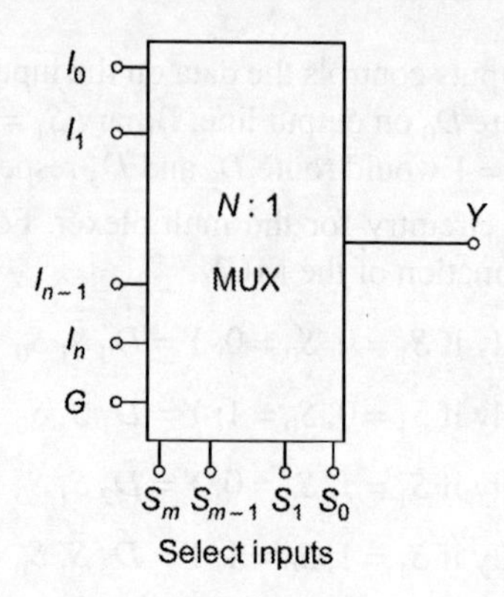

Fig. 9.2 Block Diagram of a N:1 Multiplexer

9.5 COMBINATIONAL LOGIC DESIGN USING MULTIPLEXERS

A multiplexer can easily and efficiently be used in a combinational logic design. For this even standard ICs are available as given in Table 9.3 for 2:1, 4:1, 8:1 and 16:1 multiplexers.

When a multiplexer is used for combinational logic design, the following advantages are obtained:

1. There is no need for simplification of logic expressions.
2. The number of IC package count is reduced.
3. Logic design is simplified.

Table 9.3 Various multiplexer ICs

IC No.	*Description*	*Output*
74157	Quad 2:1 Multiplexer	Same as input
74158	Quad 2:1 Multiplexer	Inverted input
74153	Dual 4:1 Multiplexer	Same as input
74352	Dual 4:1 Multiplexer	Inverted input
74151A	8:1 Multiplexer	Complementary output
74152	8:1 Multiplexer	Inverted input
74150	16:1 Multiplexer	Inverted input

9.6 DESIGN METHODOLOGY FOR COMBINATIONAL LOGIC DESIGN USING MUX

For designing a logic element using MUX, either a truth table or one of the standard forms of logic expression must be available. The design procedure is as given below:

Step I Corresponding to each minterm in the expression, identify the input lines and corresponding to these numbers, the lines are connected to logic 1 level.

Step II Connect all other input lines to logic 0 level.

Step III The inputs are applied to the select inputs.

The following example illustrates the above procedure.

Example 9.4 Implement the expression using a multiplexer

$$f(P, Q, R, S) = \Sigma m\,(0, 1, 4, 7, 8, 10, 11, 13)$$

Solution There are four variables P, Q, R & S and hence a MUX with four select inputs is required. The number of input lines will be $n = 2^4 = 16$. So we require a 16:1 MUX. Only one IC package is needed for implementation. In case the O/P of the MUX is active-low, the logic 0 and logic 1 input needs to be interchanged.

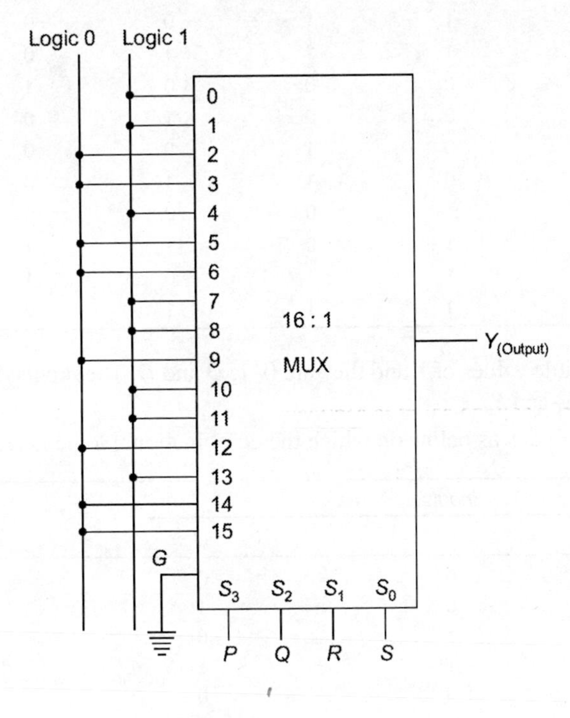

Example 9.5 Implement the problem using 8:1 MUX.

$$f(A\ B\ C\ D) = \Pi\, M\,(2, 4, 6, 7, 9, 10, 11, 12, 15)$$

Solution Simplify the expression so that only 3-select lines may suffice the requirement of the problem. Take another input D and find the relationship between D and Y, the output.

$$f(A, B, C, D) = \Pi M\,(2, 4, 6,7, 9, 10, 11, 12, 15)$$

$$= \Sigma m\,(0, 1, 3, 5, 8, 13, 14)$$

[Since ΠM and Σm terms are complementary.]

Now make the truth table

	Inputs				Outputs	
I/D	A	B	C	D	Y	
0	0	0	0	0	1	
1	0	0	0	1	1	1
2	0	0	1	0	0	
3	0	0	1	1	1	D
4	0	1	0	0	0	
5	0	1	0	1	1	D
6	0	1	1	0	0	
7	0	1	1	1	0	0
8	1	0	0	0	1	
9	1	0	0	1	0	$\overline{D}$
10	1	0	1	0	0	
11	1	0	1	1	0	0
12	1	1	0	0	0	
13	1	1	0	1	1	D
14	1	1	1	0	1	
15	1	1	1	1	0	$\overline{D}$

There are four *X* possible values of *Y* and they are 0, 1, *D* and $\overline{D}$. The inputs *A*, *B* and *C* are connected with the select lines S_2, S_1 and S_0.

The simplified truth table is as below on which the combinational logic is realized using 8:1 MUX.

Inputs			Outputs
A	B	C	Y
0	0	0	1
0	0	1	D
0	1	0	D
0	1	1	0
1	0	0	$\overline{D}$
1	0	1	0
1	1	0	D
1	1	1	$\overline{D}$

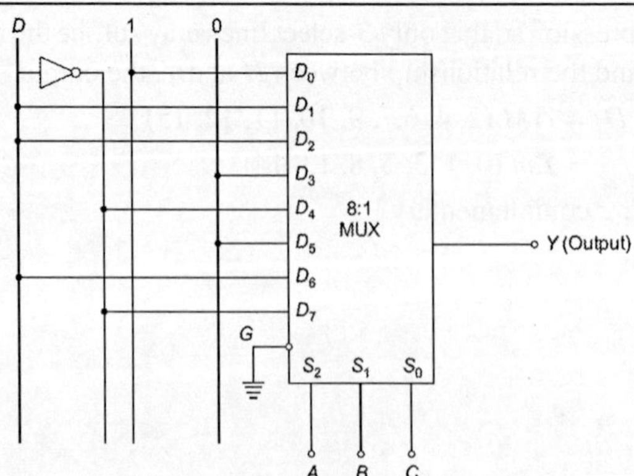

Realization of 4-variable truth table using 8:1 MUX

9.7 MULTIPLEXER TREE

Multiplexers are available in ICs. To meet the larger input needs there must be provision for expansion, so that two or more lower input MUX can be employed to analyze and realize larger input applications. For example, if we want to realize an application (system) that has 16 inputs but only 4-input MUX (4:1 MUX) are available to us, in that case Multiplexer-Trees are used.

There are two general techniques that are used to an *n* input multiplexer without any difficulty.

Technique 1 Using multiplexers and OR gate.

- Realize 32:1 MUX using two 16:1 MUX and an OR gate.

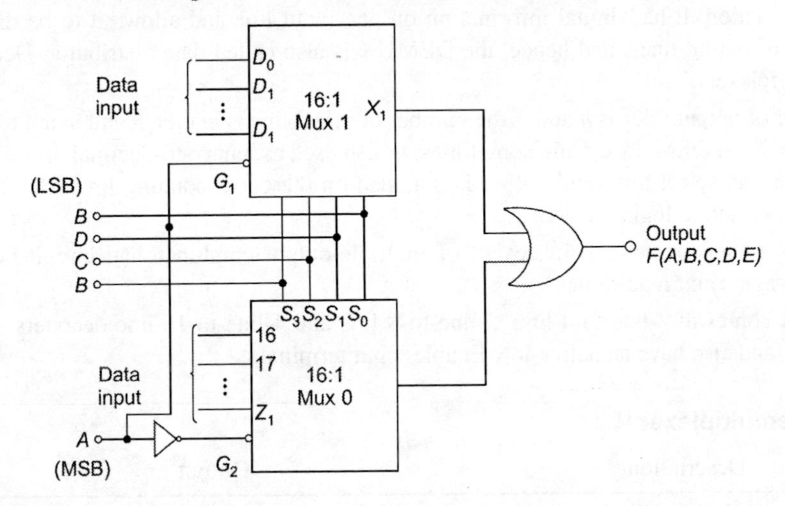

When $A = 01$, MUX 0 is selected.

When $A = 0$, MUX 1 is selected.

Technique 2 Using multiplexers only.

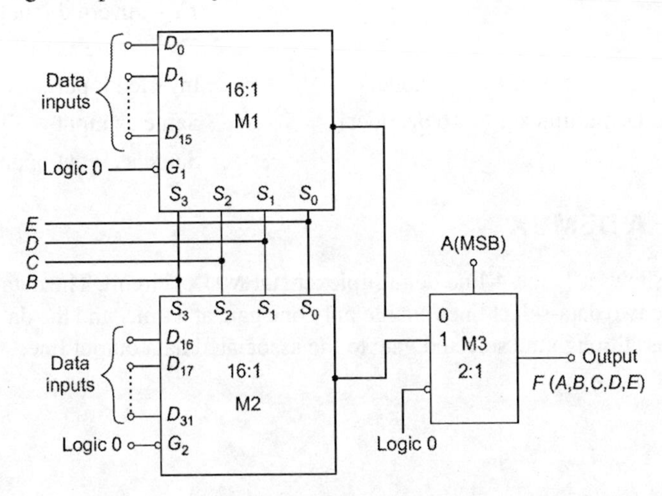

When $A = 0$, M_1 will be selected.

When $A = 0$, M_2 will be selected.

- Realize 32:1 MUX using two 16:1 MUX and one 2:1 MUX.
  When $A = 0$, M_1 will be selected.
  When $A = 1$, M_2 will be selected.

9.8 DEMULTIPLEXERS/DECODERS

The reverse operation of a multiplexer is performed by a demultiplexer. It has a single input line and it can be routed on any one of the many output lines. The select input lines decide on which output line the data input will be routed. A demultiplexer (also called DEMUX) basically performs (reverses) the multiplexing function. It has digital information on one input line and allows it to be distributed to a given number of output lines, and hence, the DEMUX is also called data distributor. Decoders can be used as demultiplexers.

The number of output lines is n and if the number of select line is m then n and m are associated with the relation $n = 2^m$. A DEMUX circuit, sometimes, is also used as binary-to-decimal decoder with binary input applied at the select-lines and output is obtained on the corresponding line. The data input line needs to be connected to logic 1 level.

A DEMUX is very beneficial in design of multiple-output combinational circuit because of the minimum package count requirement.

A DEMUX comes in 2-line-to-4 line, 3-line-to-8-line and 4-line-to-16 line decoders. Their outputs are active-low, and also have an active-low Enable input terminal.

Available Demultiplexer ICs

IC No.	Description	Output
74139	Dual 1:4 Demultiplexer (2:4 decoder)	Inverted input
74155	Dual 1:4 Demultiplexer (2:4 decoder)	1Y – Inverted input 2Y – Same as input open collector
74156	do	1Y – Inverted input 2Y – Same as input
74138	1:8 Demultiplexer (3:8 Decoder)	Inverted input
74154	1:16 Demultiplexer (4:16 decoder)	Same as input
74159	do	Same as input open collector.

9.9 DESIGN OF A DEMUX

The following figure shows a 1-line-4-line demultiplexer (DEMUX) circuit. The data-input line goes to all the AND gates. The two data-select lines enable only one gate at a time, and the data appearing on the data-input line will pass through the selected gate to the associated data output line.

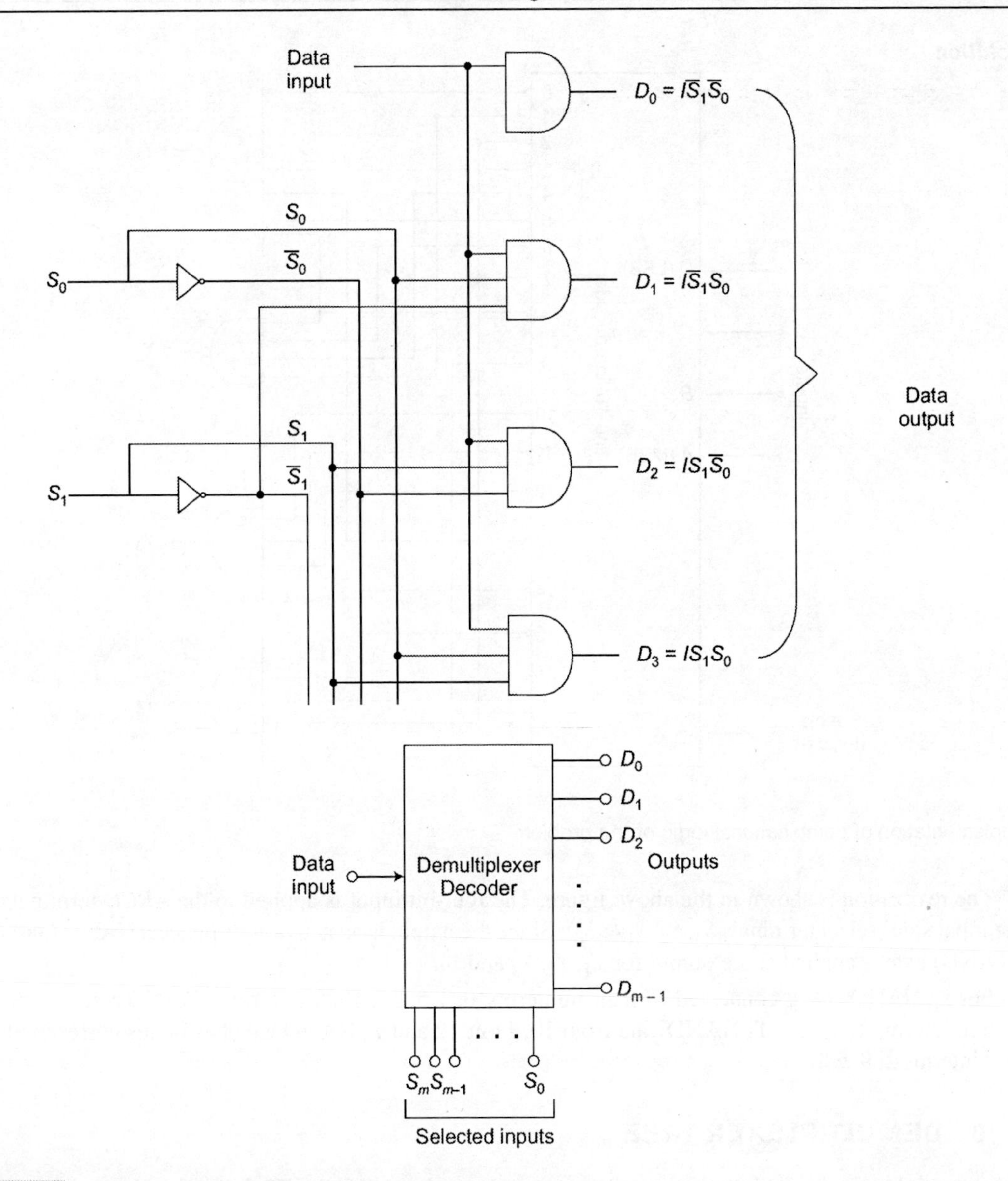

Fig. 9.3 Block diagram of a demultiplexer

Example 9.6 Implement the following multi-output logic circuit using a 4-to-16 line decoder.

$$Y_1 = \Sigma m\ (0, 3, 5, 8, 13)$$
$$Y_2 = \Sigma m\ (1, 4, 6, 7)$$
$$Y_3 = \Sigma m\ (10, 11, 12)$$
$$Y_4 = \Sigma m\ (2, 9, 14)$$

Solution

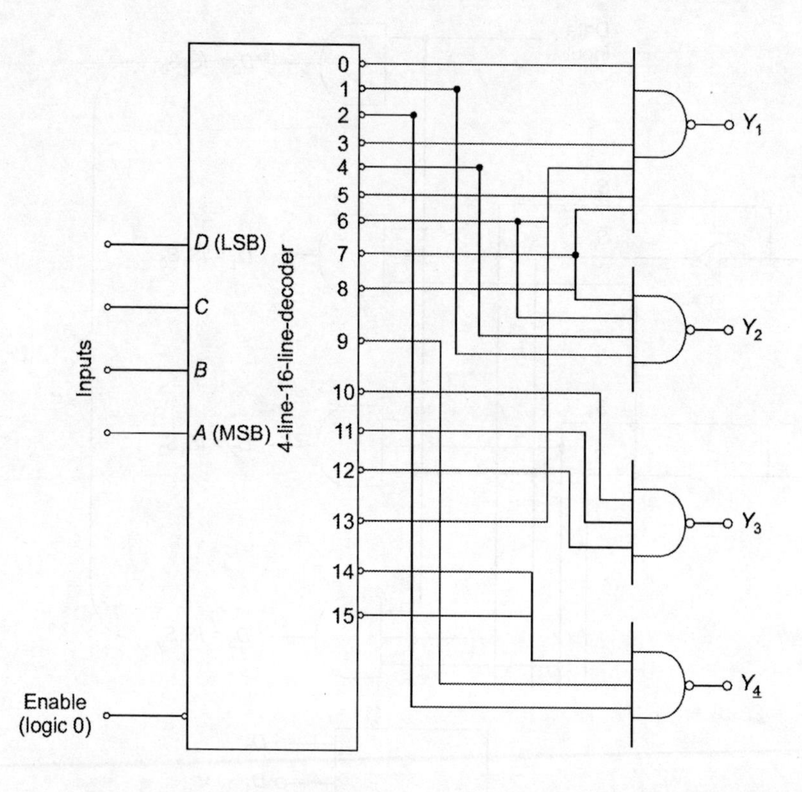

Implementation of combinational logic of the problem

The realization is shown in the above figure. The four-bit input is applied to the *ABCD* terminals in the input side (select terminals S_3, S_2, S_1 & S_0). Since the output is an active-low, hence a NAND (and not an AND) gate is applied at the output for Y_1, Y_2, Y_3 and Y_4.

For Y_1, NAND gate is connected with the minterms, 0, 3, 5, 8 & 13. Similarly Y_2 NAND gate has inputs from minterms 1, 4, 6, 7, Y_3 NAND gate from 10, 11 & 12 and Y_4 NAND gate has inputs corresponding to minterms 2, 9 & 14.

9.10 DEMULTIPLEXER TREE

Like multiplexers, demultiplexers can also be employed in tree form. Their need and requirement is same as explained for multiplexer tree, i.e., they are meant as the provision for expansion. This is made possible by ENABLE input terminal. The following figure shows 5-line-to-32 line decoder using a two 4-line-16-line decoder. Similarly, an 8-line-to-256 line decoder can be realized using four 4-line-to-16 line decoders.

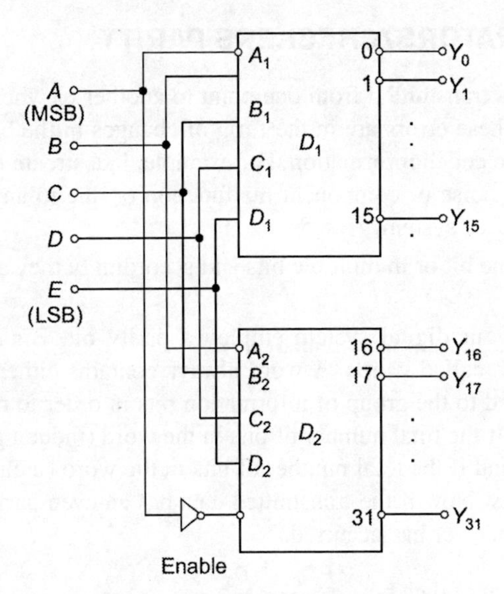

Fig. 9.4 5-line-to-32-line decoder using two 4-line-to-16-line decoders.

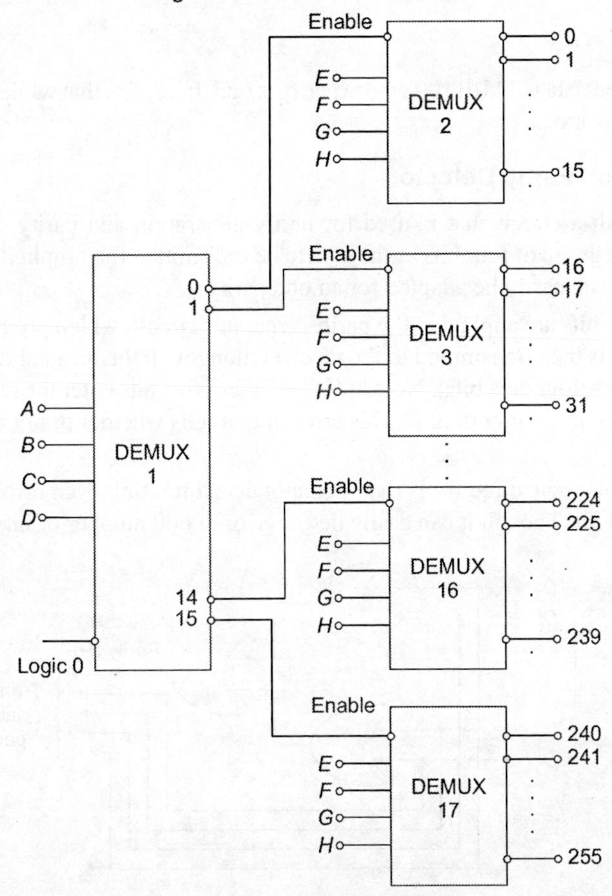

Fig. 9.5 8-line-to-256-line decoder using 4-line-to-16-line decoders tree.

9.11 PARITY GENERATORS/CHECKERS PARITY

Whenever any digital data is transmitted from one point to another or when the data is within a digital system, errors may occur. These errors are in the form of changes in the bits which is undesirable and it changes the contents of the coded information. For example, in a stream of data 0 may change to 1 or vice versa, due to electrical noise or component malfunction or due to any other reason. No practical system can ever be an error-free system.

The error may occur in one bit or in multiple bits and accordingly they are termed single-bit error or multibit error.

To detect errors, we, in our digital system employ a parity bit. Binary informations are always transmitted in groups of bits called words. A word always contains either an even or an odd number of 1's. A parity bit is attached to the group of information bits in order to make the total number of 1's always even or always odd. If the total number of bits in the word (inducing parity bit) is even, then the parity is called even parity and if the total number of bits in the word (including parity bit) is odd, then the parity is called odd parity. Now if the transmitted data has an even parity and the received data has odd parity, this means that an error has occurred.

For example, let the stream of data be: $\overset{P}{\boxed{1}}\overset{D_6}{1}001\overset{D_0}{100}$.

here parity is even.

Suppose the received data is 11101100 whose parity is odd. It implies that while the data is transmitted, there occurred some error.

Parity Generator and Parity Detector

Now we will deal with a circuit that is used for parity generation and parity checking. Here, in this example, we will use a group of four bits as the data to be transmitted for simplicity, and we will use even parity bit. The system can easily be adapted for an odd parity.

First of all, the data-bits are applied to the parity-generator circuit, which produces an even parity bit, *P*, at its output, which is then transmitted to the receiver along with the original data bits, making a total of five bits (1 parity bit + four data bits). Now in Fig. 9.6 these five bits enter the receiver's parity-checker circuit, which produces an error output, *E*. This error output tells whether or not any single-bit-error has occurred.

The limitation of this logic circuitry is that it cannot detect multibit even errors, neither it can locate the position of errored bit. Though it can easily detect error in odd number of bits but not its location.

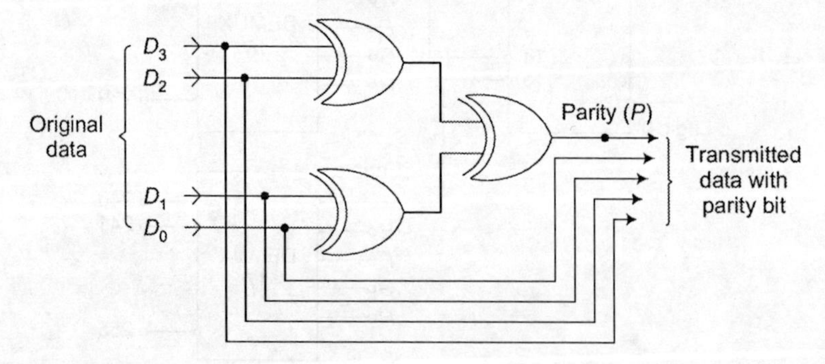

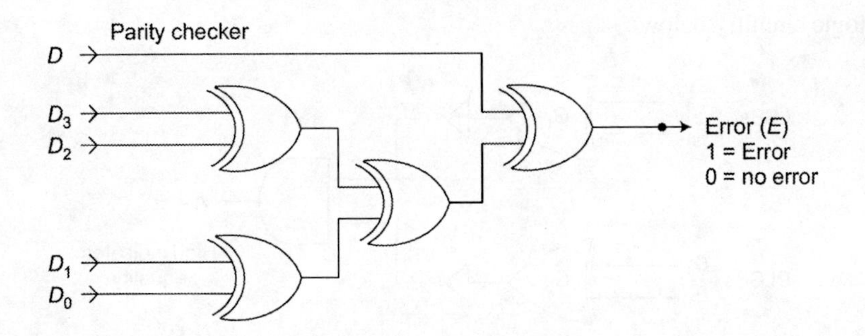

Fig. 9.6 Even parity checker circuitry.

In order to check for or to generate the proper parity in a given code, the basic principle used is:

The sum (disregarding carries) of an even number of 1's is always zero, and the sum of an odd number of 1's is always 1.

So in order to find the parity (even or odd) all the bits of the bit-stream (data) are added by exerting them one by one. The sum of four bits can be done by using three XOR gates connected as shown below. When the number of 1's on the inputs is even, the OP X is 0 (low), when the number of 1's is odd, output X is 1 (high).

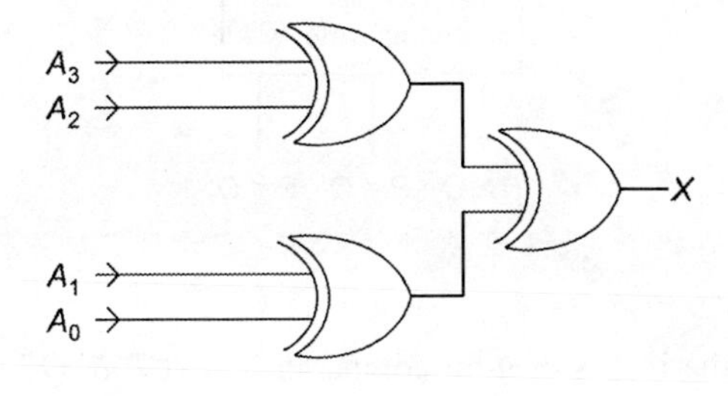

9.12 MAGNITUDE COMPARATORS

A magnitude comparator is a logic circuitry that is used to compare the magnitude (quite clear from the name) of two quantities and then determine the relationship of the quantities compared.

The gate that acts as the basic comparator is exclusive-OR gate owing to the fact that a XOR-gate has always a high (1) output when the input bits are unequal and low (0)

Suppose we have two numbers $A \rightarrow D_1D_0$ and

$B \rightarrow E_1E_0$

Refer to the logic circuitry below:

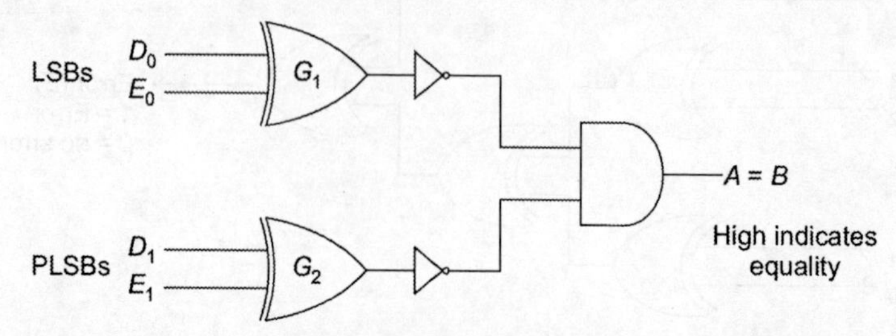

Here the LSBs (D_0 & E_0) of the numbers are applied to the inputs of XOR-gate G_1, i.e., the LSBs are compared by gate G_1, and the two MSBs are compared by Gate G_2. Now if the two numbers are same, the outputs of both the XOR-gates (G_1 & G_2) would be low and hence the overall output will be high. In case any of the bits of the two numbers differ, XOR-gate outputs will be high and hence overall output will be low, and thus, it can be concluded that the two numbers are unequal.

9.13 DIGITAL COMPARATOR

Comparators can be designed for comparing multibit numbers. Given below is the block diagram of logic circuitry for comparing a multibit number.

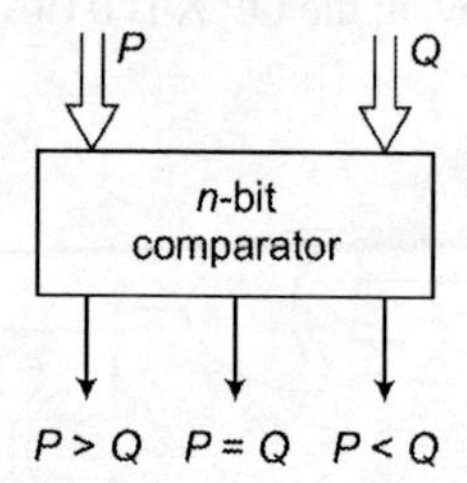

Fig. 9.7 Digital comparator.

Two n-bit numbers form the inputs of n-bit comparators. Let P & Q be the two m-bit numbers (in example we will deal with 2-bit number for simplicity). The expressions for P and Q, $R < Q$ and $P = Q$ are simplified by using K-map and then realized using gates.

Truth-Table of 2-bit Comparator

Inputs				*Outputs*		
P_1	P_0	Q_1	Q_0	$P > Q$	$P = Q$	$P < Q$
0	0	0	0	0	1	0
0	0	0	1	0	0	1
0	0	1	0	0	0	1
0	0	1	1	0	0	1
0	1	0	0	1	0	0
0	1	0	1	0	1	0
0	1	1	0	0	0	1
0	1	1	1	0	0	1

1	0	0	0	1	0	0
1	0	0	1	1	0	0
1	0	1	0	0	1	0
1	0	1	1	0	0	1
1	1	0	0	1	0	0
1	1	0	1	1	0	0
1	1	1	0	1	0	0
1	1	1	1	0	1	0

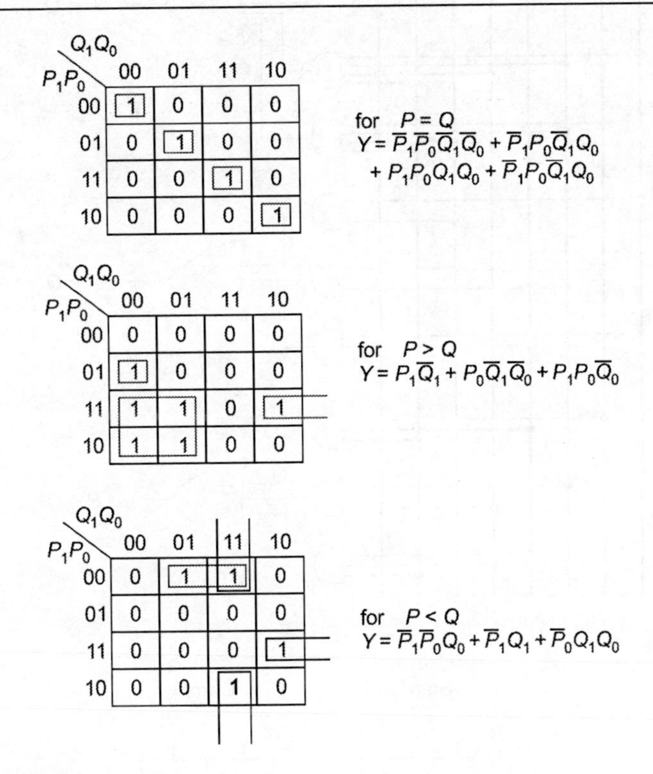

9.14 HALF ADDER

Half adder is a combinational logic that adds two single-bit numbers and output sum and carry. Truth table of half adder, K-map and gate realization of the same is given below:

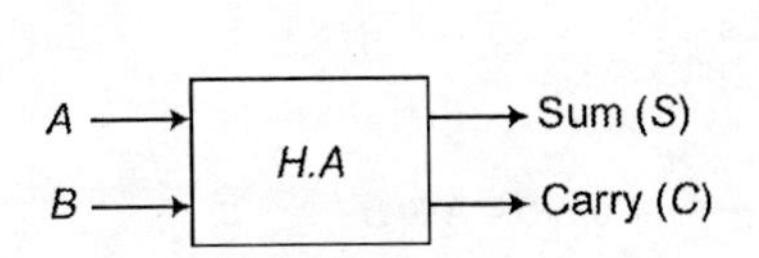

Fig. 9.8 Block diagram of half adder.

Input		Output	
A	B	S	C
0	0	0	0
0	1	1	0
1	0	1	0
1	1	0	1

Table 9.3 Truth table of half adder.

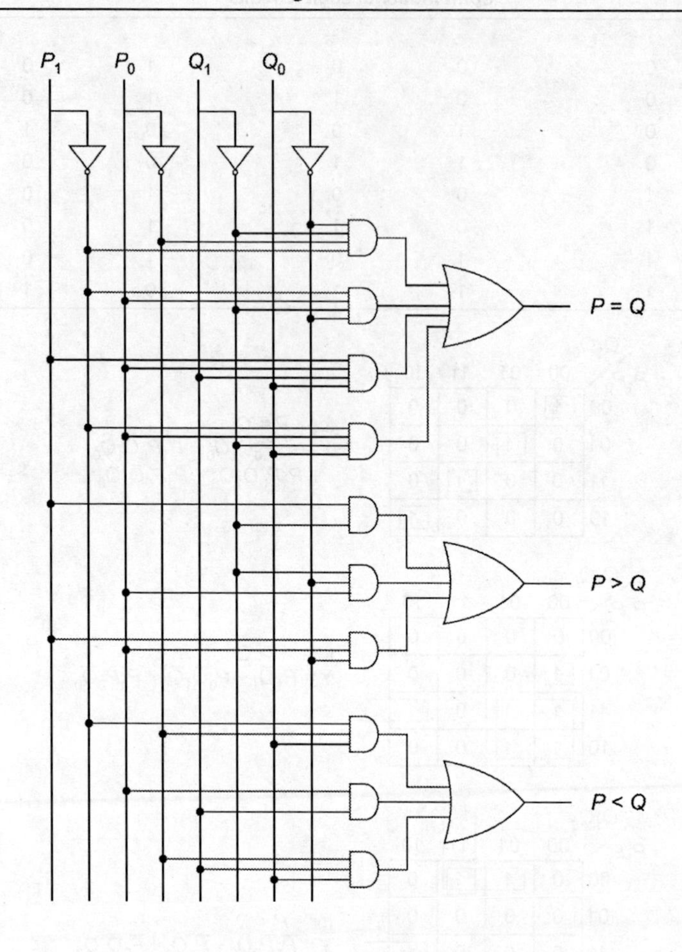

K-maps

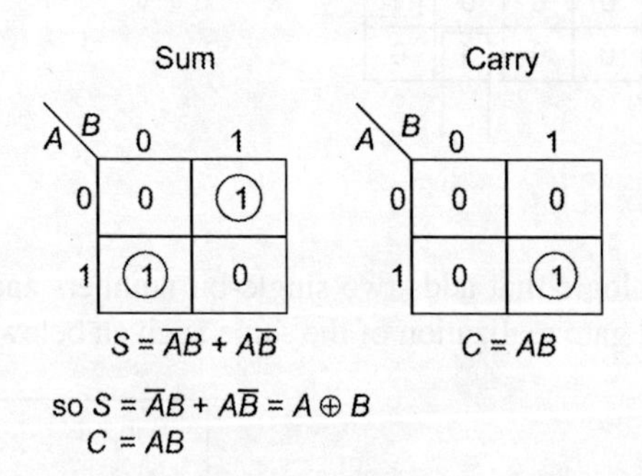

so $S = \overline{A}B + A\overline{B} = A \oplus B$

$C = AB$

Gate implementation

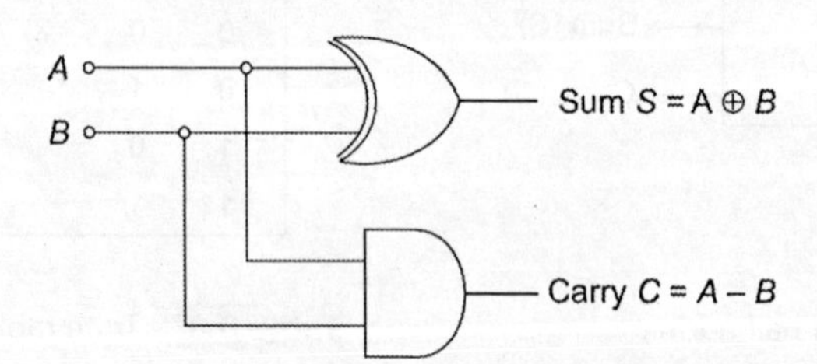

Half adder using NAND gates only

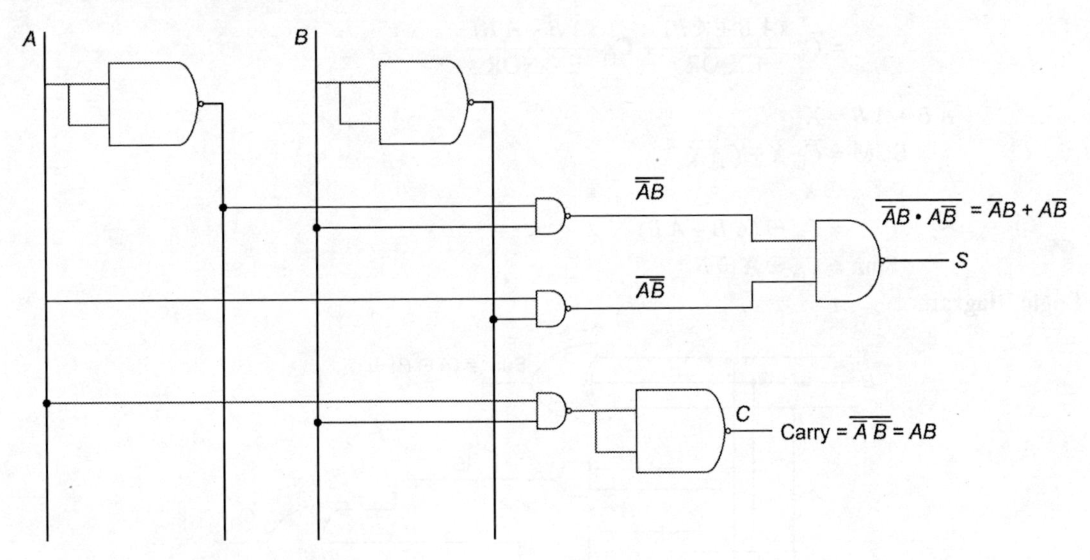

9.15 FULL ADDER

If full adder can add two one-bit numbers A & B and carry C_{in}. The full adder is a three-input and two-output combinational circuit.

(a) Block diagram

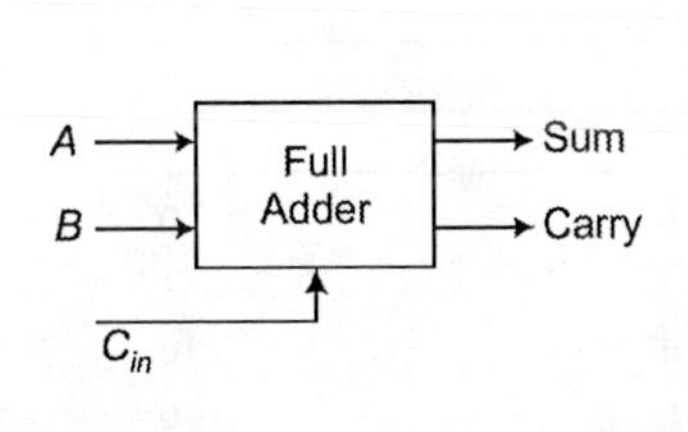

Fig. 9.9 Block diagram of full adder circuit.

Table 9.3 Truth table of full adder.

Inputs			*Outputs*	
A	*B*	C_{in}	*Sum*	*Carry*
0	0	0	0	0
0	0	1	1	0
0	1	0	1	0
0	1	1	0	1
1	0	0	1	0
1	0	1	0	1
1	1	0	0	1
1	1	1	1	1

K-map simplification

1. Sum output

C_{in} \ AB	00	01	11	10
0	0	1	0	1
1	1	0	1	0

2. Carry output

C_{in} \ AB	00	01	11	10
0	0	0	1	0
1	0	1	1	1

$$\text{Sum} = \overline{A}\,B\,\overline{C}_{in} + A\,\overline{B}\,\overline{C}_{in} + \overline{A}\,\overline{B}\,C_{in} + ABC_{in} \qquad \text{Carry} = AB + BC_{in} + AC_{in}$$

$$= \overline{C}_{in}\underbrace{(\overline{A}\,B + A\,\overline{B})}_{\text{EX-OR}} + C_{in}\underbrace{(\overline{A}\,\overline{B} + A\,B)}_{\text{EX-NOR}}$$

Let $\overline{A}\,B + A\,\overline{B} = X$

$\therefore$ $\text{SUM} = \overline{C}_{in}\,X + C_{in}\,\overline{X}$

$C_{in} \oplus X$

$= C_{in} \oplus (\overline{A}\,B + A\,\overline{B})$

$\text{Sum} = C_{in} \oplus A \oplus B$

Logic Diagram

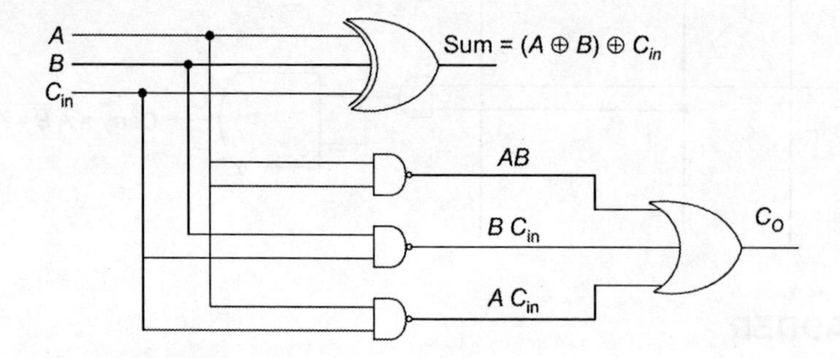

9.16 HALF SUBTRACTOR

Half subtractor provides the difference and borrow output of two single bit inputs.

If we have to perform P-Q, then P is called minuend and Q is called subtrahend.

Truth table, K-map and gate realization of half subtractor are given below.

(a) Truth Table

Inputs		*Outputs*	
A	*B*	*A-B = D D = Difference*	*Borrow B_0*
0	0	0	0
0	1	1	1
1	0	1	0
1	1	0	0

(b) K-maps

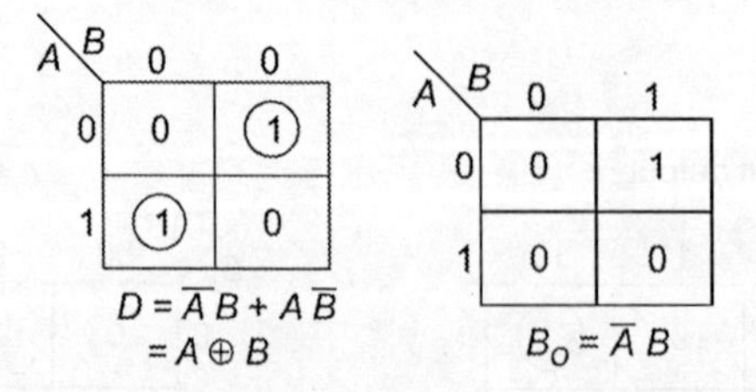

(c) Gate implementation

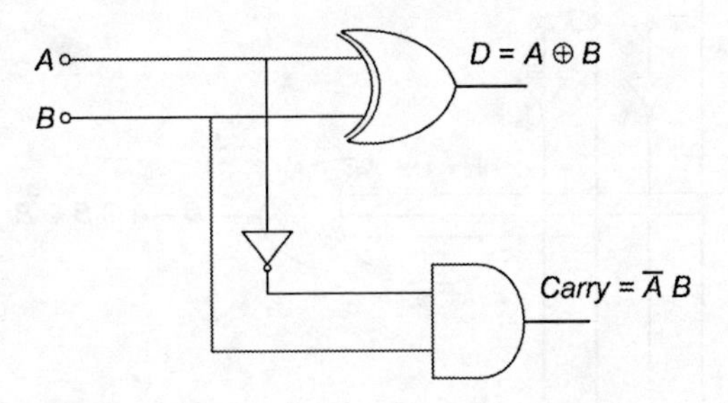

Half subtractor using NAND gates

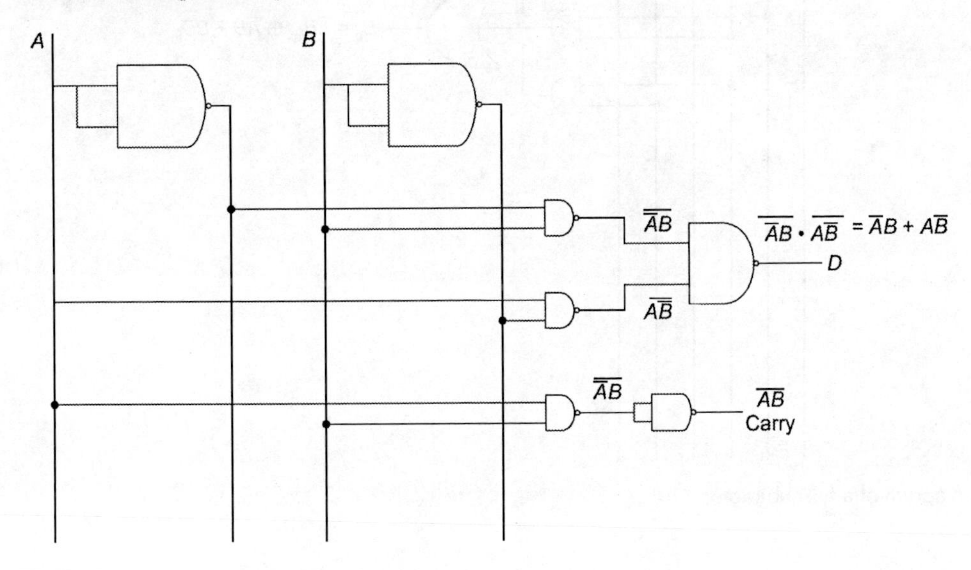

$$D = \overline{A}\,\overline{B}\,B_{in} + \overline{A}\,B\,\overline{B}_{in} + A\,\overline{B}\,\overline{B}_{in} + A\,B\,B_{in}$$
$$= B_{in}\,(\overline{A}\,\overline{B} = A\,B) + \overline{B}_{in}\,(\overline{A}\,B + A\,\overline{B})$$
$$= B_{in}\,(A \oplus B) + \overline{B}_{in}\,(A \oplus B)$$
$$= B_{in}\,\overline{X} + \overline{B}_{in}\,X$$
$$= B_{in} \oplus X$$
$$D = B_{in} \oplus A \oplus B$$

and $$B_0 = \overline{A}\,B_{in} + \overline{A}\,B + B\,B_{in}$$

9.17 FULL SUBTRACTOR

A full subtractor is a combinational circuit with inputs A, B and B_{in} and output D and B_0, where A is minuend, B is subtrahend, B_{in} is borrow in and D is difference output m, (A-B) and B_0 is borrow output.

A full subtractor performs the difference between two single-bit numbers and input borrow bit produces their difference and output borrow.

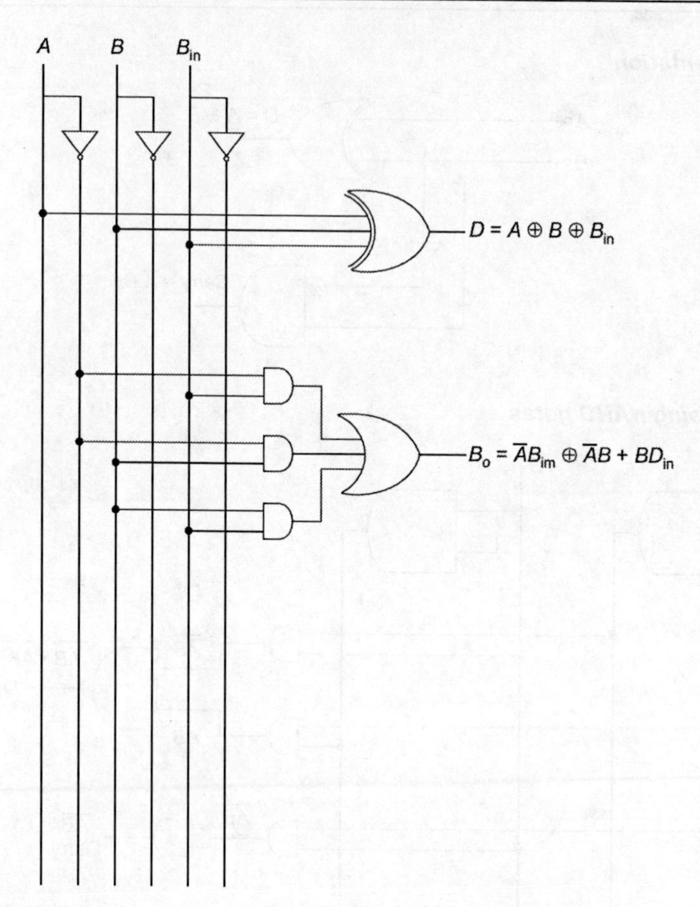

Logic diagram of a full subtractor

The truth table, K-maps and gate realization of the same are given below.

(a) Truth Table

Input			Output	
A	*B*	B_{in}	$D = A\text{-}B\text{-}B_{in}$	B_0
0	0	0	0	0
0	0	1	1	1
0	1	0	1	1
0	1	1	0	1
1	0	0	1	0
1	0	1	0	0
1	1	0	0	0
1	1	1	1	1

K-maps

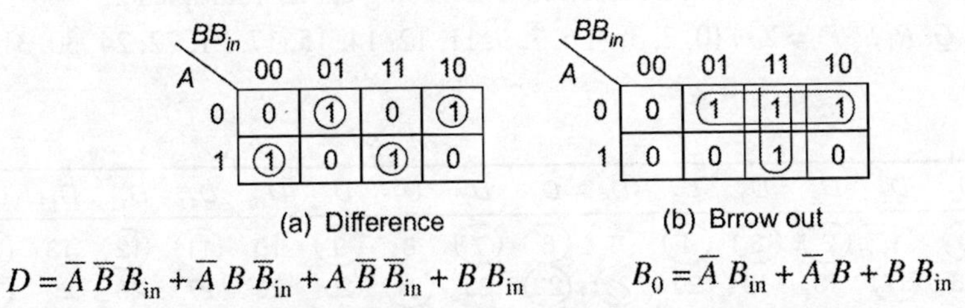

$$D = \overline{A}\,\overline{B}\,B_{in} + \overline{A}\,B\,\overline{B}_{in} + A\,\overline{B}\,\overline{B}_{in} + B\,B_{in} \qquad B_0 = \overline{A}\,B_{in} + \overline{A}\,B + B\,B_{in}$$

Example 9.7 Implement the following function using 16:1 multiplexer.

$f(P, Q, R, S, T) = \Sigma m\,(3, 5, 7, 11, 15, 16, 17, 18, 25, 27, 30, 31)$

Solution Design Table

Inputs	D_0	*1*	*2*	*3*	*4*	*5*	*6*	*7*	*8*	*9*	*10*	*11*	*12*	*13*	*14*	*15*
$\overline{P}$	0	1	2	3	4	5	6	7	8	9	10	11	12	13	14	15
P	16	17	18	19	20	21	22	23	24	25	26	27	28	29	30	31
input to MUX	P	P	P	$\overline{P}$	0	$\overline{P}$	0	$\overline{P}$	0	P	0	1	0	0	0	1

2. Realization

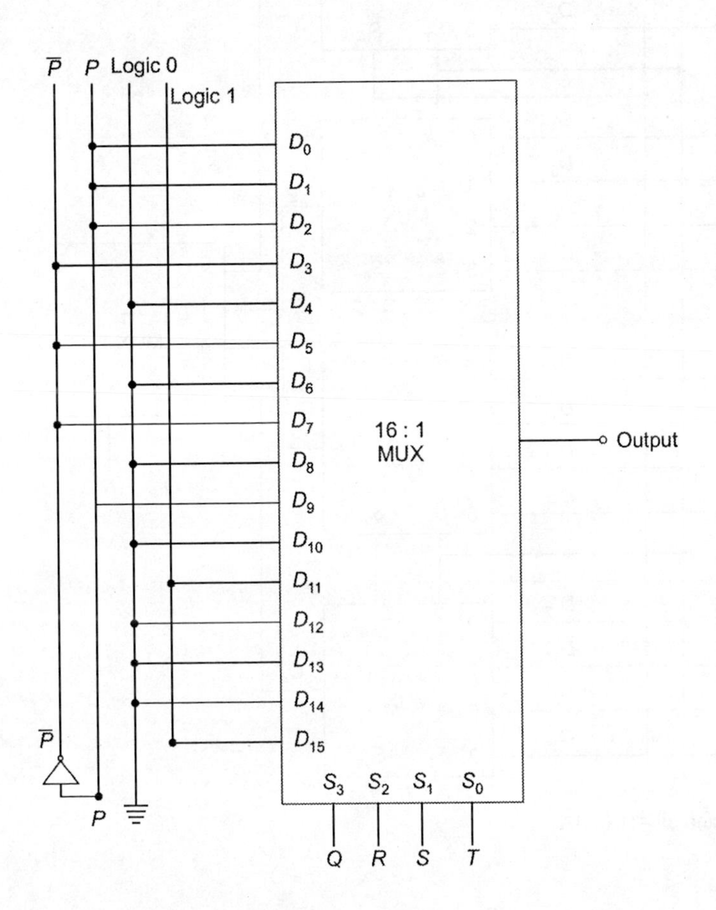

Example 9.8 Implement the following Boolean function using all 4:1 multiplexers

$$f(P, Q, R, S, T) = \Sigma m\,(0, 2, 3, 4, 6, 7, 9, 11, 12, 14, 15, 17, 21, 22, 24, 30, 31)$$

Solution

1. Design table

Inputs	D_0	D_1	D_2	D_3	D_4	D_5	D_6	D_7	D_8	D_9	D_{10}	D_{11}	D_{12}	D_{13}	D_{14}	D_{15}
$\overline{P}$	(0)	1	(2)	(3)	(4)	5	(6)	(7)	8	(9)	10	(11)	(12)	13	(14)	(15)
P	16	(17)	18	19	20	(21)	(22)	23	(24)	25	26	27	28	29	(30)	(31)
Input to MUX	$\overline{P}$	P	$\overline{P}$	$\overline{P}$	$\overline{P}$	P	1	$\overline{P}$	P	$\overline{P}$	0	$\overline{P}$	$\overline{P}$	0	1	1

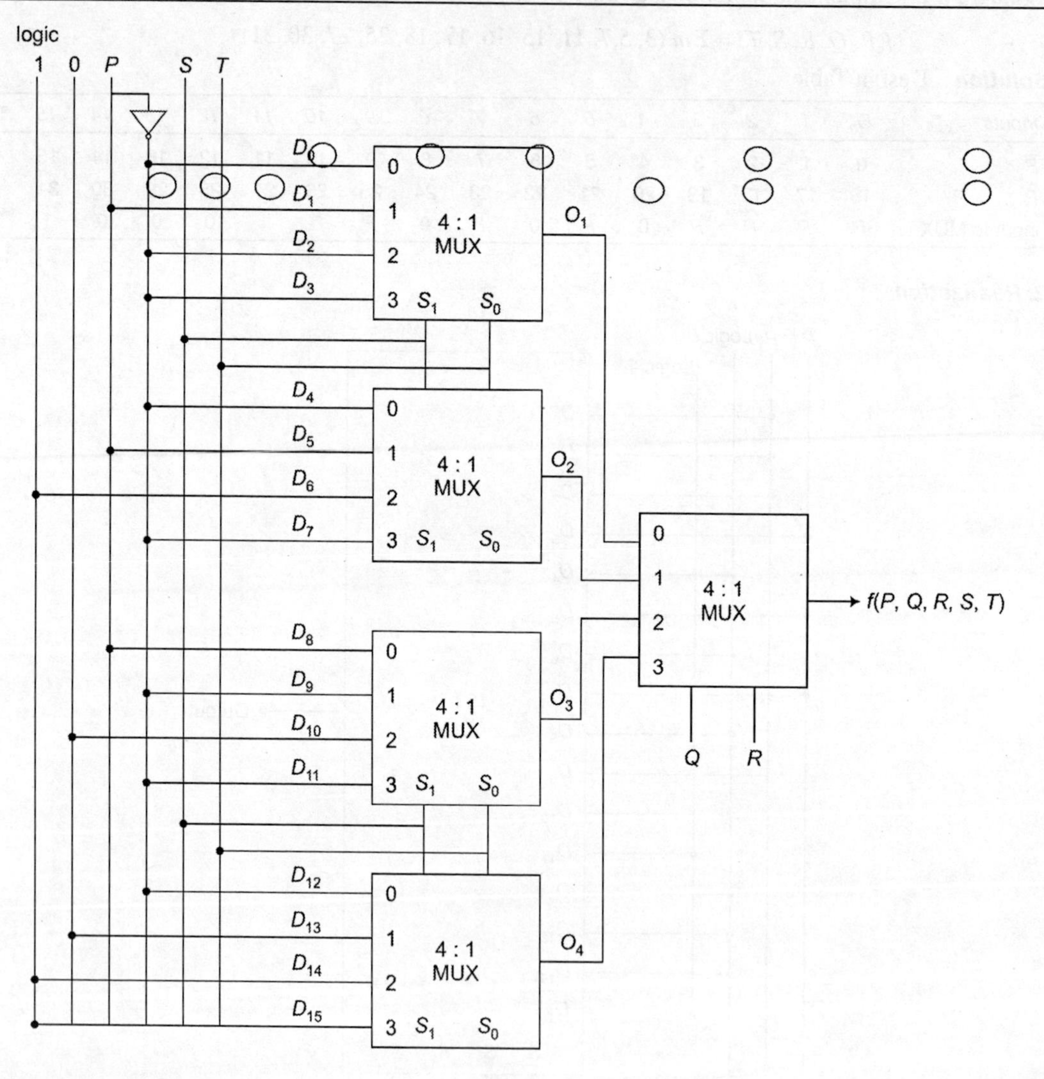

Implementation using all 4:1 MUX

Example 9.9 Implement using an 8:1 multiplexer

$$f(P, Q, R, S) = \Pi M\,(1, 3, 5, 7, 9, 10, 12, 14)$$

Solution

1. Design table.

	D_0	D_1	D_2	D_3	D_4	D_5	D_6	D_7
$\overline{P}$	(0)	1	(2)	3	(4)	5	(6)	7
P	(8)	9	10	(11)	12	(13)	14	(15)
input to MUX	1	0	$\overline{P}$	P	$\overline{P}$	P	$\overline{P}$	P

2. Implementation using 8:1 MUX:

Example 9.10 Implement the full subtractor using a 1:8 demultiplexer.

Solution 1. Truth Table

Inputs			*Outputs*	
A	*B*	B_{in}	*D*	B_{out}
0	0	0	0	0
0	0	1	1	1
0	1	0	1	1
0	1	1	0	1
1	0	0	1	0
1	0	1	0	0
1	1	0	0	0
1	1	1	1	1

$$D = f(A, B, B_{\text{in}}) = \Sigma m\,(1, 2, 4, 7)$$

$$B_{\text{out}} = f(A, B, B_{\text{in}}) = \Sigma m\,(1, 2, 3, 7)$$

2. Realization: D_{in} is connected to logic 1 permanently and A, B and B_{in} to the select input S_2, S_1, S_0 respectively.

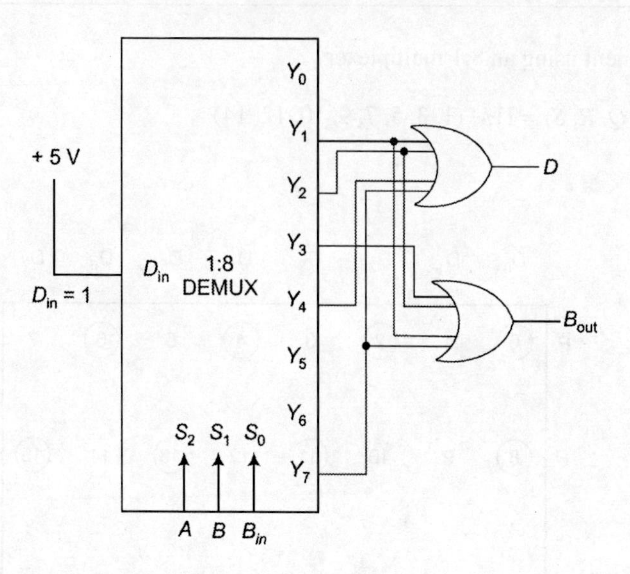

Example 9.11 Implement a full adder using demultiplexer.

Solution Truth table of a full adder

Inputs			*Outputs*	
A	*B*	C_{in}	*Sum*	*Carry*
0	0	0	0	0
0	0	1	1	0
0	1	0	1	0
0	1	1	0	1
1	0	0	1	0
1	0	1	0	1
1	1	0	0	1
1	1	1	1	1

2. From the truth table, we can conclude that:

$$\text{Sum} = \Sigma m\ (1, 2, 4, 7)$$

and

$$\text{Carry} = \Sigma m\ (3, 5, 6, 7)$$

3. Realization.

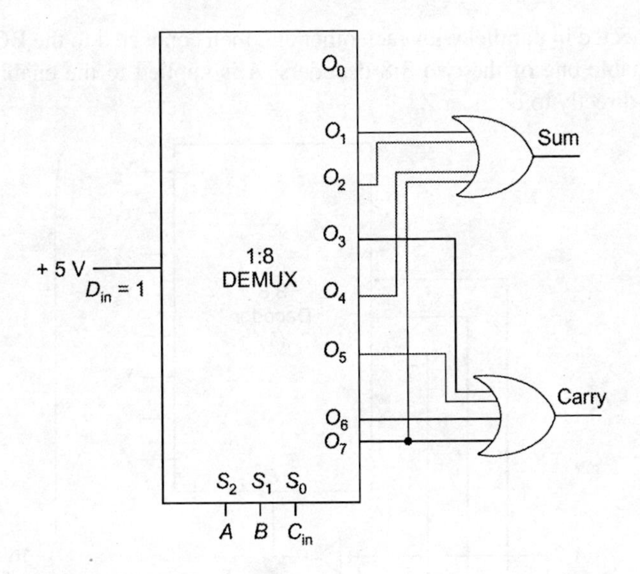

DECODER

Example 9.12 Implement the following Boolean function using a 3:8 decoder and external gates.

$$f(P, Q, R) = \Sigma\ (1, 2, 3, 6, 7)$$

Solution The decoders produce minterms. The outputs Y_1, Y_2, Y_3, Y_6 and Y_7 are required to produce the required output logic implementation.

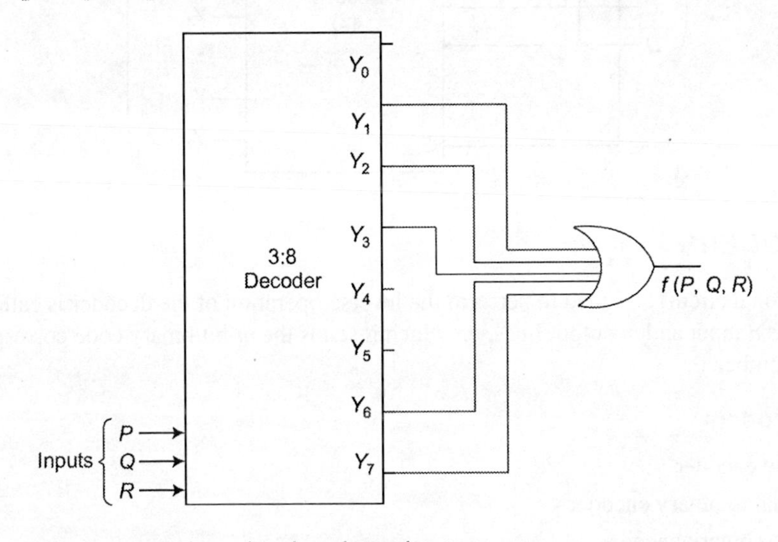

Implementation of Boolean equation using decoder and gate

Example 9.13 Implement a 4-line-to 16-line Decoder using 3:8 decoder.

Solution Connection diagram is shown below:

Three inputs are connected in parallel with each other and then connected to the BCD inputs. The fourth input A is used to enable one of the two 3:8 decoders. $\bar{A}$ is applied to the enable input of decoder 1 whereas A is applied directly to decoder Z.

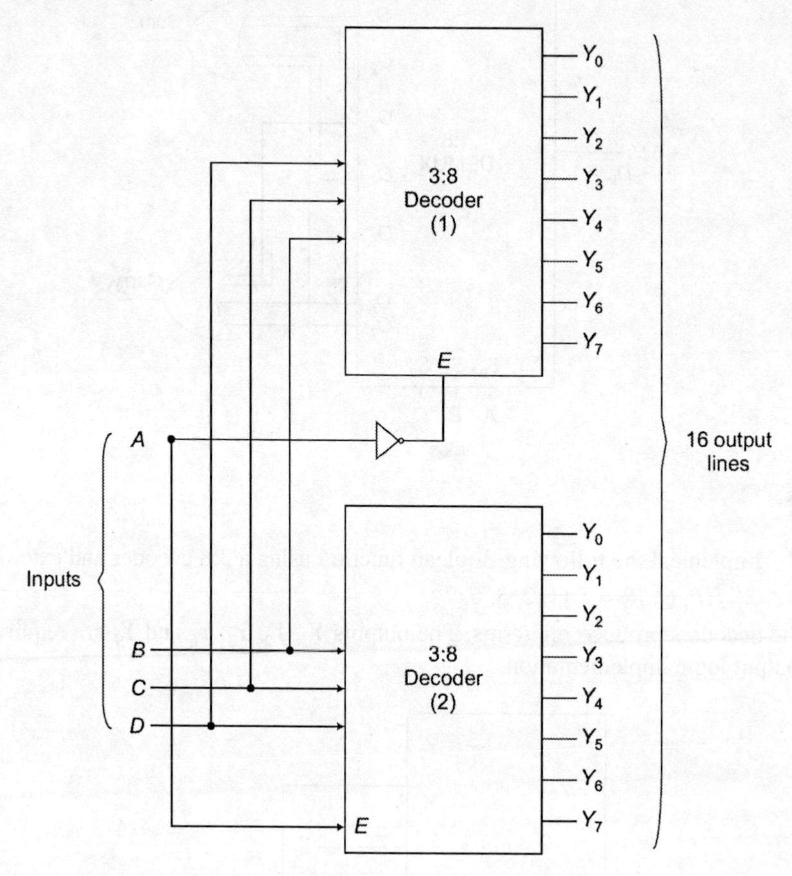

9.18 ENCODERS

The combinational circuit designed to perform the inverse operation of the decoder is called an encoder. An encoder has n input and m output lines, on which it sends the m-bit binary code corresponding to the digital input number.

Types of Encoders

1. Priority encoder
2. Decimal to binary encoder
3. Octal to binary encoder
4. Hexadecimal to binary encoder

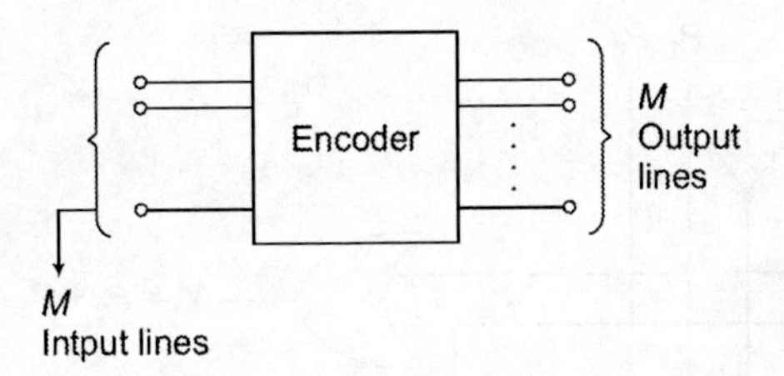

Fig. 9.10 Block diagram at encoder.

1. Priority Encoder This encoder prioritizes between the inputs if two or more input lines simultaneously go high.

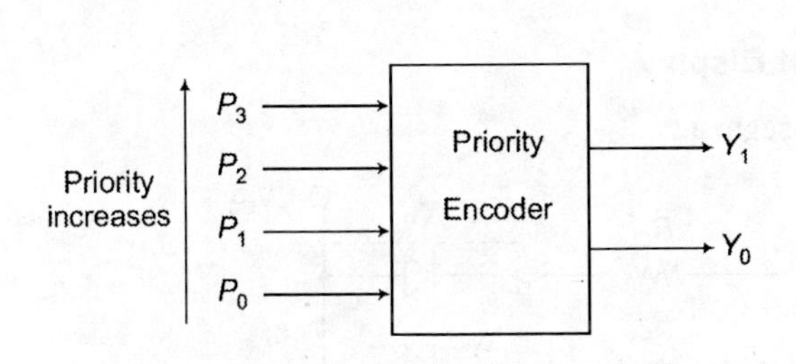

(a) Block diagram of a priority encoder.

Input				Outputs	
P_3	P_2	P_1	P_0	Y_1	Y_0
0	0	0	0	X	X
0	0	0	1	0	0
0	0	1	X	0	1
0	1	X	X	1	0
1	X	X	X	1	1

(b) Truth table of a priority encoder.

K-map

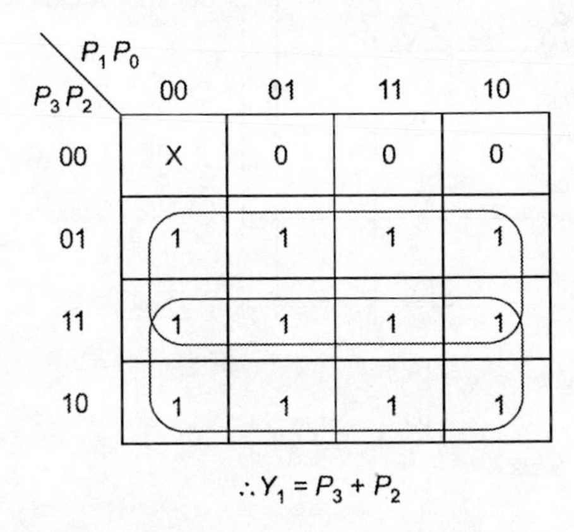

$P_1 P_0$
$P_3 P_2$
00
01
11
10
00
X
0
0
0
01
1
1
1
1
11
1
1
1
1
10
1
1
1
1

$\therefore Y_1 = P_3 + P_2$

$P_3P_2 \backslash P_1P_0$	00	01	11	10
00	X	0	1	1
01	0	0	0	0
11	1	1	1	1
10	1	1	1	1

$\therefore Y_0 = P_3 + \overline{P}_2 P_1$

III Gate Implementation

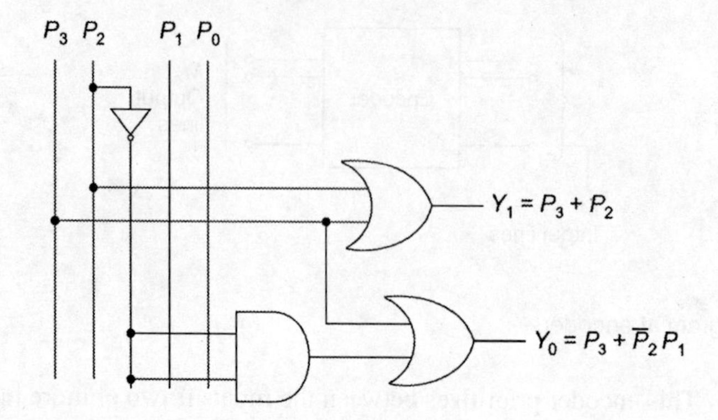

Priority Encoder Logic Diagram.

Use of Decoder for Driving 7-Segment Display

1. Circuit set-up for common-anode 7-segmant.

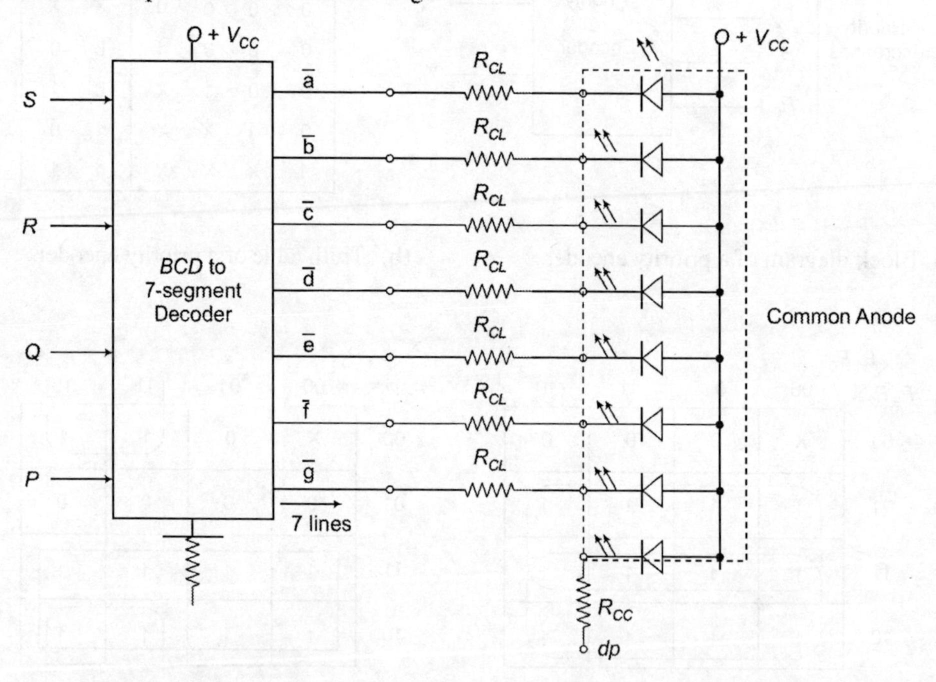

R_{cl} = Current Limiting Resistors.

2. Circuit set-up for a common cathode 7-segment display

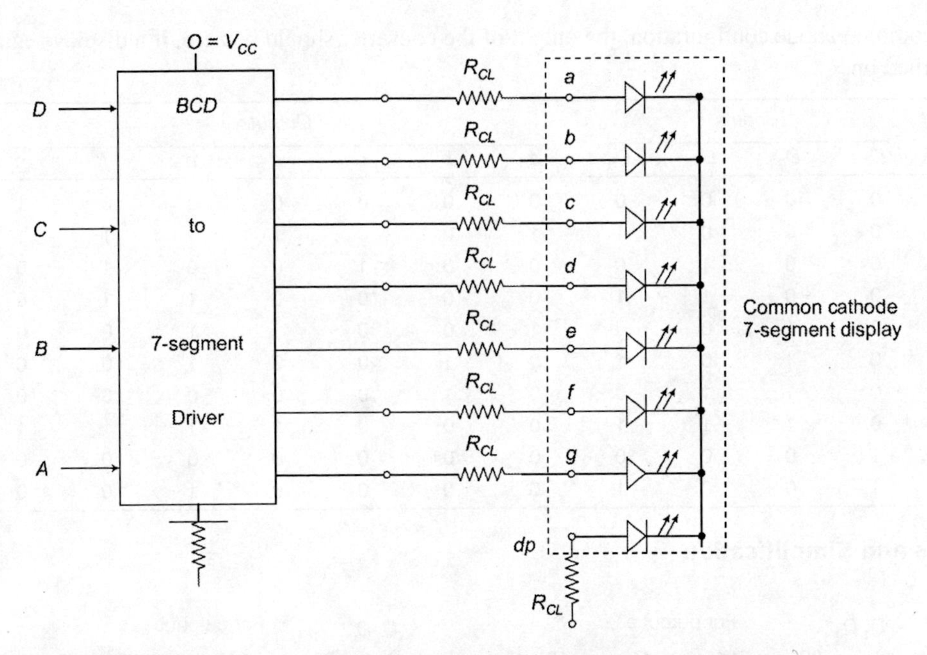

9.19 BCD TO SEVEN-SEGMENT DISPLAY CODE CONVERTER

1. Common Anode

Construction of a seven-segment display.

Each segment (a - g and dp) consists pf an λeD

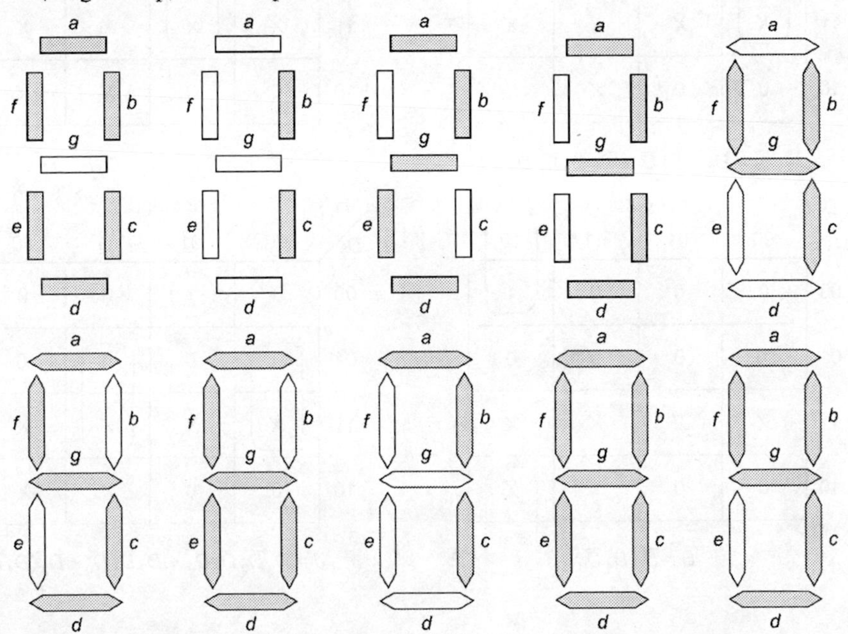

Fig. 9.11 7-segment display of various digits

For common anode configuration, the output of the converter should be zero, if a display segment is to be turned on.

Decimal	Inputs				Outputs						
	D_3	D_2	D_1	D_0	a	b	c	d	e	f	g
0	0	0	0	0	0	0	0	0	0	0	1
1	0	0	0	1	1	0	0	1	1	1	1
2	0	0	1	0	0	0	1	0	0	1	0
3	0	0	1	1	0	0	0	0	1	1	0
4	0	1	0	0	1	0	0	1	1	0	0
5	0	1	0	1	0	1	0	0	1	0	0
6	0	1	1	0	0	1	0	0	0	0	0
7	0	1	1	1	0	0	0	1	1	1	1
8	1	0	0	0	0	0	0	0	0	0	0
9	1	0	0	1	0	0	0	0	1	0	0

K-maps and Simplification

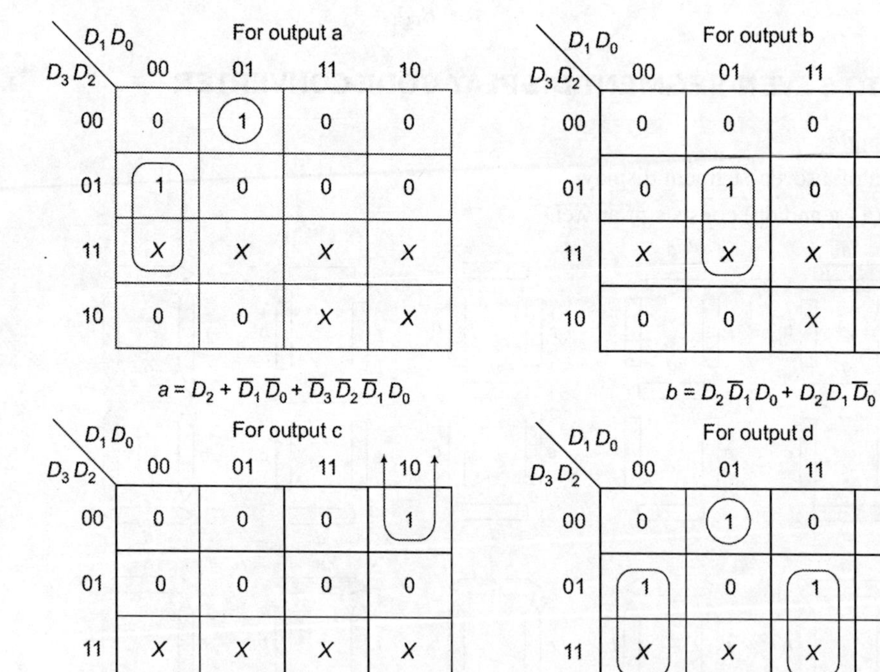

$a = D_2 + \overline{D}_1\overline{D}_0 + \overline{D}_3\overline{D}_2\overline{D}_1 D_0$

$b = D_2\overline{D}_1 D_0 + D_2 D_1\overline{D}_0$

$C = \overline{D}_2\ D_1\overline{D}_0$

$d = D_3\overline{D}_2\overline{D}_1 D_0 + D_2\overline{D}_1\overline{D}_0 + D_2(D_1 D_0)$

For output e

$D_3D_2 \backslash D_1D_0$	00	01	11	10
00	0	1	1	0
01	1	1	1	0
11	X	X	X	X
10	0	1	X	X

$e = \overline{D}_0 + \overline{D}_2\overline{D}_1$

For output f

$D_3D_2 \backslash D_1D_0$	00	01	11	10
00	0	1	1	1
01	0	0	1	0
11	X	X	X	X
10	0	0	X	X

$f = D_1D_0 + \overline{D}_3\overline{D}_2D_0 + \overline{D}_2D_1\overline{D}_0$

For output f

$D_3D_2 \backslash D_1D_0$	00	01	11	10
00	1	1	1	0
01	0	0	1	0
11	X	X	X	X
10	0	0	X	X

$g = \overline{D}_3\overline{D}_2\overline{D}_1 + D_2D_1D_0$

Realization See figure on page 208.

K-map Simplification for Outputs

1. Output for a

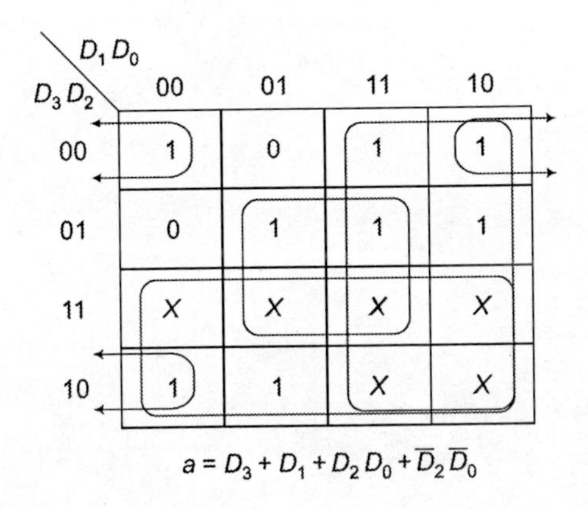

D1 D0
D3 D2
00 01 11 10
00 1 0 1 1
01 0 1 1 1
11 X X X X
10 1 1 X X

$a = D_3 + D_1 + D_2D_0 + \overline{D}_2\overline{D}_0$

2. Output for b

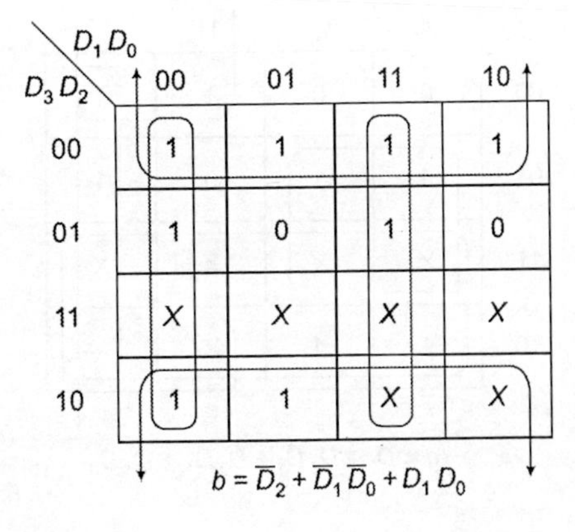

D1 D0
D3 D2
00 01 11 10
00 1 1 1 1
01 1 0 1 0
11 X X X X
10 1 1 X X

$b = \overline{D}_2 + \overline{D}_1\overline{D}_0 + D_1D_0$

3. Output c.

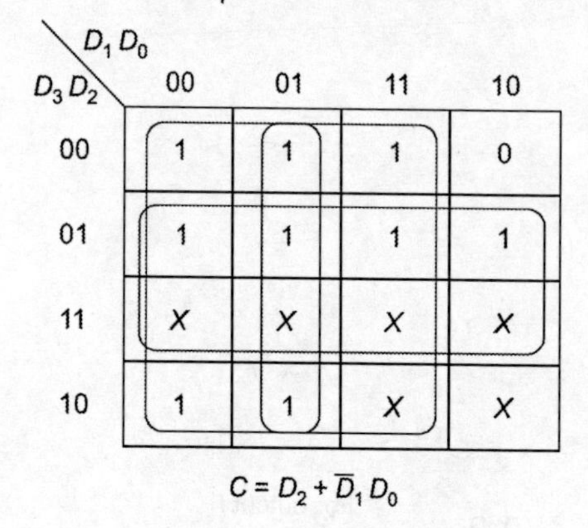

$C = D_2 + \overline{D}_1 D_0$

4. Output d.

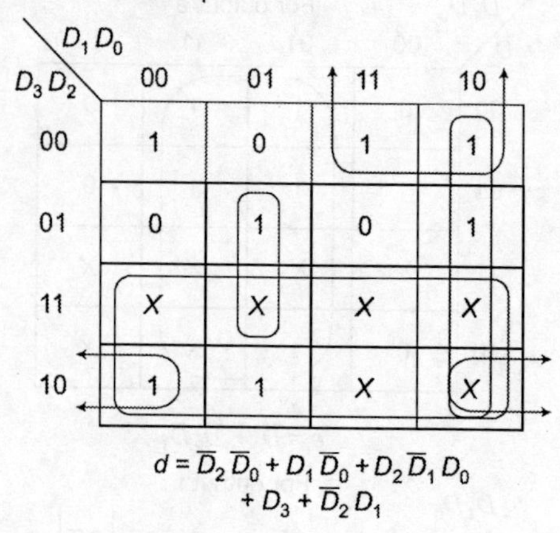

$d = \overline{D}_2\,\overline{D}_0 + D_1\overline{D}_0 + D_2\overline{D}_1 D_0 + D_3 + \overline{D}_2 D_1$

5. Output e.

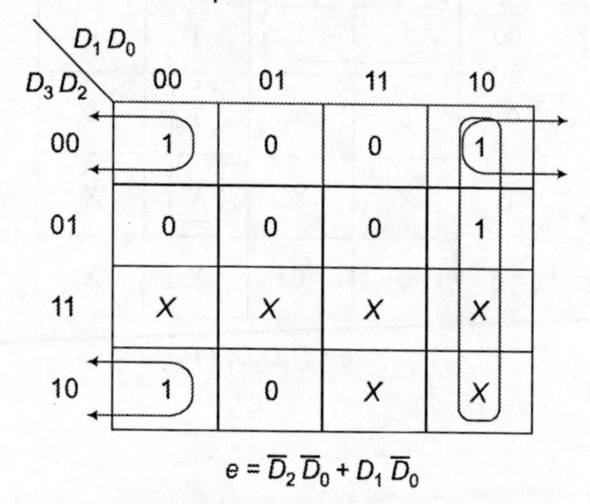

$e = \overline{D}_2\,\overline{D}_0 + D_1\overline{D}_0$

6. Output f.

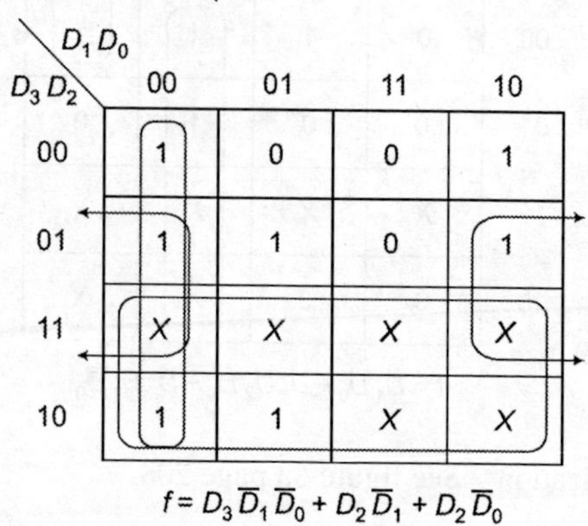

$f = D_3\overline{D}_1\overline{D}_0 + D_2\overline{D}_1 + D_2\overline{D}_0$

7. Output g.

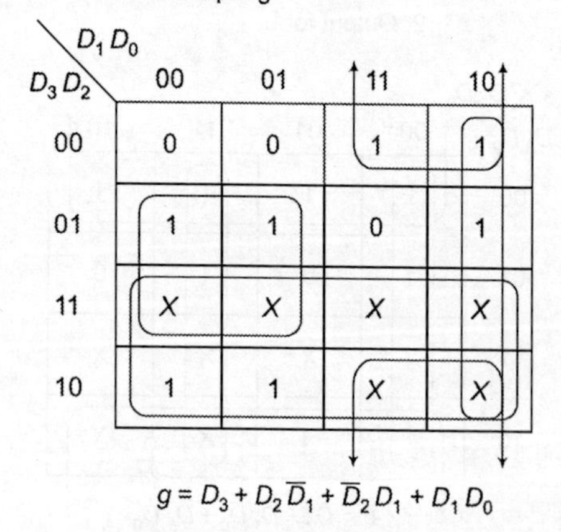

$g = D_3 + D_2\overline{D}_1 + \overline{D}_2 D_1 + D_1 D_0$

9.20 BCD-TO SEVEN-LIGAMENT DECODER USING COMMON CATHODE DISPLAY

- If the common cathode type display is used, then we need a high (1) as the output to turn the segament ON.
- Truth Table

Decimal	Inputs				Outputs						
	D_3	D_2	D_1	D_0	*a*	*b*	*c*	*d*	*e*	*f*	*g*
0	0	0	0	0	1	1	1	1	1	1	0
1	0	0	0	1	0	1	1	0	0	0	0
2	0	0	1	0	1	1	0	1	1	0	1
3	0	0	1	1	1	1	1	1	0	0	1
4	0	1	0	0	0	1	1	0	0	1	1
5	0	1	0	1	1	0	1	1	0	1	1
6	0	1	1	0	1	0	1	1	1	1	1
7	0	1	1	1	1	1	1	0	0	0	0
8	1	0	0	0	1	1	1	1	1	1	1
9	1	0	0	1	1	1	1	1	0	1	1

Gate implementation for various outputs of common cathode 7-segment driver.
See figure on page 210.

1. Design of a Four-bit Parallel Adder

The schematic for design of four-bit parallel adder is shown below. It uses four full-adders. P_0 & Q_0 show the LSB of two numbers (P & Q) to be added which are of 4-bit length.

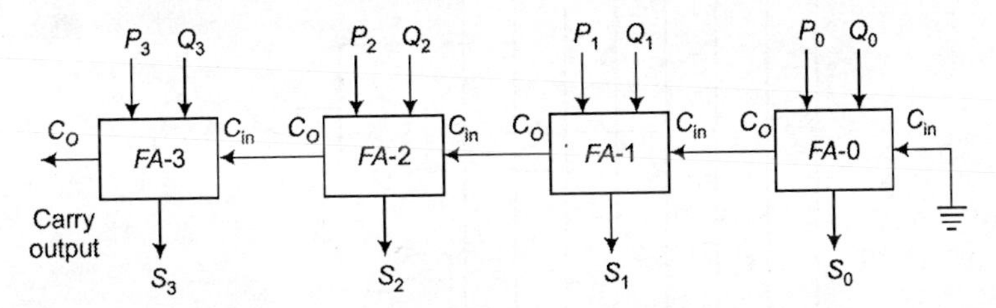

Block-diagram of a 4-bit parallel adder.

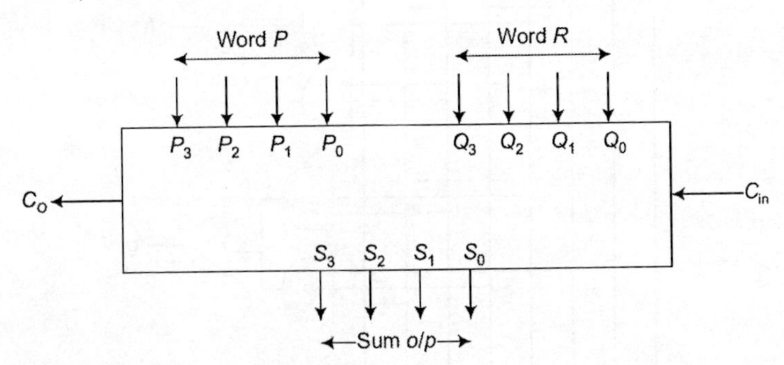

Schematic for 4-bit parallel adder

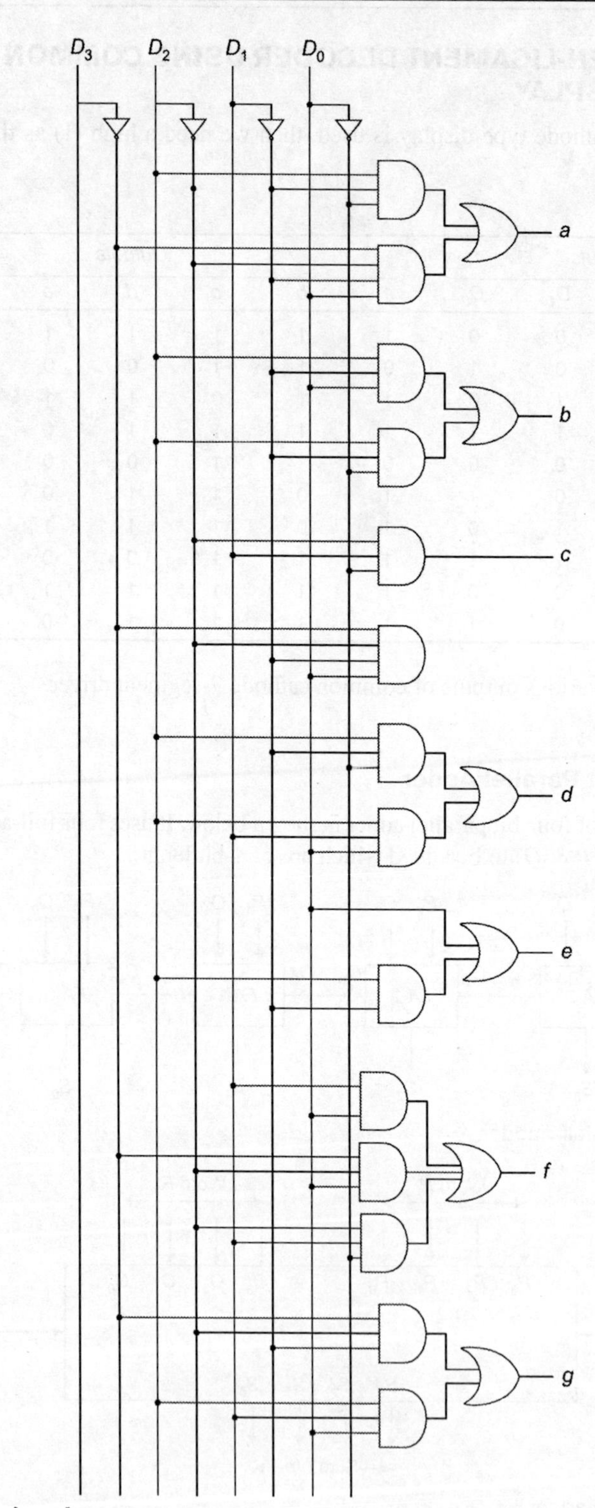

BCD-to 7-segment decoder.

9.21 N-BIT PARALLEL ADDER

The greatest limitations of the full adder discussed earlier was their ability to add only two-single bit binary numbers along with carry input. But practically we require more than this, i.e., we deal with numbers which are much longer than just one bit. To add two n-bit binary numbers we requirer n-bit parallel adder, the block diagram of which is shown below:

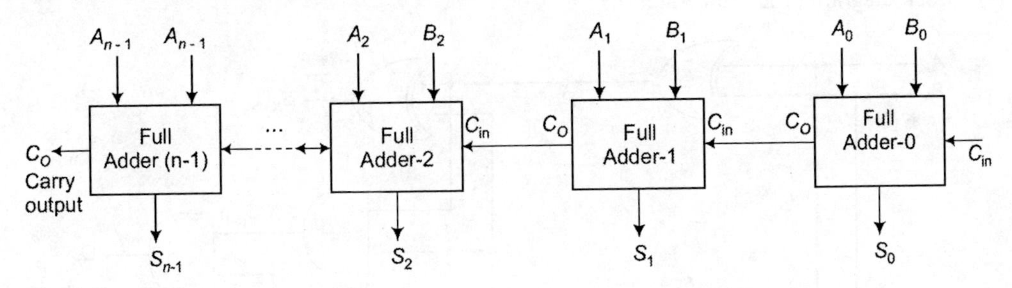

Block diagram of an n-bit parallel adder.

Here we use n-full adders in cascade where the carry output of the first stage serves as the carry input of the next stage, and so on until we confront with the final, i.e. n^{th} full-adder [or $(n-1)^{th}$ if we start with 0th F.A. The system is quite clear from the block diagram shown above.

The first stage (F. A-0) provides the LSB of sum and the last stage (F. A-$(n-1)^{th}$) provides the MSB of the sum as well as the output carry.

Limitations of Parallel Adder

The greatest limitation of a parallel adder is the propagation delay involved in the transmission of carry O/P from 1st to last stage. The carry of first stage propagates to the 2nd stage, then to the 3rd stage and hence-forth until it reaches the last stage. Such type of propagation is called ripple carry propagation, which results in time delay (also called the propagation delay) in addition process.

Suppose we have a 4-bit parallel adder. So S_3 will be MSB of the sum. Again suppose each stage has a propagation delay of 15 min, the final bit of S_3 will reach after a delay of 3×15 min = 45 min from the instant when LSB carry is generated. So the total time required to perform the addition will be 4×15 = 60 min.

Now again suppose we need to add two words of length Z_0 bit. The problem worsens now. The total time required for the addition process will be $Z_0 \times 15$ min = 300 min, which is a significant time duration in digital circuits. Thus, the O/P comes more and more delayed as the number of bits increases in the words to be added.

To counter this problem, there is another technique which we can use for adding two n-bit numbers, called Look-Ahead-Carry Addition, wherein there is very small propagation delay.

9.22 LOOK-AHEAD-CARRY ADDITION

Look-Ahead-Carry Addition is the technique that eliminates the interstage carry delay problem found in n-bit parallel adder. This technique, though requires additional hardware but the speed of the adder is increased and is independent of the number of bits to be added.

Given below is the block diagram of full adder.

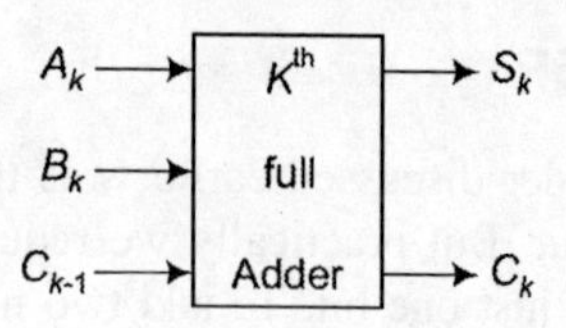

(a) Block diagram of Kth full adder.

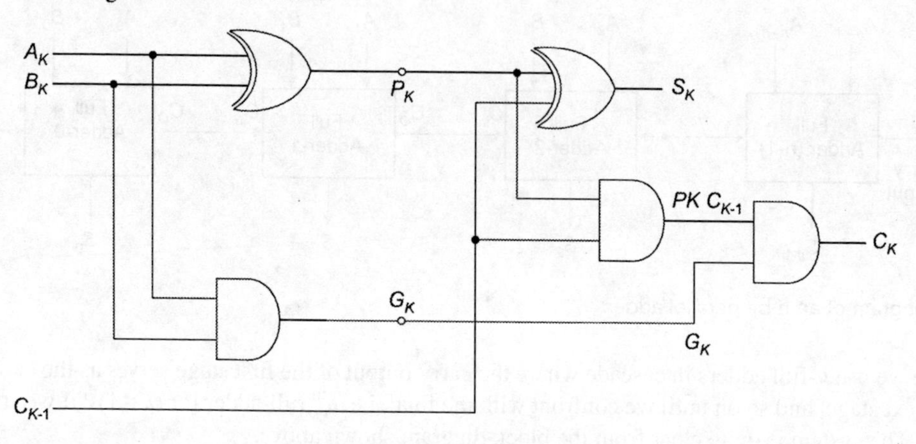

(b) Realization using gates.

Realization The variables to be used are G_0, G_1, G_2, G_3, P_0, P_1, P_2, P_3 and C_{-1}. See figure on page 215.

Reduction of Expressions for Carry Outputs of Various Stages

Refer to the Fig. (b)

$$P_K = A_K \oplus B_K$$

and $$G_K = A_K B_K$$

$$S_K = P_K \oplus C_{K-1} = A_K \oplus B_K \oplus C_{K-1}$$

and $$C_K = G_K + P_K C_{K-1}$$

where the variable G_K is carry generate having value independent on carry. P_K is called carry propagate as it is linked with the propagation of carry from $(K-1)^{th}$ stage to K^{th} stage.

The relation between carry of Kth stage and carry generate and propagate is:

$$C_K = G_K + P_K C_{K-1}$$

Using the above relation, the expression for carry output of each stage can be formulated as:

$$C_0 = G_0 + P_0 C_{-1}$$

$$C_1 = G_1 + P_1 C_0 = G_1 + P_1 (G_0 + P_0 C_{-1})$$

$\Rightarrow$ $$C_1 = G_1 + P_1 G_0 + P_0 P_1 C_{-1}$$

Similarly, $$C_2 = G_2 + P_2 C_1 = G_2 + P_2 (G_1 + P_1 G_0 + P_0 P_1 C_{-1})$$

$\therefore$ $$C_2 = G_2 + P_2 G_1 + P_2 P_1 G_0 + P_2 P_1 P_0 C_{-1}$$

And $$C_3 = G_3 + P_3 C_2 = G_3 + P_3 (G_2 + P_2 G_1 + P_2 P_1 G_0 + P_2 P_1 P_0 C_{-1})$$

$\therefore$ $$C_3 = G_3 + P_3 G_2 + P_3 P_2 G_1 + P_3 P_2 P_1 G_0 + P_3 P_2 P_1 P_0 C_{-1}$$

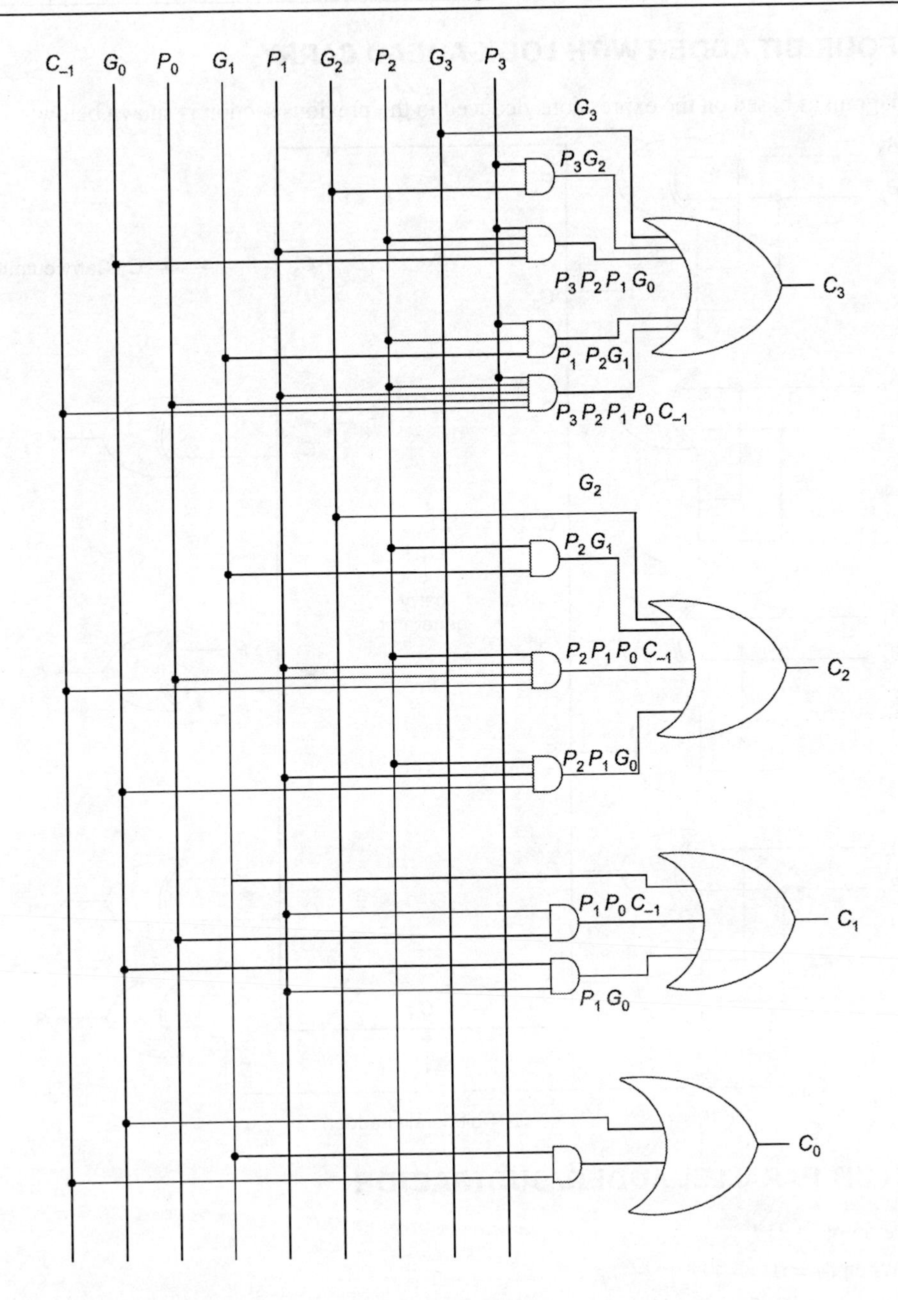
C_{-1}
G_0
P_0
G_1
P_1
G_2
P_2
G_3
P_3
G_3
P_3G_2
$P_3P_2P_1G_0$
C_3
$P_1P_2G_1$
$P_3P_2P_1P_0C_{-1}$
G_2
P_2G_1
$P_2P_1P_0C_{-1}$
C_2
$P_2P_1G_0$
$P_1P_0C_{-1}$
C_1
P_1G_0
C_0

9.23 FOUR-BIT ADDER WITH LOOK-AHEAD CARRY

Block-diagram as based on the expressions deduced in the previous section is shown below:

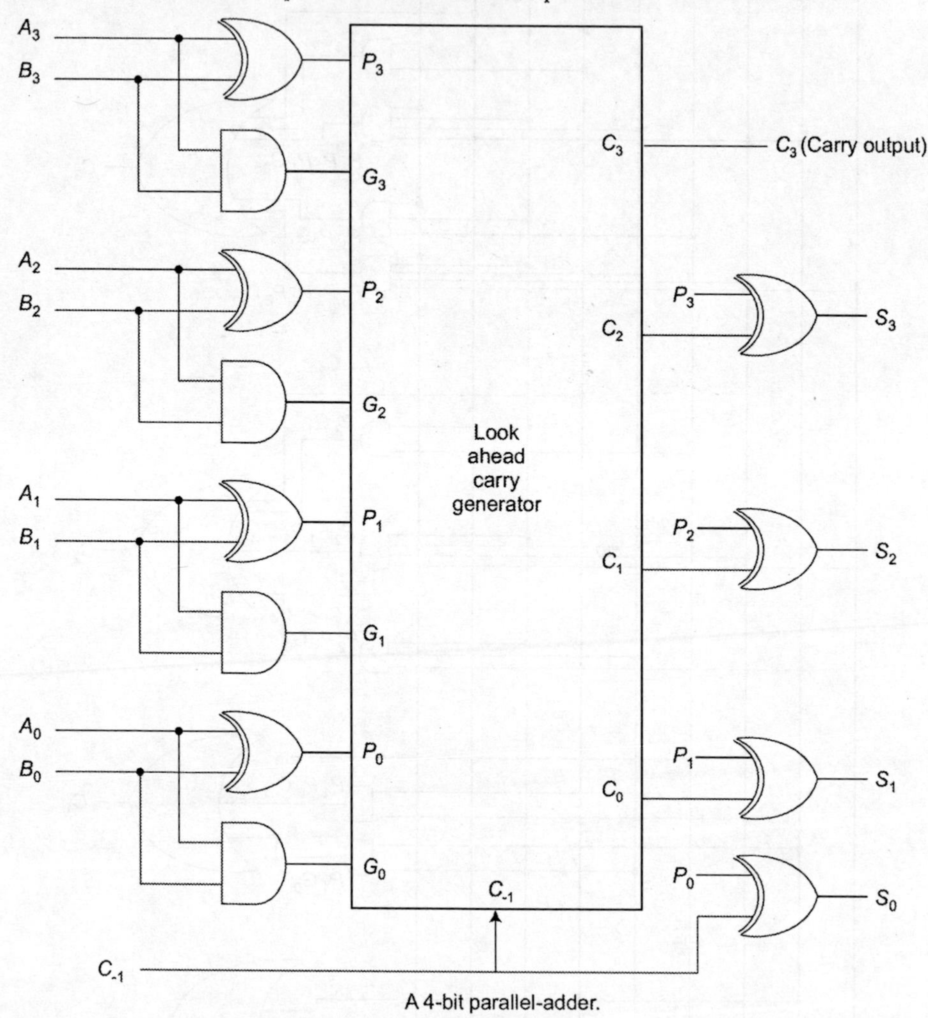

A 4-bit parallel-adder.

9.24 N-BIT PARALLEL ADDER/SUBTRACTOR

See figure on page 217.

1. When $M = 0$.

   Here $C_{in} = 0$. Output of input Q applied to EX-OR gates will be same as Q_3, Q_2, Q_1, Q_0 because $0 \oplus 0 = 0$ and $0 \oplus 1 = 1$. So, the second word goes to the adder/subtractor without getting complemented.

   So, addition will be performed and output will be as per $P + Q + C_{in} = P + Q$ ($\therefore C_{in} = 0$)

   So with $M = 0$, addition will take place.

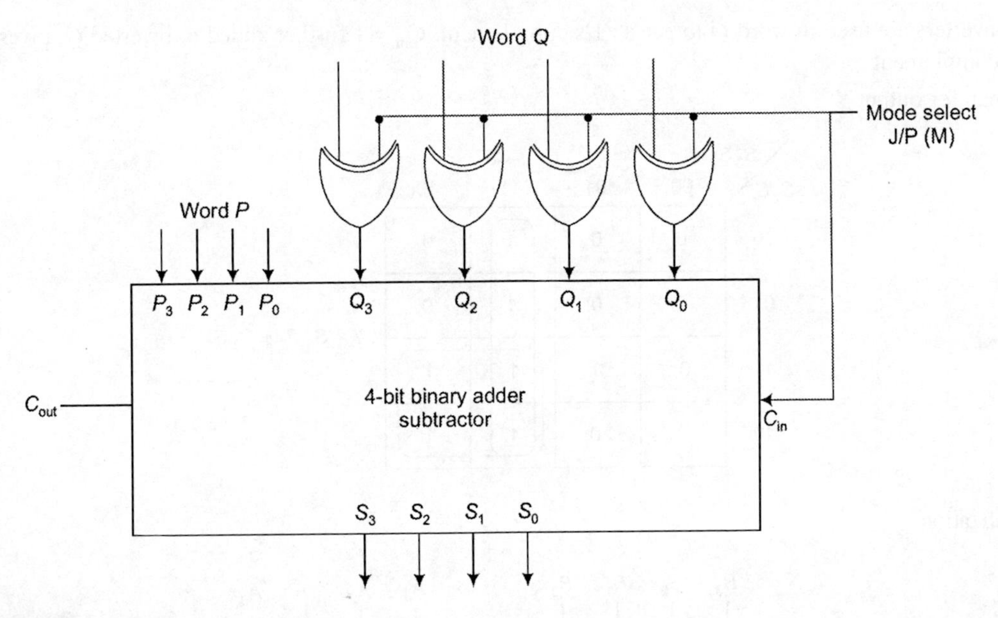

Fig. 9.12 4-bit Binary adder/subtractor

2. When $M = 1$.

   $C_{in} = 1$ and EX-OR gate will complement each of the a-bits (Q_3, Q_2, Q_1, Q_0) because $1 \oplus 0 = 1$ and $1 \oplus 1 = 0$ further C_{in} is added to inverted Q to give its 2's complement which is then added to P.

   Hence, P + 2's complement of $Q = P - Q$ is the function performed here, so with $M = 1$, it acts as a subtractor.

n-bit Parallel Subtractor

Subtraction can be carried out by taking 1's or 2's complements of the number to be subtracted. Refer to the section of n-bit parallel adder. The same circuit (design) with slight manipulation can be used as an n-bit parallel subtractor. Moreover, same can also be modified as n-bit parallel adder or subtractor by providing a mode select input M.

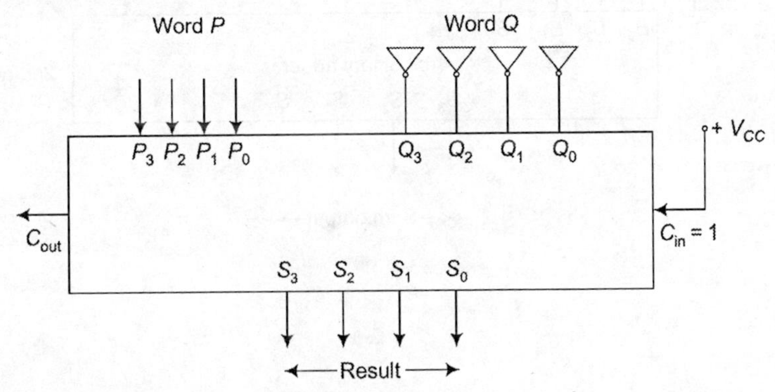

Fig. 9.13 4-bit binary subtractor using 2's complement

Inverters are used in word Q to get its 1's complement. $C_{in} = 1$ further added to inverted Q, gives its 2's complement.

K-map for output Y

$S_1 S_0$ \ $S_3 S_2$	00	01	11	10
00	0	0	1	0
01	0	0	1	0
11	0	0	1	1
10	0	0	1	1

$$Y = S_3 S_2 + S_3 S_1$$

Realization:

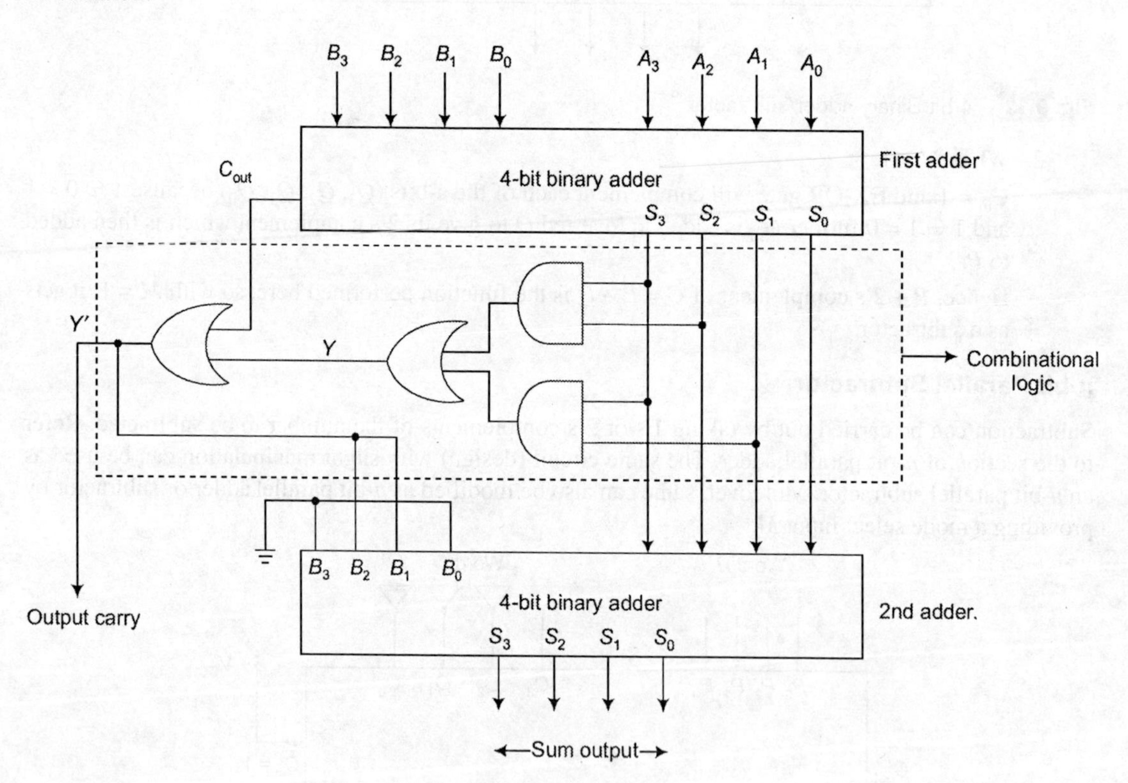

A-bit BCD adder

Case I If sum ≤ 9, carry = 0; $y' = 0$. O/P of adder 2 will be same as output of adder 1.

Case II If sum > 9, carry = 0, $y' = 1 \Rightarrow B_3\, B_2\, B_1\, B_0 = 0110$

Hence, $(6)_{10}$ will be added to the O/P of adder 1 and adder 2 will give convicted BCD sum.

Case III Sum ≤ 9, C = 1 ⇒ $y' = 1$ and hence $B_3\, B_2\, B_1\, B_0 = 0110$.

O/P of the adder two will be sum of 0110 and O/P of adders to provide corrected (valid) BCD result.

9.25 BCD-ADDER

A BCD adder adds two BCD digits and produces a BCD digit. A BCD digit can never be greater than 9. Six (0110) is added whenever the sum exceeds 9 or carry generated is 1.

Block diagram of BCD Adder

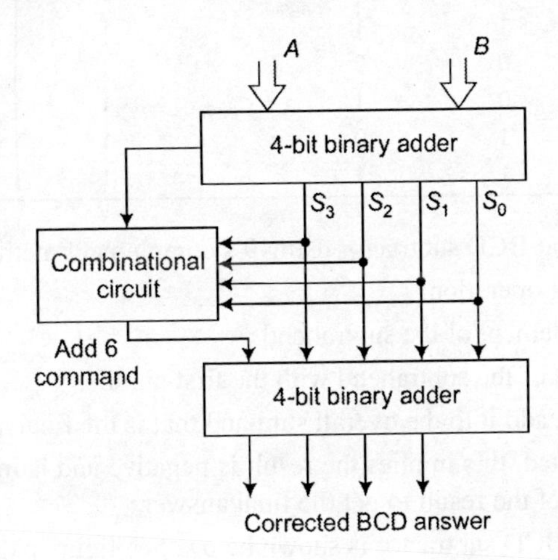

Fig. 9.14 Block diagram of BCD Adder.

A four-bit binary adder adds the two given numbers *A* & *D*. The combinational circuit checks if the sum is greater than 9 or carry = 1. If the sum exceeds 9 or carry is 1, the second binary adder adds six (0110) to the incorrect sum.

Design of Computational Circuit

1. Truth Table

Input				Outputs	
S_3	S_2	S_1	S_0	Y	
0	0	0	0	0	Sum is correct (Valid BCD number)
0	0	0	1	0	
0	0	1	0	0	
0	0	1	1	0	
0	1	0	0	0	
0	1	0	1	0	
0	1	1	0	0	
0	1	1	1	0	
1	0	0	0	0	
1	0	0	1	0	
1	0	1	0	1	incurred sum (or invalid BCD)
1	0	1	1	1	
1	1	0	0	1	
1	1	0	1	1	
1	1	1	0	1	
1	1	1	1	1	

Example 9.14 Design a 4-bit BCD subtractor using 9's complement method.

Solution Basic principle of operation:

1. Obtain the 9's complement of the subtrahend.
2. Add 9's complement of the subtrahend with the first number.
3. If carry is generated, add it to the overall sum and that is the final answer.
4. If no carry is generated, this implies the result is negative and is in 9's complement form. Take the 9's complement of the result to get the final answer.

Block diagram of a 4-bit BCD subtractor is shown below. See figure page 219.

Details The first adder produces the 9's complement of the subtrahend (2nd number).

To the 2nd and 3rd 4-bit binary adder add Q & 9's complement of CD. and also perform the required BCD correction to get the final result as a valid BCD. The combinational logic used between the adders & $D3$, is same as that in BCD adder and same design Mule and K-map follows here also.

The fourth adder performs the check whether the result of subtraction is positive or negative (in 9's complement form). If there is a carry generated at the output of 4-bit binary adder 3, it is supplied as C_{in} to 4th adder and the same in inverted form is supplied to the EX-OR gates, so as to perform simple addition of result with 1 ($C_{in} = 1$) [as $A_3 A_2 A_1 A_0 = 0\,0\,0\,0$].

If carry = 0, $C_{in} = 0$ and $A_3 A_2 A_1 A_0 = 1001$, also Q is inverted and thus 9's complement of the result is obtained to get the final answer.

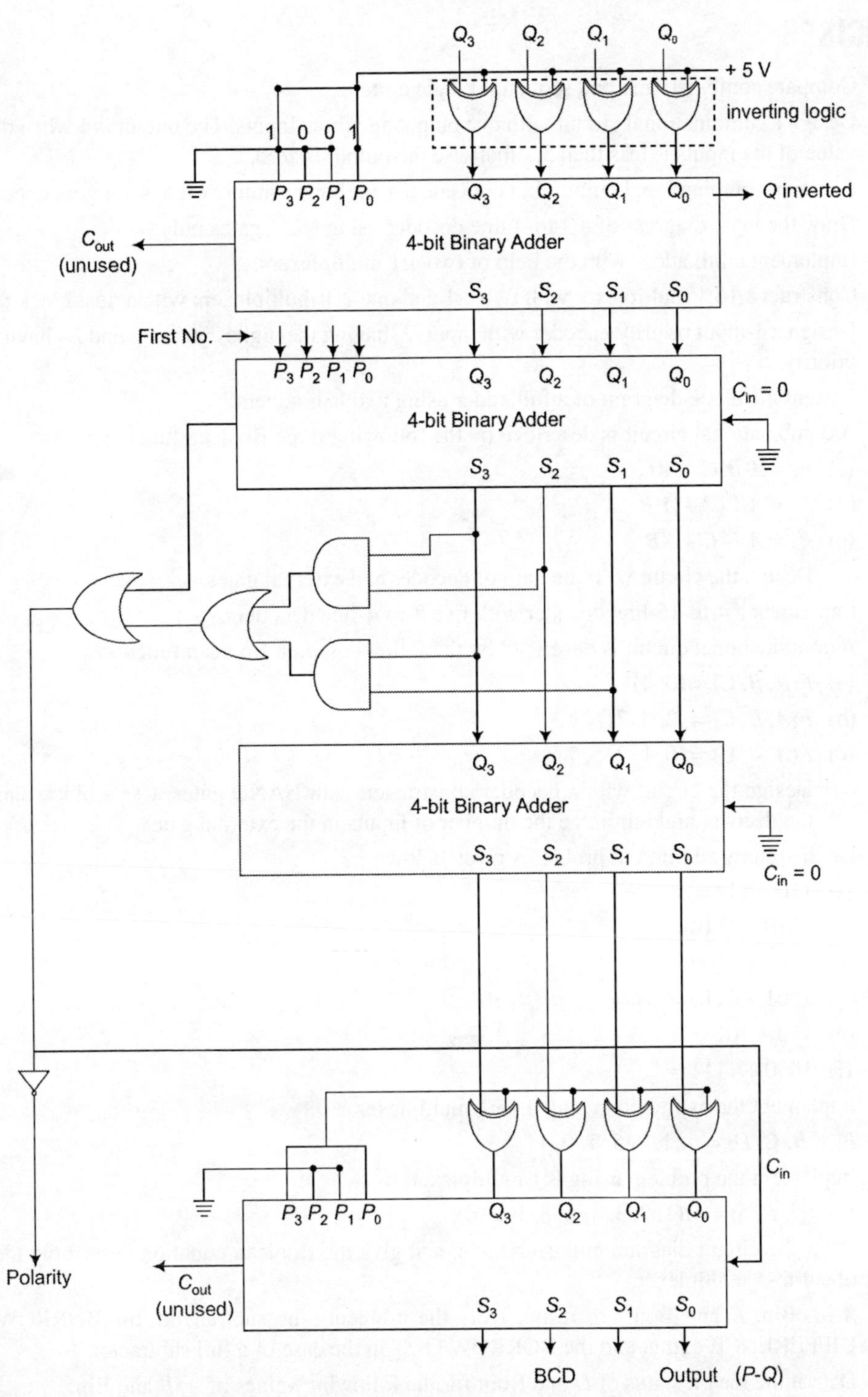

Fig. 9.15 BCD subtractor using 9's comlement methods.

EXERCISES

1. Compare combinational and sequential logic circuits.
2. Design a combinational circuit with one output and three inputs. The output is 1 when the binary value of the inputs is less than 3, otherwise the output is zero.
3. Design a combinational circuit that converts a 4-bit binary number to a 4-bit gray code.
4. Draw the logic diagram of a 2-to-4 line decoder using NOR gates only.
5. Implement a full adder with the help of two 4:1 multiplexers.
6. Construct a 16:1 multiplexer with two 8:1 and one 2:1 multiplexers with a neat block diagram.
7. Design a 4-input priority encoder with input D_0 having the highest priority and D_3 having lowest priority.
8. Present the block diagram of a full adder using two half adders.
9. A combinational circuit is described by the following three Boolean functions:
   (a) $F_1 = A'B'C' + AC$
   (b) $F_2 = AB'C' + A'B$
   (c) $F_3 = A'B'C + AB$
   Design the circuit with the help of decoder and external gates.
10. Implement a 4-to-16-line decoder with five 2-to-4-line decoders.
11. A combinational circuit is described by the following three Boolean functions:
    (a) $F(A, B, C) = (0, 4)$
    (b) $F(A, B, C) = (2, 3. 7)$
    (c) $F(A, B, C) = (0, 1, 2, 5, 7)$
    Design the circuit with a decoder constructed with NAND gates. Use a block diagram for the decoder and minimize the number of inputs in the external gates.
12. Do the binary addition in problems given below:
    (a) 110 + 111 =
    (b) 1010 + 1110 =
    (c) 1001 + 1100 =
    (d) 1001 + 0110 =
    (e) 110 + 101 =
    (f) 1000 + 1111 =
13. Implement the expression using a 16:1 multiplexer
    $F(A, B, C, D) = \text{m}(1, 3, 5, 7, 9, 11, 13)$
14. Implement the problem using 8:1 multiplexer
    $F(P, Q, R, S) = M(1, 2, 3, 4, 6, 8, 10, 14)$
15. Draw the circuit diagram and truth table, and give the Boolean equation describing the output, of a 4-to-1 multiplexer.
16. *A*, *B*, Bin, *D* and Bout are respectively the minuend, the subtrahend, the BORROW-IN, the DIFFERENCE output and the BORROW-OUT in the case of a full subtractor.
    Determine the bit status of *D* and Bout for the following values of *A*, *B* and Bin:

(a) $A = 0, B = 1$, Bin = 1
(b) $A = 1, B = 1$, Bin = 0
(c) $A = 1, B = 1$, Bin = 1
(d) $A = 0, B = 0$, Bin = 1

17. Implement the product-of-sums Boolean function expressed by (1, 2, 5) by a suitable multiplexer.
18. Figure 8.13 shows the use of an 8-to-1 multiplexer to implement a certain four-variable Boolean function. From the given logic circuit arrangement, derive the Boolean expression implemented by the given circuit.

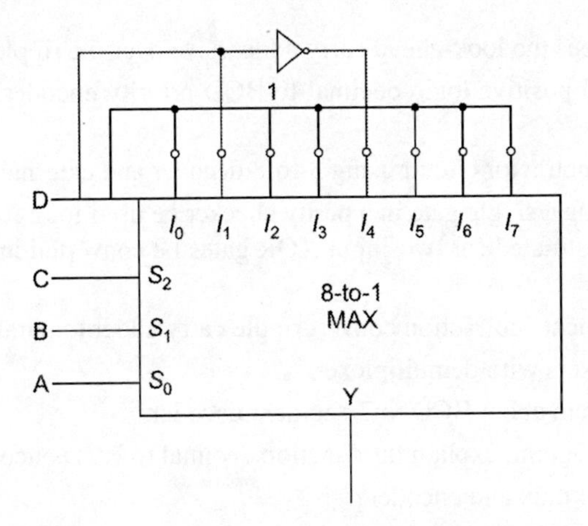

19. Design a 16-to-1 multiplexer using two 8-to-1 multiplexers having an active LOW ENABLE input.
20. Draw the truth table of a BCD-to-decimal decoder, and show how the generation of the first four codes could be achieved using a demultiplexer.
21. We have an eight-line to three-line priority encoder circuit with D0_D1_D2_D3_D4_D5_D6 and D7 as the data input lines, the output bits are A(MSB), B and C (LSB). Higher-order data bits have been assigned a higher priority, with D7 having the highest priority. If the data inputs and outputs are active when LOW, determine the logic status of output bits for the following logic status of data inputs:
    (a) All inputs are in logic '0' state.
    (b) D1 to D4 are in logic '1' state and D5 to d7 are in logic '0' state.
    (c) D7 is in logic '0' state. The logic status of the other inputs is not known.
22. Implement a full adder circuit using a 3-to-8-line decoder.
23. What is a multiplexer circuit? Briefly describe one or two applications of a multiplexer?
24. What is an encoder? How does a priority encoder differ from a conventional encoder?
25. Write out the truth table for the 4-line-to-2-line encoder that takes a four-line decimal signal and convert it to binary code. Design and draw the circuit to implement this encoder.

26. A combinational circuit is defined by F = 0, 2, 5, 6, 7. Implement the Boolean function F with a suitable decoder and an external OR/NOR gate having the minimum number of inputs.
27. Give the applications of encode and decoders.
28. Design, using two-input XOR gates, a comparator which will give an active-LOW output if two four-bit words, A and B, are the same
29. A two-bit comparator gives an active-HIGH output, Y, if two two-bit words, A and B are the same. Give Y in fundamental sum of products form and then use Boolean algebra to show that $Y = (A_0 \oplus B_1) + (A_1 \oplus B_0)$
30. What type of circuit is a ripple carry adder, what basic unit is it built from, and what is its major disadvantage?
31. What advantage does the look-ahead carry adder have over the ripple carry adder?
32. Design a two-level positive logic decimal-to-BCD priority encoder for decimal inputs from 0 to 4.
33. Implement a full subtractor circuit using 3-to-8 decoder and external NOR gates.
34. How could changing a single gate in a parity checker be used for a four-bit word (three data and one parity bit) constructed for two-input XOR gates be converted into a comparator for use on two-bit words?
35. Using 2's complement subtraction convert ripple carry adder to parallel subtractor.
36. Compare multiplexers with demultiplexer.
37. Explain with diagram about BCD to 7-segment decoder.
38. With the help of diagram, explain the function decimal to BCD encoder.
39. What is the need of mux and encoder?
40. What is meant by priority encoder? Give one example.
41. Draw the circuit diagram of 1:16 demux and explain its operation in detail.
42. What are the disadvantages of half adder?
43. Discuss the working of parity generators and parity checkers.
44. Implement the following Boolean function using all 4:1 mux
    $f(A, B, C, D, E) = m(1, 3, 5, 7, 9, 12, 13, 14, 15, 22, 24, 26, 30, 31)$
45. Design a full adder using 1:8 demux.

10

D/A and A/D Conversions

INTRODUCTION

Digital to Analog and Analog to Digital conversions form two very important aspects of digital data processing. Digital to Analog conversion involves translation of digital information into equivalent analog information. Similarly, as the process of changing an analog signal to an equivalent digital signal is accomplished by the use of an A/O converter. Some of the examples where, A/D and D/A converters are used are:

1. A digital signal system can be used to monitor the ambient temperature of on oven and if it exceeds a certain limit, it should reduce the fuel input. Here an A/D converter is required to convert the output of the sensor (which converts temperature to an analog electrical signal) to digital form. If the temperature exceeds the specific limit, some digital output is produced which is to be converted to analog form in order to control the device which reduces the fuel input.
2. A digital voltmeter is used to measure an analog voltage and display the voltage in numerical form. In this an A/D converter is required to correct the analog voltage into a digital signal. The required processing consists of determining its value. The output in this case is not required to be converted back to the analog form and hence D/A converter is not required.
3. A digital communication is used to transmit messages which are in the form of analog electrical signals. This requires an A/D converter at the transmitting end and a D/A converter at the receiving end. There are two types of commonly used D/A converters. These are:
   (i) Weighted Register D/A converter
   (ii) R-2R Ladder DAC

10.1 D/A CONVERTER

The input to D/A converter is an N-bit binary signal available in parallel form. Normally, digital signals are available at the output of the latches or registers and the voltages corresponding to logic 0 and logic 1, available to drive the converter are in general not precisely fixed voltages. Therefore, these voltages are not applied directly to the converter but are used to operate digitally controlled switches. The switch is thrown to one of the two positions depending upon the digital signal (1 or 0) which connects precisely fixed voltages v(1) or v(0) to the converter input corresponding to 1 or 0 respectively.

The analog output voltage V_0 of an N-bit state binary D/A converter is related to the digital input of the equation.

$$V_0 = K(2^{N-1} b_{N-1} + 2^{N-2} b_{N-2} + ... + 2^2 b_2 + 2b_1 + b_0)$$

where K is proportionality factor

b_n = 1 if nth bit of the digital input is 1

= 0 if nth bit of the digital input is 0

10.2 D/A CONVERTER WITH BINARY-WEIGHTED RESISTORS

D/A converter using an Op-Amp and binary weighted registers is shown in Fig. 11.1. Although in this figure the Op-Amp is connected in the inverting mode, it can also be connected in the non-inverting mode.

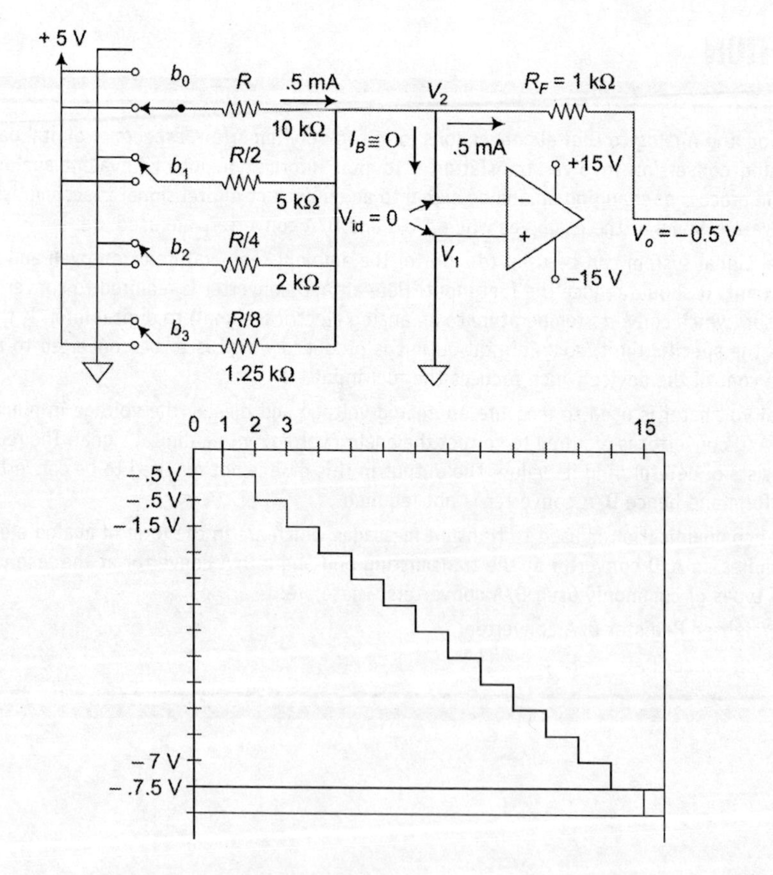

Since the number of binary inputs is 4, the converter is called a 4-bit converter.

Because there are 16 (2^4) combinations of binary input for b_0 through b_3, an analog output should have 16 possible corresponding values. When switch b_0 is closed (connected to +5 V), the voltage across R is 5 V because $V_2 = V_1 = 0$ V. Therefore, the current through R is 5 V/10 kΩ = 0.5 mA. However, the input bias current I_B is negligible hence, the current through feedback register R_f is also 0.5 mA which, in turn produces an output voltage of –(1 kΩ) (0.5 mA) = – 0.5 V.

Now suppose that switch b_1 is closed and b_0 is opened. This action connects $R/2$ to the positive supply of +5 V, causing twice as much current (1 mA) to flow through R_F, which in turn doubles the output voltage. Thus, the output voltage V_0 is –1V. When switch b_1 is closed. Similarly, if both switches b_0 and b_1 are closed, then the current through R_f will be 1.5 MA, which will be converted to an output voltage of –(1 kΩ) (1.5 mA) = –1.5 V.

Thus, depending on whether switches b_0 to b_3 are open or closed, the binary-weighted currents will be set up in input resistors. The sum of these currents is equal to the current through R_F, which in turn is converted to a proportional output voltage. When all the switches are closed, obviously the output will be maximum. The output voltage equation is given by

$$V_0 = -R_F\left(\frac{b_0}{R} + \frac{b_1}{R/2} + \frac{b_2}{R/4} + \frac{b_3}{R/8}\right)$$

where each of the inputs b_3, b_2, b_1 and b_0 may either be high (+ 5 V) or low (0 V).

10.3 D/A CONVERTER WITH R AND 2R RESISTORS

D/A converter with R and $2R$ resistors are shown in Fig. 11.2. The binary inputs are simulated by switches b_0 through b_3, and the output is proportional to the binary inputs. Assume that the MSB switch b_3 is connected to + 5 V and other switches are connected to ground.

Thevenizing the circuit to the left of switch b_3, Thevenin's equivalent resistance R_{th} is

$$R_{TH} = [\{[2R \parallel 2R + R) \parallel 2R]. + R\} \parallel 2R] + R$$
$$= 2R = 20\ \text{k}\Omega$$

In the figure below the (–) input is at virtual ground ($V_2 \cong 0$ V); therefore the current through R_{TH} (= 2 R) is zero. However, the current through $2R$ connected to + 5 V is 5V/20 kΩ = 0.25 WA.

The same current flows through R_F and in turn produces the output voltages.

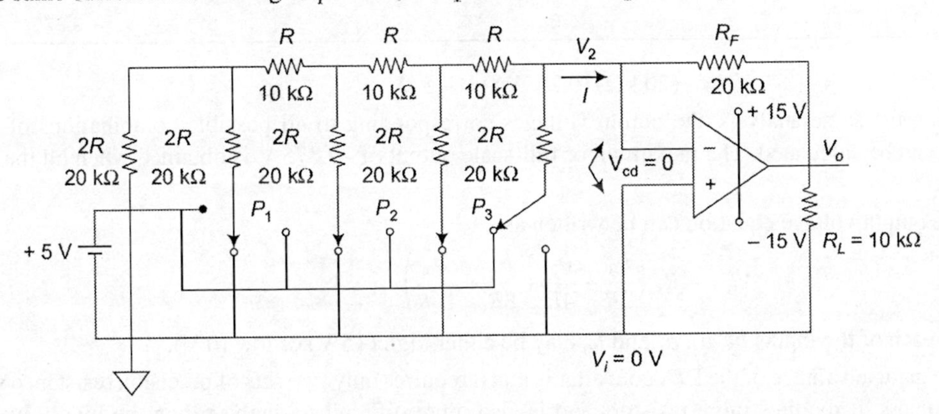

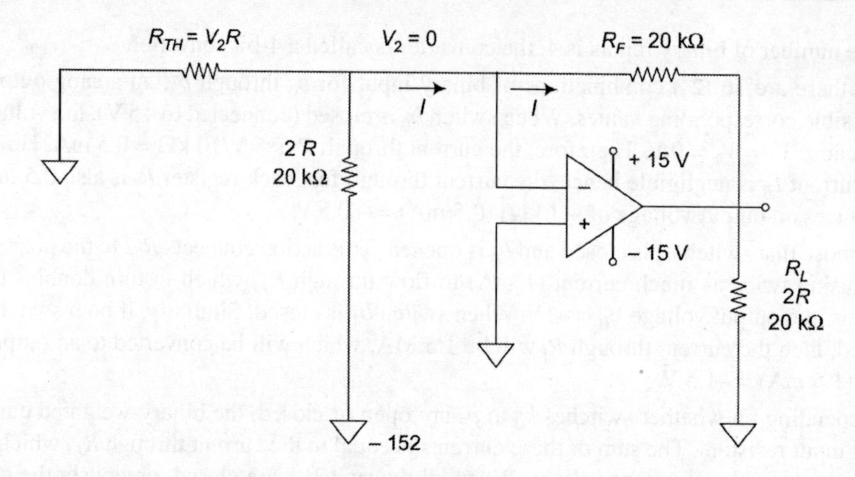

Decimal equivalent of binary in bits	Input (V) b_3	b_2	b_1	b_0	Output voltage (V)
0	0	0	0	0	0
1	0	0	0	5	−0.625
2	0	0	5	0	−1.25
3	0	0	5	5	−1.875
4	0	5	0	0	−2.50
5	0	5	0	5	−3.125
6	0	5	5	0	−3.750
7	0	5	5	5	−4.375
8	5	0	0	0	−5.0
9	5	0	0	5	−5.625
10	5	0	5	0	−6.25
11	5	0	5	5	−6.875
12	5	5	0	0	−7.50
13	5	5	0	5	−8.125
14	5	5	5	0	−8.875
15	5	5	5	5	− 9.375

$$V_0 = -(20\ k\Omega)\ (0.25\ WA) = -5\ V$$

Using the same analysis, the output voltages corresponding to all possible combinations of binary inputs can be calculated. The maximum or full-scale output of –9.375 V is obtained when all the inputs are high.

The output voltage equation can be written as

$$V_0 = -R_F\left(\frac{b_3}{2R} + \frac{b_2}{4R} + \frac{b_1}{8R} + \frac{b_0}{16R}\right)$$

where each of the inputs b_3, b_2, b_1 and b_0 may be either high (+5 V) or low (0 V).

The great advantage of the D/A converter is that it requires only two sets of precision resistance values, nevertheless, it requires more resistors and is also more difficult to analyse than the binary weighted resistor type. As the number of binary inputs is increased beyond four, both D/A converter circuits get complex and their accuracy degenerates.

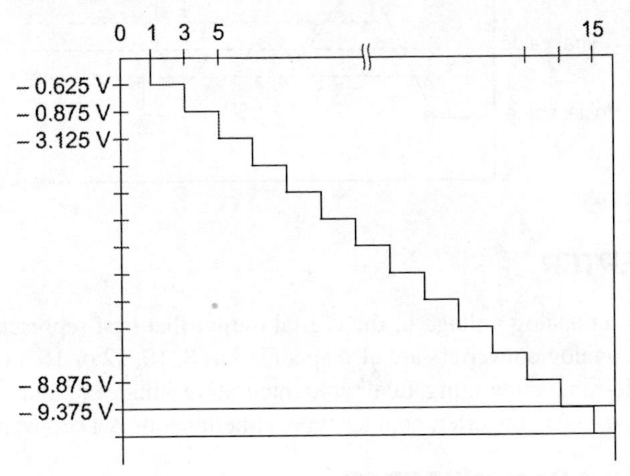

Therefore, in critical applications an integrated circuit specially designed as a D/A converter should be used.

10.4 SPECIFICATIONS FOR D/A CONVERTERS

The characteristics of a D/A converter which are generally specified by the manufacturers are:

1. Resolution
2. Linearity
3. Accuracy
4. Settling time
5. Temperature sensitivity

1. ***Resolution*** This is the smallest possible change in output voltage as a fraction or percentage of the full-scale output range. For example for an 8-bit converter, there are 2^8 or 256 possible values of analog output voltage, hence the smallest change in the output voltage is 1/255 of the full-scale output range.
2. ***Linearity*** The linearity of a converter is a measure of the precision with which the linear input-output relationship is satisfied.
3. ***Accuracy*** The accuracy of a D/A converter is a measure of the difference between the actual output voltage and the expected output voltage. It is specified as a percentage of full-scale or maximum output voltage.
4. ***Settling Time*** The time required for the analog output to settle to within ±1/2 LSB of the final value after a change in the digital input is usually specified by the manufacturers and is referred to as settling time.
5. ***Temperature Sensitivity*** The analog output voltage for any fixed digital input varies with temperature. This is due to the temperature sensitivities of the reference voltage source, resistors, Op-Amp, etc. It is specified as ± ppm°/°C.

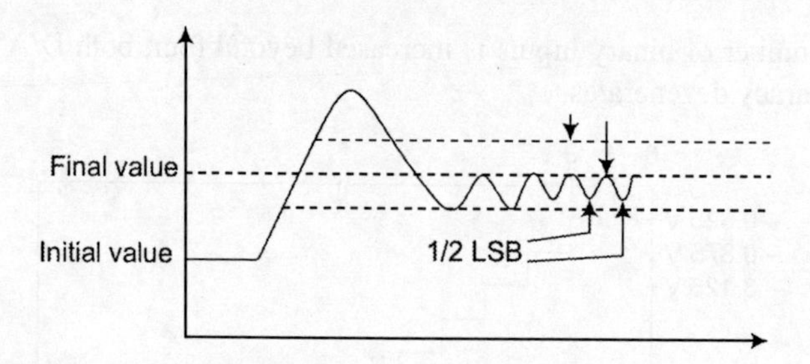

10.5 A/D CONVERTER

A/D converters convert an analog voltage to the digital output that best represents the input. As in the case of D/A converters, analog converters are also specified as 8, 10, 12 or 16 bit. There are many types of A/D converters: single-ramp integrating, dual ramp integrating, single counter, tracking and successive approximation continuous A/D converter, counter type, simultaneous A/D converter.

10.6 DUAL SLOPE A/D CONVERTER

The major four blocks used in dual slope A/D converter are:

(a) An integrator
(b) A comparator
(c) A binary counter
(d) A switch driver

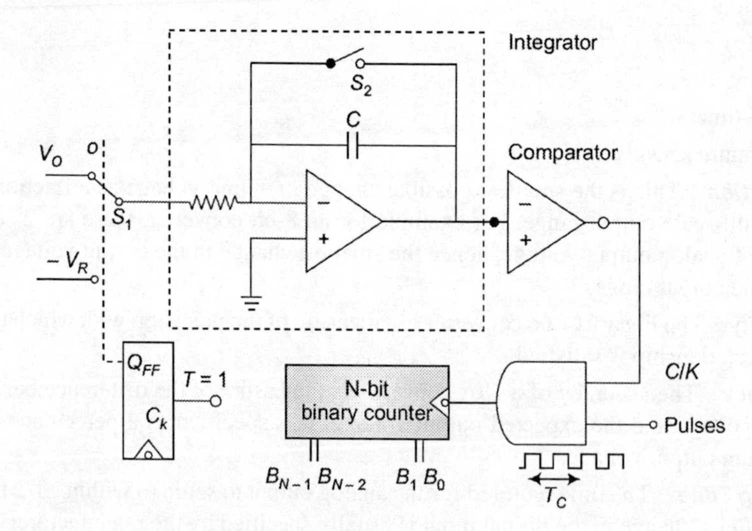

The conversion process begins at 1 = 0 with the switch S_1 in position 0 thereby connecting the analog voltage V_a to the input of the integrator.

The integrator output

$$V_0 = \frac{-1}{\tau} \int_0^t V_a \, dt$$

$$= -\left(\frac{V_a}{\tau}\right)^t$$

This results is high V_C, thus enabling the AND gate and the clock pulses reach the clock input terminal of the counter which was initially clear. The counter counts from 00 .... 00 to 111.11 when $(2^N - 1)$ clock pulses are applied. At the next clock pulse (2^N) the counter is cleared and Q becomes 1. This controls the state of S_1 which now moves to position 1 at T_1. Thereby connecting $-V_R$ to the input of the integrator. The output of the integrator now starts to move in the positive direction. The counter continues to count until V_0. As soon as V_0 goes +ve at T_2, V_C goes low disabling the AND gate. The counter will stop counting in the absence of the clock pulses. The waveforms of voltages V_0 to V_C are shown in Fig. 11.8.

The heart of the given successive approximation A/D converter is an 8-bit successive approximation register (SAR) whose output is applied to an 8-bit D/A converter. The analog output (V_0) of the D/A converter is then compared to an analog input signal V_{in} by the comparator.

The output of the comparator is a serial data input to the SAR. The SAR then adjusts its digital output (8-bit) until it is equivalent to the analog input V_{in}. The 8-bit latch at the end of conversion holds onto the resultant digital data output. The CKT works as follows and shown in Fig. 11.9.

At the start of a conversion cycle, the SAR is reset by holding the start signal high. On the first clock pulse LOW to HIGH transition, the most significant output bit Q_7 of the SAR is set. The D/A converter then regenerates an analog equivalent to the Q_7 bit, which is compared with the analog input V_{in}. If the comparator O/P is low, the D/A output > V_{in} and the SAR will clear its MSB Q_7. On the other hand, if the comparator output is high, the D/A output < V_{in} and the SAR will keep the MSB Q_7 set. In any case on the next clock pulse, low to high transition the SAR will set the next MSB Q_6. Depending upon the O/P of comparator, the SAR will then either keep or reset the bit Q_6. This process is continued until the SAR tries all the bits. As soon as LSB Q_0 is tried, the SAR forces the conversion complete (CC) Signal HIGH to indicate the parallel output lines contain valid data. The CC signal in turn enables the latch, and digital data appear at the output at the latch. Digital data is also available serially as the SAR determines each bit. To cycle the converters continuously the CC signal may be connected to the start conversion input. The advantage of successive approximation A/D converter is its high speed and excellent resolution. For example an 8-bit successive approximation A/D converter of the given figure requires only eight clock pulses.

10.7 COUNTER TYPES OR PULSE WIDTH TYPE A/D CONVERTERS

This converter is sometimes also called a staircase type A/D converter. It is also a type of analog to time to digital conversion. The block diagram is shown below.

The analog input voltage to be measured or to be converted to digital form is fed to the comparator, which is essentially an operational amplifier. The logic control circuit on receiving the signal from the comparator resets the counter to zero and starts input clock pulses to the counter. The output of the counter, giving digital display is fed to the D/A converter whose analog outforms the reference voltage of the comparator.

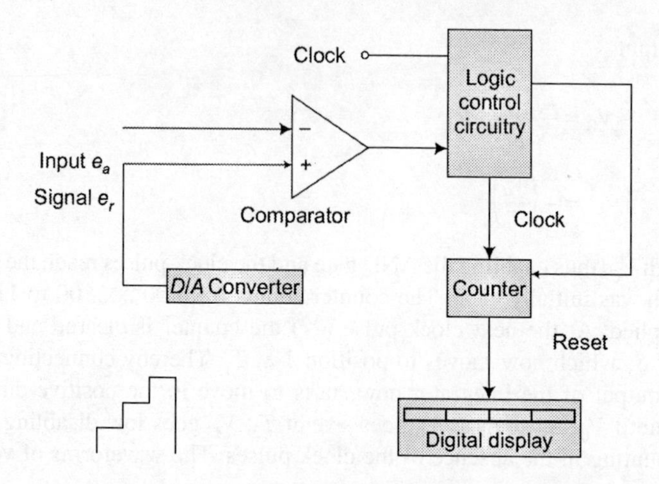

When the output of a D/A converter or reference voltage K of comparator exceeds the analog input voltage e_0, the output of comparator becomes zero, and thus the logic controls circuitry stops the counter. The binary number in the counter at that stage represents the digital output which is shown on the digital display. The conversion time depends on the amplitude of the input voltage. This conversion is the cheapest method, but has slow conversion rate, rather it is the slowest method of converting analog data to digital form.

10.8 CONTINUOUS A/D CONVERTERS

In this type of converter, a direct and continuous comparison of an unknown analog voltage, e_a and a variable voltage, e_r is performed as shown below.

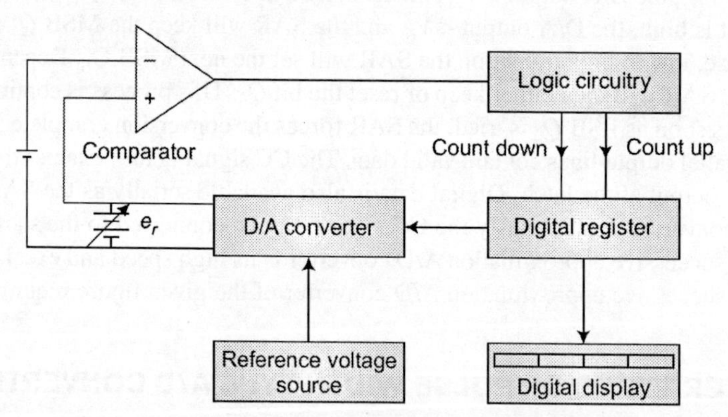

The variable voltage e_r is generated by a D/A converter, which receives information from a digital register. When analog voltage is greater than e_r, the comparator and the logic circuitry enables the digital register to count up. Thus, the digital number increases as shown by digital display, corresponding to which e_r also increases through the D/A converter until it is equal to e_a. In a similar manner, when $e_a < e_r$ the digital register counts down thus decreasing e_r until it becomes equal to e_a. The conversion of the unknown analog voltage e_a to a digital output equivalent to e_r is thus performed by using the digital circuitry to produce an analog reference voltage e_r, which is continuously balanced with the unknown analog voltage e_a. The circuit is thus a form of digital potentiometer in which the digital output tracks a slowly varying analog input. It is similar to an electromechanical system having a closed loop feedback control. This closed loop feedback control always ensures that the digital output is equal to the analog input. This type of converter is capable of following the input voltages which change at a faster rate.

10.9 SIMULTANEOUS CONVERTERS

The method or process of converting analog voltage to digital voltage is called analog to digital conversion (ADC). A number of methods have been developed. In a simultaneous method the A/D conversion is based on the use of number of comparators. One such type of circuit is shown in figure below. The analog signal to be digitized serves as one of the inputs to each comparator. The input is a standard voltage. In this case the voltages used are +V/4, + V/2 and +3V/4 and V. Then the system is capable of accepting an analog input voltage between 0 volt and V volts.

If analog input exceeds the Reference voltage at any comparator that comparator turns on. Now, if all the comparators are turned off, the analog signal must be between 0 and +V/4. If C_1 is high and C_2 and C_4 are low, then the input must be between +V/4 and +V/2 volts. So by using this process we can convert an analog signal to digital form.

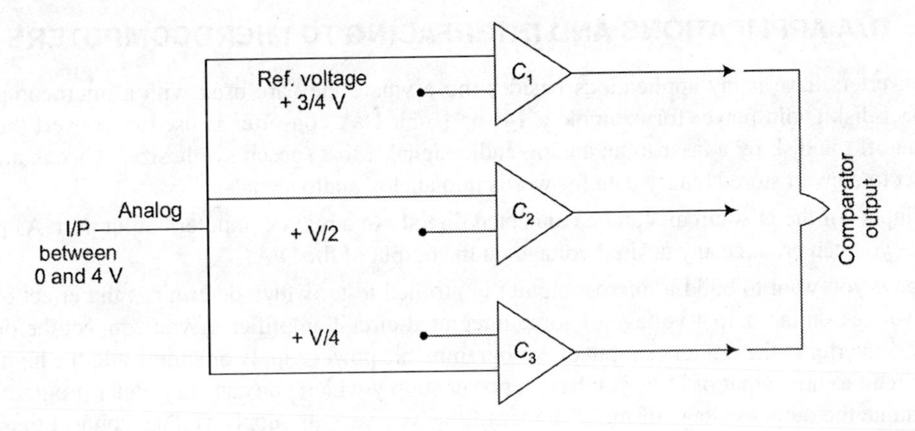

Input Voltage	C_1	C_2	C_3
0 to +V/4	0	0	0
+V/4 to +V/2	1	0	0
+V/2 to +3V/4	1	1	0
+3V/4 to V	1	1	1

10.10 A/D CONVERTER SPECIFICATIONS

A wide variety of A/D converters with different specifications are available. Suitable and proper selection has to be made, depending upon the particular requirement for the specific application. These specifications are as follows:

1. ***A/D Resolution*** It is defined as the change in the input voltage required for 1-bit change in the output. The resolution of A/D converter is equal to the resolution of the D/A converter. It depends upon the number of bits in O/P digital code. It can also be expressed as a percentage.

2. ***A/D Gain*** It is defined as the equivalent voltage or the digital output divided by the analog input voltage at the linearity reference line.
3. ***A/D Drift*** Drift of an A/D quality of a CKT to change the parameters with time. Drift errors of 1.5 LSB can cause a maximum error of 1 LSB from first the transition to the last transition. It is difficult to achieve a very low drift as it increases the cost of A/D converters.
4. ***A/D Speed*** It is the time required to perform one conversion or as the time taken between successive conversions at the highest possible rate. It depends upon the settling time of the various components and the internal speed of the logic.
5. ***Aperture Rate*** It is the rate discrete points along which the waves are analysed and expressed in conversion/second. The higher the aperture rate, the better the fidelity.
6. ***Accuracy*** The maximum difference between these two analog voltages is expressed as a fraction of full-scale output voltage and is defined as the accuracy of A/D converters.
7. ***Quantization Error*** Quantization is the process of reducing series of complex waveforms to reasonably accurate straight line approximations. Resolution can also be taken as built-in error which is referred to as quantization error. This can be reduced by increasing the number of bits in the digital counter and the D/S converter.

10.11 D/A APPLICATIONS AND INTERFACING TO MICROCOMPUTERS

D/A converters have many applications besides those where they are used with a microcomputer, in a compact-disk audio player for example a 14- or 16-bit D/A converter is used to convert the binary data read off the disk by a laser to an analog audio signal. Most speech synthesizer ICs contain a D/A converter to convert stored binary data for words into analog audio signals.

The inputs of the D/A circuit can be connected directly to a microcomputer output port. As part of a program, you can produce any desired voltage on the output of the D/A.

Suppose you want to build a microcomputer-controlled tester which determines the effect of power supply voltage on the output voltage of some integrated-circuit amplifier. If you connect the output of the D/A converter to the reference input of a programmable power supply or simply add the high current buffer circuit to the output of D/A, you have a power supply which you can vary under program control to determine the output voltage of the IC under test as you vary its supply voltage connect the input of an A/D converter to the IC output, and connect the output of the A/D converter to an input port of your microcomputer. You can then read in the value of the output voltage on the IC.

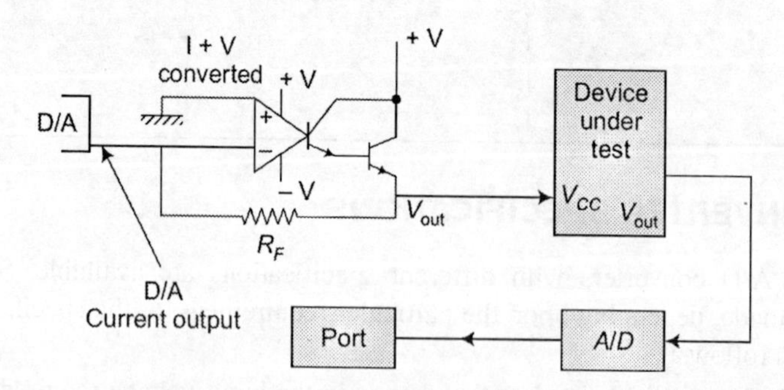

Another application you might use is a D/A and a power buffer to vary the voltage supplied to a small resistive heater under program control. Also, the speed of small d.c. motors is proportional to the amount of current passed through them, so you could connect a small d.c. motor to the output of the power buffer and control the speed of the motor with the value you output to the D/A.

Interfacing different types of A/F converters to the microcomputers.

1. ***Interfacing to Parallel – Comparator A/D Converters*** In any application where a parallel comparator converter is used, the converter is most likely going to producing digital output values much faster than a microcomputer could possibly read them. Therefore, separate circuitry is used to bypass the microprocessor and load a set of samples from the converter directly into a series of memory locations. The microprocessor can later perform the desired operation on the samples. Bypassing the microprocessor in this way is called direct memory access or DMA.
2. ***Interfacing a Successive Approximation A/D Converter*** Successive approximation A/D converters usually have outputs for each bit. The code output on these lines is usually straight binary or offset binary. You can simply connect the parallel outputs of the converter to the required number of input port pins and read in the converter output under program control. In addition to the data lines, there are two other successive approximation A/D converter signal lines you need to interface to the microcomputer for the data transfer. The first of these is a START CONVERT signal which you output from the microcomputer to the A/D to tell it to do a conversion for you. The second signal uses the FOC signal which the A/D converter outputs to indicate that the conversion is complete and the word on outputs is valid.

Example 10.1 Find the binary equivalent weight of each bit in a 4-bit system.

Solution

The LSB has a weight of $\frac{1}{24-1} = \frac{1}{16-1} = \frac{1}{15}$

The second LSB has a weight of $\frac{2 \times 1}{15} = \frac{2}{15}$

The third LSB has a weight of $\frac{4 \times 1}{15} = \frac{4}{15}$

The MSB has a weight of $\frac{8 \times 1}{15} = \frac{8}{15}$

$\therefore$ as a check the sum of all the weights must be equal to 1.

is
$$\frac{1}{15} + \frac{2}{15} + \frac{4}{15} + \frac{8}{15} = 1.$$

The following table gives you detail

B_3	B_2	B_1	B_0	Weight values	Equivalent value
0	0	0	0	0 + 0 + 0 + 0	0 V
0	0	0	1	0 + 0 + 0 + 1/15	1/15 V
0	0	1	0	0 + 0 + 2/15 + 0	2/15 V
0	0	1	1	0 + 0 + 2/15 + 1/15	3/15 V

0	1	0	0	0 + 4/15 + 0 + 0	4/15 V
0	1	0	1	0 + 4/15 + 0 + 1/15	5/15 V
0	1	1	0	0 + 4/15 + 2/15 + 0	6/15 V
0	1	1	1	0 + 4/15 + 2/15 + 1/15	7/15 V
1	0	0	0	8/15 + 0 + 0 + 0	8/15 V
1	0	0	1	8/15 + 0 + 0 + 1/15	9/15 V
1	0	1	0	8/15 + 0 + 2/15 + 0	10/15 V
1	0	1	1	8/15 + 0 + 2/15 + 1/15	11/15 V
1	1	0	0	8/15 + 4/15 + 0 + 0	12/15 V
1	1	0	1	8/15 + 4/15 + 0 + 1/15	13/15 V
1	1	1	0	8/15 + 4/15 + 2/15 + 0	14/15 V
1	1	1	1	8/15 + 4/15 + 2/15 + 1/15	1 V

Example 10.2 For a four-input resistive divider (0 = 0 V, 1 = + 10 V). Find

(a) the full-scale output voltage

(b) the output voltage change due to +0 the LSB

(c) the analog output voltage for a digital input of (1011)

Solution

(a) The maximum output voltage occurs when all the inputs are at +10 volts. If all the inputs are at +10 V, the output must also be +10 V.

(b) For a 4-bit digital number, there are 16 possible states. There are 15 steps between these 16 states, and the LSB must be equal to 1/15 of full-scale output voltage.
Therefore, the change in output voltage due to LSB is +10 × 1/15 = 2/3 volts.

(c) To calculate the analog output voltage for a digital input, we have to use Millers theorem which is given as

$$V_A = \frac{V/R_0 + V/(R_0/2) + V/(R_0/4) + V/R_0/8)}{1/R_0 + 1/(R_0/2) + 1/(R_0/4) + 1/(R_0/8)}$$

$$V_A = \frac{10\,(R_0) + 10\,(R_0\;2) + 10\,(R_0\;4) + 10\,(R_0\;8)}{1/(R_0) + 1/(R_0/2) + 1/(R_0/4) + 1/(R_0/8)}$$

$$= \frac{110}{15} = \frac{22}{3} = +7\frac{1}{3}\,\text{V}$$

Example 10.3 For a 5-bit resistive divider, determine the following:

(a) the weight assigned to LSB

(b) the weight assigned to the second and third LSB.

(c) the change in output voltage due to a change in the LSB, the second LSB, and the third LSB

(d) the output voltage of a digital input of 10101 assume 0 = 0 V, 1 = 10 volts

Solution

(a) The LSB weight is $1/2^5 - 1 = 1/31$

(b) The second LSB weight is 2/31, the third LSB is 4/31.

(c) The LSB causes a change in output voltage of 10/31 V. The second LSB causes an output voltage change of 20/31 V, the third LSB causes an output voltage of 40/31 V.

(d) The output voltage for a digital input at 10101 is

$$V_A = \frac{10 \times 2^{\circ} + 0 \times 2^1 + 10 \times 2^2 + 0 \times 2^3 + 10 \times 2^4}{2^5 - 1}$$

$$= \frac{10(1 + 0 + 4 + 0 + 16)}{31}$$

$$= \frac{210}{31} = +6.77 \text{ volts.}$$

EXERCISES

1. Explain the principle of variable resistor divider and binary ladder D/A converter and why are they considered decoders?
2. What are the advantages of binary ladder network over resistive divider network?
3. Write a short note on D/A specifications.
4. Find out the binary equivalent weights of each bit in a 5-bit variable resistive divider D/A converter.
5. Find out the output voltage of 5-bit binary ladder with the digital input (10 – 110) assume the input levels as 0 = 0 V and 1 = 10 V.
6. Explain the principle of working of a dual slope integrator A/D converter. Also mention its advantages.
7. Explain the working of successive approximation A/D converter, and what are the advantages of successive approximation method.
8. Explain the application of A/D and D/A converters.
9. What are the output voltages caused by each bit in a 4-bit binary ladder if the input levels are taken as '0' = 0 volts and 1 = 10 volts.
10. What is the resolution in volts of a 12-bit binary ladder D/A converter? If full-scale output voltage is +10 volts find the percentage solution.
11. Define and explain the working principle of a continuous A/D converter.
12. What is counter type A/D converter? Explain the working principle in detail with the help of diagram.
13. How simultaneous conversion is used for A/D conversion?
14. Compare the working advantages of A/D converter and which one is the most accurate.
15. How many comparators are required to build a 5-bit simultaneous A/D converter?
16. Find the output voltage of a 6-bit binary ladder in which the following inputs:
    (a) 101001 (b) 111011 (c) 110001
17. Discuss the overall acceptable accuracy of 10-bit ADC in terms of a quantization error, ladder accuracy, comparator accuracy, converter accuracy, etc.
18. What is the conversion time of a 12-bit successive approximation type A/D converter using a 1 MHz clock?

11

Semiconductor Memories

INTRODUCTION

Semiconductor memory is an electronic data storage device used in computers. It is fabricated on semiconductor based integrated circuits. In our daily life we are familiar with the internet and communication, entertainment devices like mobile phones, camcorders, ipods. In this chapter, we discuss the Random Access Memory (RAM), Read-Only Memory (ROM), Static RAMs (SRAMs), Dynamic RAMs (DRAMs), Volatile, Non-volatile, Programmable Read-Only Memories (PROMs), Erasable Programmable Read-Only Memories (EPROMs), Electrically Erasable Programmable Read-Only Memories (EEPROMs), Flash Memories, Memory sticks and cache memories.

A major advantage of digital technology over analog systems is its ability to store large quantity of digital information. Before discussing the different types of memories used in computer system, it is necessary to study the terminology used.

Memory Cell: Basically a flip-flop used to store a single bit.

Memory Word: A group of cells which represents the data. Modern computers typically range in word sizes from 8 to 64 bits.

Capacity: It shows how many bits can be stored in a particular memory.

Address: A location of word in memory.

Read Operation: The operation by which the data are read from the memory location.

Write Operation: The operation by which the data are written onto the memory location.

Access time: It is the operating speed od a memory device.

Volatile Memory: Memory that requires electrical power to store information.

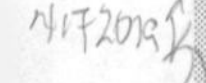

Non-volatile Memory: Memory that does not require electrical power to store information.

11.1 RANDOM ACCESS MEMORY (RAM)

RAM is the best known form of computer memory. It is known as Random Access Memory because you can access any memory location randomly if you know the row and column that intersect the cell. In a typical RAM the access time is independent of the location of the data within the memory device. RAM is a volatile memory, which means that the data is lost when power is removed. Another types of memory is the serial access memory (SAM) in which the memory store data sequentially. If the data is not in the current location, then each memory cell is checked until the required data is found. SAM is usually used for memory buffers, where the data is stored in the order in which it will be used as a buffer memory on a video card. SAM is a non-volatile memory, i.e., the stored data is retained even through the power is removed.

RAM typically comes with the word capacities of 1 K, 4 K, 8 K, 16 K, 64 K, 128 K, 256 K, 1024 K upto 2048 M and with the word sizes of one, four or eight bits. The basic internal architecture of 64 × 4 RAM is shown in Fig. 11.1. These words can address from 0 to $(63)_{10}$. To access 64 address locations for reading or writing, a binary code of 6-bit (26 = 64) is required. For example, if 001001 is applied then the decoder ouput 9 will be accessed for read or write operation.

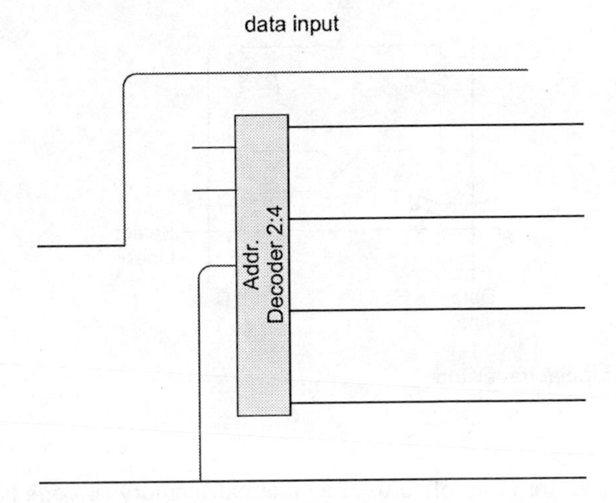

Fig. 11.1 Random access memory

Read Operation

The read operation will read the contents of the selected register when the READ/$\overline{\text{WRITE}}$ (R/$\overline{\text{W}}$) input must be 1 and the Chip Select ($\overline{\text{CS}}$) must be 0.

Write Operation

The write operation will read the contents of the selected register when the READ/$\overline{\text{WRITE}}$ (R/$\overline{\text{W}}$) input must be 1 and the Chip Select ($\overline{\text{CS}}$) must be 0.

Chip Select

Each memory chip has one chip select signal, which is used to enable the entire chip or disable it completely. In disabled mode, all data inputs and outputs are disabled, i.e., neither a read nor a write operation can be performed.

11.2 STATIC RAMs

The basic RAM that we have discussed which can store data as long as power is applied to the chip is known as static RAM. In static RAM, the memory cells are flip-flops which stay in the given state and retain data indefinitely as long as the power is applied to the flip-flops.

Static RAMs are available in bipolar, MOS and Bi-CMOS technologies. Each technology having its own advantages, as bipolar RAMs are high speed and the MOS devices have greater capacities and lower power consumption. Figure 11.2 shows the typical bipolar static memory cell. The bipolar cell is made up of two bipolar transistors and two resistors. Figure 11.3 shows NMOS static RAM memory cell which is made up of four N-channel MOSFETs. The bipolar RAM cell requires more chip area than MOS because a bipolar transistor is more complex than MOSFET.

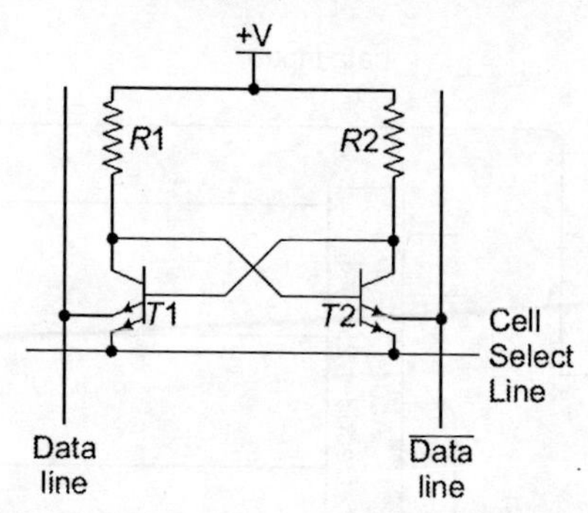

Fig. 11.2 Static RAM bipolar transistors

Static RAM Timing

Random Access Memories are more often used as internal memory of computer which continuously performs read and write operations on this memory at a very fast rate. The computer designer must be concerned with RAMs timing characteristics as the memory chips that are interfaced to CPU must be synchronized to the CPU read/write commands.

All types of RAMs have not the same timing characteristics but most of them are similar which varies from one manufacturer to another. Figure 11.4 shows the memory timing diagram for the read and the write cycle of a RAM chip.

READ Cycle

The waveform of Fig. 11.4 shows the Read cycle of a typical RAM cell, how the address, READ/$\overline{\text{WRITE}}$ and CS inputs behave during memory write cycle. Before the beginning of read cycle, the address inputs hold the address of its proceeding operation. At the beginning of read cycle, at time T_0 the R/$\overline{\text{W}}$ line is high and stays high throughout the read cycle. At $T = T_0$, the processor applies a new address to RAM inputs.

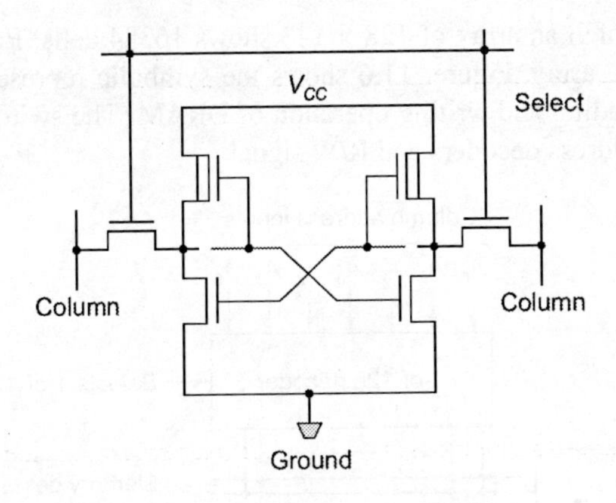

Fig. 11.3 Static RAM (N-MOS cell)

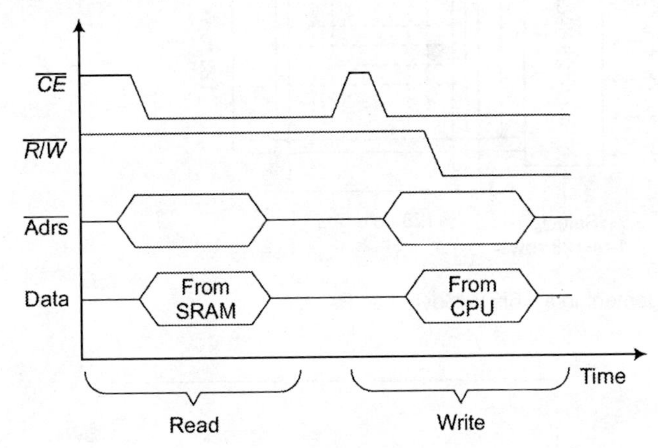

Fig. 11.4 Timing diagram.

Write Cycle

The waveform of Fig. 11.4 shows the Write cycle of a typical RAM cell, how the address, READ/$\overline{\text{WRITE}}$ and CS inputs behave during the memory write cycle. Before the beginning of the write cycle, the address inputs hold the address of its prceding operation. At the beginning of the write cycle, at time T_0 the R/$\overline{\text{W}}$ line is low and stays low throughout the write cycle. At $T = T_0$, the processor applies a new address to RAM inputs.

11.3 DYNAMIC RAMs

Dynamic RAM uses MOS technology which stores data as charges on capacitors. They are known for high capacity, low power requirement and moderate operating speed but dynamic RAM's major disadvantage is that the capacitors cannot hold the charge for long, so the data will gradually and slowly disappear due to discharge of a capacitor. For the applications where the speed and reduced complexity are more critical than cost, space, power considerations, static RAMs are the best. Static RAMs are generally faster and do not require any refresh operation but they cannot compete with higher capacity and lower power requirement of dynamic RAM. The basic cell arrangement of 16 K × 1 dynamic RAM is shown in Fig. 11.5 in which an array of 128 × 128 shows 16384 cells. Each cell occupies a unique row and column within the array. Figure. 11.6 shows the symbolic representation of dynamic RAM cell which describes the reading and writing operation of DRAM. The switches S1, S2, S3 and S4 are MOSFETs controlled by address decoders and R/$\overline{W}$ signal.

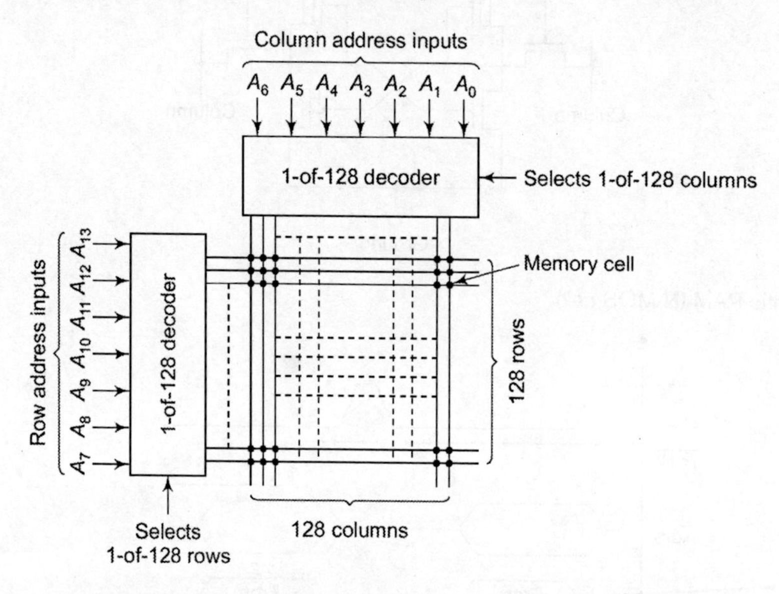

Fig. 11.5 Cell arrangement in a 16K × 1 dynamic RAM.

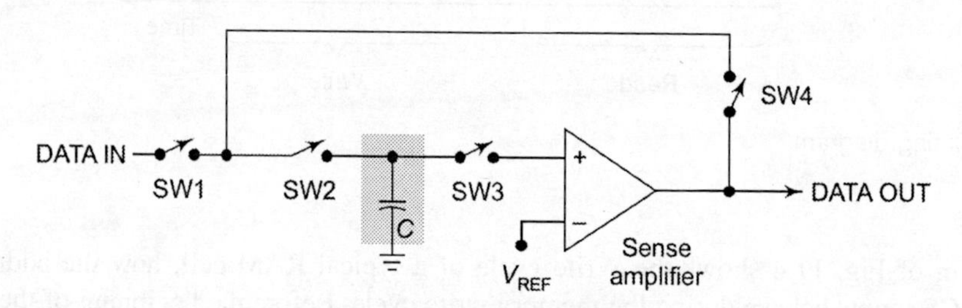

Fig. 11.6 Symbolic respresentation of a dynamic memory cell.

To write data in the cell signals from the address decode circuit and R/$\overline{W}$ logic will close switches S1 and S2 while keeping S3 and S4 open. This will charge the capacitor C. To read data from the cell switches S2, S3 and S4 are closed and S1 is kept open which connects the capacitor voltage to the sense amplifier to read the data.

11.4 READ ONLY MEMORIES

The Read Only Memory (ROM) is a semiconductor memory which holds data permanently. During the normal operation of the computer, no data can be written or altered into a ROM but can be read from it. ROM are used for the storage of data and information that are not to change during the normal operation of the system. All ROMs are non-volatile in which the programs are not lost when the electric power is turned off. ROMs are basically used in computers, microprocessor based equipment such as ATMs, security systems etc. Fig. 11.7 shows the basic block diagram of ROM which has three sets of signals, address inputs, control inputs and data inputs.

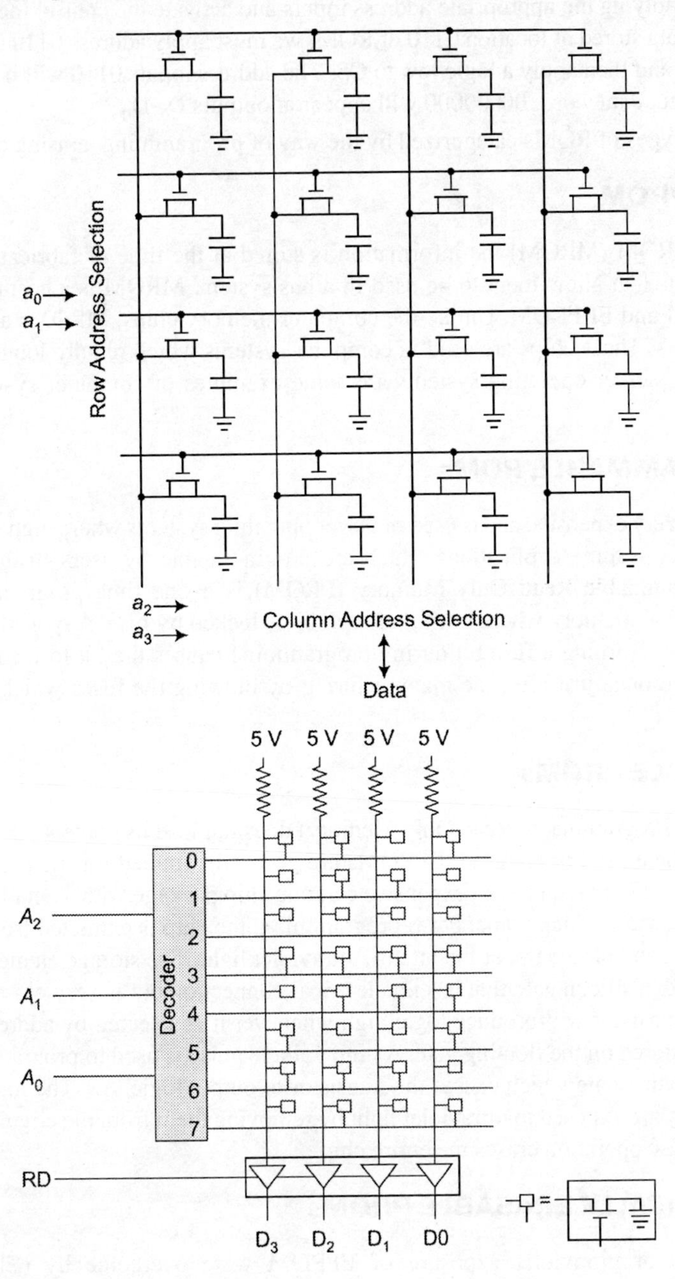

Fig. 11.7 .

From the four address lines we can address upto $2^4 = 16$ possible addresses. The data outputs of ROM are tristate outputs which are used to connect many ROM chips for memory expansion on the same data bus. The control input CS used for chip select which enables or disables the chip. The Read operation can be performed by applying the appropriate address inputs and activate the control inputs. For example, if you want to read data stored at location 0110 of ROM, we must apply address 0110 to the address inputs $A_3A_2A_1A_0 = 0110$ and then apply a logic low to CS. The address inputs 0110 will be decoded inside the ROM and the correct data word, 00000000 will appear at outputs D_7-D_0.

There are various types of ROMs categorized by the way of programming, erasing and reprogramming.

11.5 MASK PROM

In Mask Program ROM (MROM) the information is stored at the time of fabrication of IC. MROMs have tristate outputs that allow them to be used in a bus system. MROM is a cost-effective alternative to PROM, EPROM and EEPROM, Unlike the computer memory chips, MROMs are manufactured by arranging transistors. These chips are used in computer systems which require long term sustainability. Operating systems, server operating systems are some examples of computer systems that use Mask ROM chips.

11.6 PROGRAMMABLE ROMs

The Mask ROM is very expensive and is used in server operating systems where high volume applications are required. Lower volume applications which are programmable by users fusible link PROMs are required. A Programmable Read Only Memory (PROM) is a one time programmable non-volatile memory. It is a digital memory where setting of each bit is locked by fuse. A typical PROM comes with all bits reading as 1. Burning a fuse bit during programming causes the bit to read as 0. The memory can be programmed once just after the manufacturing by blowing the fuses which is a non-reversible process.

11.7 ERASABLE PROMs

EPROM (Erasable Programmable Read Only Memory) does not lose its content when the power supply is cut off and can be erased or re-fused. EPROMs are generally applied for the programs designed for repeated use. EPROM chip requires an expensive ceramic chip package with a small quartz window that is covered with an opaque, sticky tape. For reprogramming, the chip is extracted from the circuit board, the tape is removed, and placed under the intense ultraviolet light. The storage element of a EPROM is a MOS transistor with a silicon gate that has no electrical connection but is very close to an electrode. In normal state, the transistor will produce logic high whenever it is selected by address decoder because there is no charge stored on the floating gate. A high voltage pulse is used to program a 0 which leaves a net charge on a floating gate which causes the transistor to output logic low. The major disadvantage of EPROM is that they are exposed to ultraviolet light by removing them from the circuit to be programmed and erased. The erase operation erases the entire chip.

11.8 ELECTRICALLY ERASABLE PROMs

The disadvantages of ultraviolet exposure of EPROM were overcome by (Electrically Erasable Programmable Read Only Memory). It is user modifiable ROM that can be erased and reprogrammed repeatedly through the application of electrical voltage. EEPROM need be removed from the computer for reprogramming and EEPROM chip has to be entirely erased and reprogrammed not selectively. The internal process storing data value in an EEPROM is quite slow.

11.9 CD-ROM

A very familiar type of read-only memory used in computer systems is Compact Disk. CD-ROM is a prepressed compact disk that contains data accessible to, but not writable by a computer for data storage. CD-ROMs are used for computer software, games, multimedia applications. Some CDs can store both video and audio that can be played on CD/DVD player. A standard 74 min. CD contains 333000 blocks. Each sector is 2352 bytes and contains 2048 bytes of PC, 2336 bytes of VCD and 2352 bytes of audio data.

EXERCISES

1. What do you mean by memory? Describe the advantages of semiconductor memories.
2. List the uses of read only memories.
3. Explain the difference between MROM and PROM.
4. Give the difference between RAM and ROM.
5. What is the basic difference between volatile and non-volatile memory?
6. What do you mean by ROM? Give its types and merits.
7. Define RAM along with its characteristics
8. Discuss the operation of Random Access Memory (RAM).
9. Differentiate between SRAM and DRAM.
10. Discuss the operation of SRAM cell and also make a neat diagram of NMOS static RAM memory cell.
11. Draw the diagram for SRAM timing for read and write cycle.
12. Make the basic block diagram of ROM which contains address inputs, control inputs and data inputs.
13. Describe various types of ROM in detail.
14. What is the use of Erasable PROMs.
15. What is the basic difference between EPROM and EEPROM.
16. If a computer has 4 Mbytes of RAM, how many bytes of read/write memory does it contain?
17. How many pins would an E2PROM chip have if it has a capacity of 256 K bit organised as 32 Kbits by 8 bit?
18. Which of the memory types uses the most transistors per bit and which uses the least?
19. Which type of memory device is used as the main semiconductor memory in PC computers?
20. Why do you have to refresh a DRAM memory device?
21. What power supply voltages are needed for EPROM, flash, SRAM, E2PROM?
22. What happens when CE is held high for an SRAM device?
23. Which of these devices must be erased before they can be written: EPROM; E2PROM; SRAM; flash?

12

Introduction to Microprocessors

INTRODUCTION

A microprocessor is a miniature electronic device that contains the arithmetic, logic and control circuitry needed to function as a digital computer's Central Processing Unit (CPU). It is an integrated circuit that can interpret and execute program instructions, used in bank ATM's, point-of-sale devices and to control instrumentation in industry, hospitals, microwave ovens, etc. Nowadays, many automobiles use microprocessor-controlled ignition and fuel systems.

Microprocessors are found in every consumer product that requires electric power such as video recorders, cellular phones, digital cameras, etc.

High performance microprocessors are used to store and distribute web content, such as streaming audio and video, desktop computers and high speed network switches which constitute Web infrastructure. More powered microprocessors are the heart of computers, laptops, etc. and low powered microprocessors provide the control and flow logic of consumer hand-held devices such as digital cameras, cellular and cordless phones and etc.

Microprocessors are generally classified according to semiconductor technology used such as TTL (Transistor-transistor logic), CMOS (Complementary Metal Oxide Semiconductor) or ECL (Emilder-couple logic), by the width of data (4-bit, 8-bit, 16-bit, 32-bit or 64-bit) they process, or by their instruction set such as CISC (Complex Instruction Set Computer) or RISC (Reduced Instruction Set Computer).

TTL logic family is most commonly used in microprocessors. Because of low power consumption of CMOS, it is used in portable computers and other low powered battery consumption devices.

High speed simulations require high speed offsets. ECL is the fastest logic family and consumes more power for high-speed processors.

12.1 HOW MICROPROCESSORS WORK

A microprocessor is an integrated circuit, divided into several different areas, each of which performs a specific function. Some areas are used to retrieve data from computer's main memory (RAM) or hard disk, while other sections translate data from English like text and machine language into binary code that the processor can understand. Some portions of processor perform millions of mathematical, logic operations. Each component of a microprocessor is vital in enabling the computer to help in composing the letters, send e-mails, managing finances or Web surfing. Each processor consists of the following units.

- ***Prefetch Unit*** The prefetch unit monitors all incoming data and decides when to order data and instructions from instruction cache or computer's main memory based on commands. Prefetch unit also decides where the data is stored in instruction cache.
- ***Data Cache Unit*** Data cache unit works very closely with ALU (Arithmetic Logic Unit) or FPU (Floating Point Unit) and the registers. Data from the Decode unit are stored in DCd for later use by ALU. Final results for any program are prepared in DCU for distribution to different parts of a computer.
- ***Control Unit*** The control unit of microprocessor makes sure everything happens in the right place at right time. It oversees the processing of instructions from instruction cache, routes instructions to decode unit for binary conversion. It also sends data to ALU or FPU for processing. Basically, it creates control signal that tells ALU and other registers how to operate, what to operate on, and what to do with the result.
- ***Decode Unit*** Decode unit translates machine language programs into a language understood by ALU and other registers which makes processing more efficient.
- ***Arithmetic Logic Unit (ALU)*** This is the last stage processing in the chip. It performs functions like Adding, Subtracting, Multiplying, Dividing, Increment, Decrement arithmetic operations and logic functions like AND, OR, NOT.
- ***Floating Point Unit (FPU)*** Known as math co-processors which handle complex mathematical operations that include floating point numbers fractional or integer numbers.
- ***Instruction Cache*** It is a warehouse of instructions where instructions from a CPU's instruction set and external software instructions wait in line to be carried out by control unit.
- ***Bus Unit*** It is the place where instructions flow in and out of the microprocessor from the computer's main memory.
- ***Instruction Set*** All CPUs are designed with preloaded set of instructions that tells them how to function and respond to external commands. These instructions are stored permanently in the circuitry that makes up this portion of a processor and need to be translated into binary by the decode unit in order to be executed.

12.2 8-BIT MICROPROCESSOR – 8085

8085 microprocessor is an 8-bit microprocessor available in 40-pin-dual-in-line package. It was introduced in 1976 by Intel. It is a modified version of 8080. It consists of various units listed below:

- Accumulator (A)
- Arithmetic & Logic Unit (ALU)
- General Purpore Registers (B, C, D, E, H, L)
- Program Counter (PC)
- Stack Pointer

- Temporary Register
- Flags
- Instruction Register & Decoder
- Timing & Control Unit
- Interrupt Control
- Serial Input/Output Control
- Address Buffer and Address-Data Buffer
- Address Bus and Data Bus.

The detailed pin diagram of 8-bit 8085 microprocessor is shown in Fig. 12.1 and schematic block diagram is shown in Fig. 12.2.

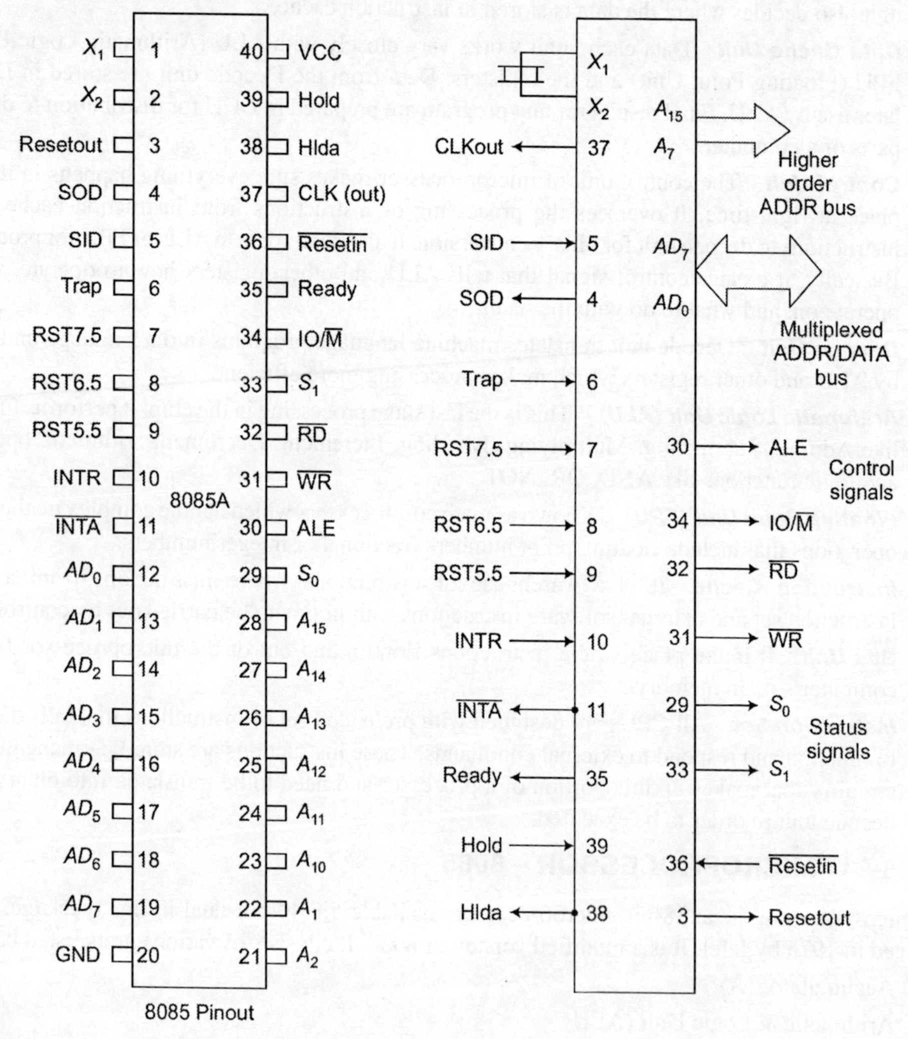

Fig. 12.1

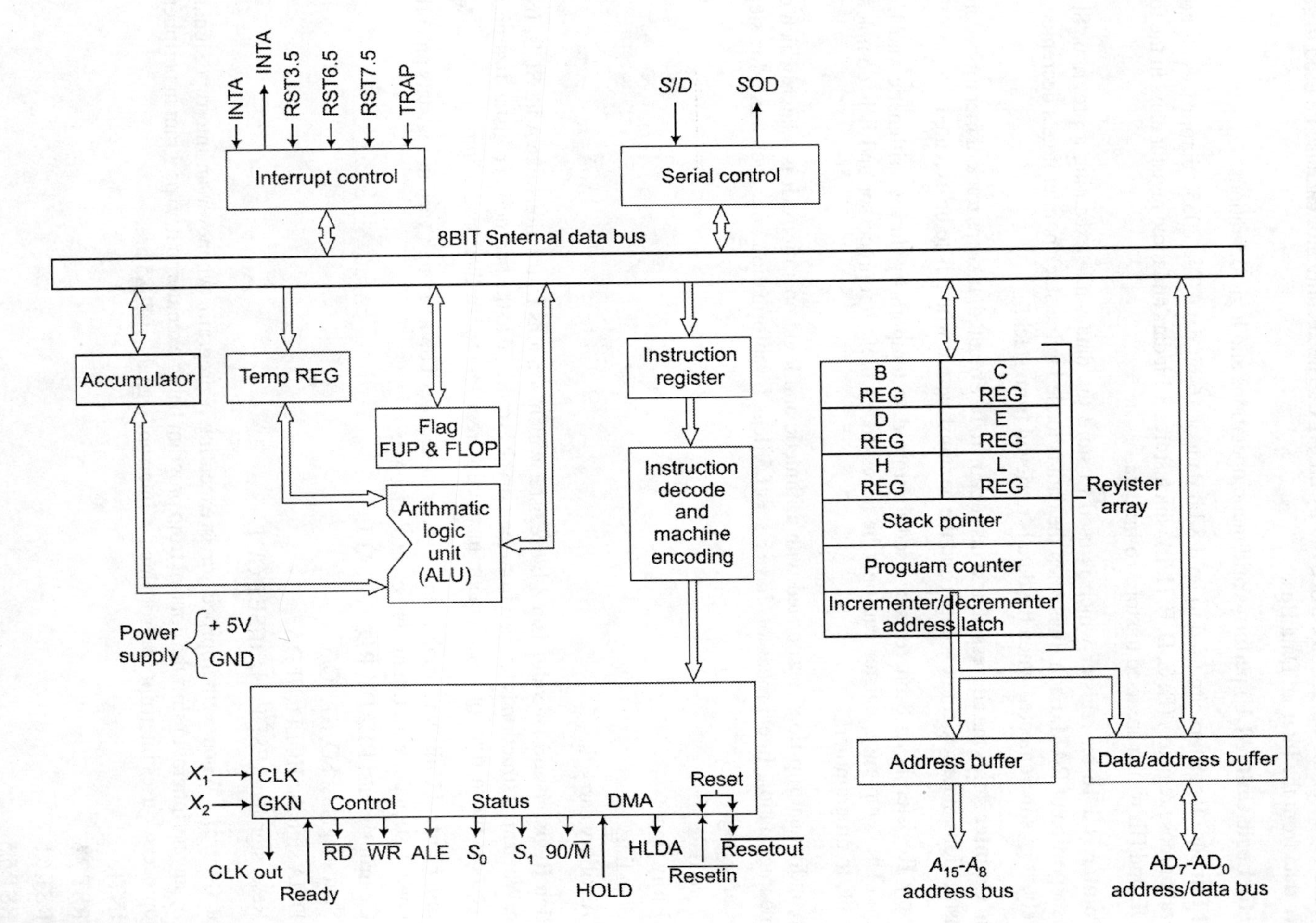

Fig. 12.3 8085 Block Diagram

Accumulator is a special kind of register of 8-bit length. It stores one of the data to be processed by ALU and the result of the operation carried out by the ALU. The accumulator can also be used to send or receive data from the Internal Data Bus.

Arithmetic Logic Unit (ALU) performs arithmetic operations and logic operations.

General Purpose Registers are used to hold 8-bit data or 16-bit data. The 8085 microprocessor holds six general purpose registers (B, C, D, E, H, L) which stores 8-bit data and three register pairs in the form of BC, DE and HL which are used to store 16-bit data.

Stack Pointer is a 16-bit register which is used to store the data while executing a program. Stack pointer is a portion of RAM. Each time when the data is loaded in the stack pointer it gets decremented. Conversely, it gets incremented when the data is retrieved from stack.

Program Counter stores the address of next instruction to be executed. It keeps track of memory address of the instructions that are being executed and the next instruction which is to be executed.

Temporary Register is an 8-bit register used to store data temporarily during arithmetic and logic operations. The temporary register can only be accessed by the microprocessor and it is completely inaccessible to programmers.

Flags are individual flip flops, associated with arithmetic and logic operations. It is a latch which can hold some information. Intel processors have a set of 5 flags for information after each execution step.

- Carry flag
- Zero flag
- Parity flag
- Sign flag
- Auxiliary carry flag

Instruction Register and Decoder is a 8-bit register which contains instructions like Add, SUB, INR, DCR, etc. When an instruction is fetched from memory, it is directed to the instruction register. Instruction decoder decodes the instructions available in instruction registers for further processing.

Timing and Control Unit synchronizes the registers and flow of data through various registers and other units. Signals associated with timing and control unit are:

- Control signals: READY, $\overline{RD}$, $\overline{WR}$, ALE
- Status signals: SO, S1, IO/$\overline{M}$
- DMA signals: HOLD, HLDA
- Reset signals: RESET IN, RESET OUT.

Interrupt Control When a microprocessor is executing a program, whenever an interrupt signal is enabled by any peripheral to shift the control from program being executor to new program in peripheral, the microprocessor uses interrupt control signals. Interrupt signals of 8085 are:

- INTR
- RST 7.5
- RST 6.5
- RST 5.5
- TRAP

Out of five, TRAP is non-maskable and others are maskable.

Serial I/O controls are used to input and output the serial data by using two instructions:

SID – Serial Input Data

SOD – Serial Output Data

Two instructions in 8085 are used to perform serial-parallel conversion

SIM – Set Interrupt Mask

RIM – Read Interrupt Mask.

Address and Address Data Buffer It stores the contents of stack pointer and program counter. These buffers are used to drive the external address bus and address-data bus. The memory and I/O chips are connected to these buses, the CPU can exchange the desired data to the memory and I/O chips. The address-data bus can send and receive data from the internal data bus.

Address Bus and Data Bus 8085 processor holds 16-bit address bus and 8-bit data bus. The address/data bus is time multiplexed which can be used to transmit address and data simultaneously by a signal ALE (Address Latch Enable).

The architecture of 8085 are explained in detail.

12.2.1 Instruction Set

Intel 8085 microprocessor consists of the following instructions:

- Data Moving Instructions
- Arithmetic Instructions
- Logic–AND, OR, XOR
- Control transfer – Conditional, Unconditional, Call, Return, etc.
- Input/output instructions
- Setting/Clearing Flag, Enabling/Disabling Interrupts, stack operations, etc.

The detailed instruction set of 8085 is out of scope of this book.

12.2.2 Addressing Modes

The outline of addressing modes of 8085 processor is given below:

Register Refers the data in register or in register pair

Register Indirect Instruction specifies register pair containing address.

Direct Direct In or Out the data to or from peripheral devices.

Immediate Instructions which contain 8-bit or 16-bit data itself.

12.3 16-BIT MICROPROCESSOR – 8086

The 8086 was the first 16-bit microprocessor introduced by Intel. The word 16-bit means that its internal registers, ALU, and most of its instructions are designed to work with 16-bit binary words. The Intel 8086 has 16-bit data bus and 20-bit address bus, so it can address any one of 2^{20} or 1,048,576 (1 MB) memory locations.

The architecture of Intel 8086 includes

- Bus Interface Unit
- Execution Unit

shown in 12.3. The pin schematic of 8086 is shown in Fig. 12.4.

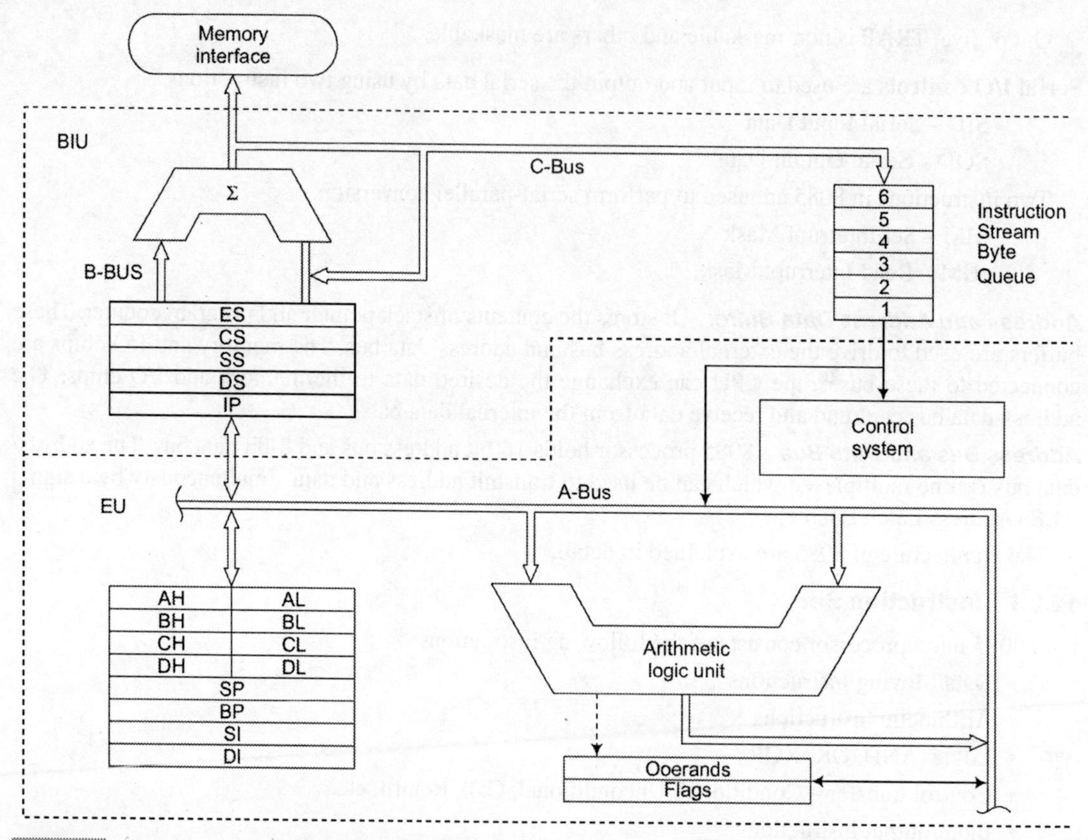

Fig. 12.3 Block Diagram of 8086 (16 bit processor)

Bus Interface Unit is made up of the address generation and bus control unit, instruction queue and instruction pointer. There are four different 64 KB segments for instructions, stack, data and extra data. Four segment registers are used in 8086 processor:

- ***Segment Registers***

  Code Segment (CS) It points to the current code segment. Instructions are fetched from this segment. **Data segment (DS)** register points the current data segment.

  Stack segment (SS) register points the current stack segment, stack operations are performed in stack segment (SS). **Extra segment (ES)** is used for extra data storage. Each segment register is 64 KB long.

- ***Address control*** is used to generate 20-bit address that gives physical or actual location of data in memory. This unit consists of instruction pointer, segment registers, and address generation.
- ***Instruction Pointer*** is a 16-bit register which is used to tell the processor, about the next instruction to be executed.
- ***Instruction Queue*** is a temporary memory storage area for data instructions that are to be executed by MPU.

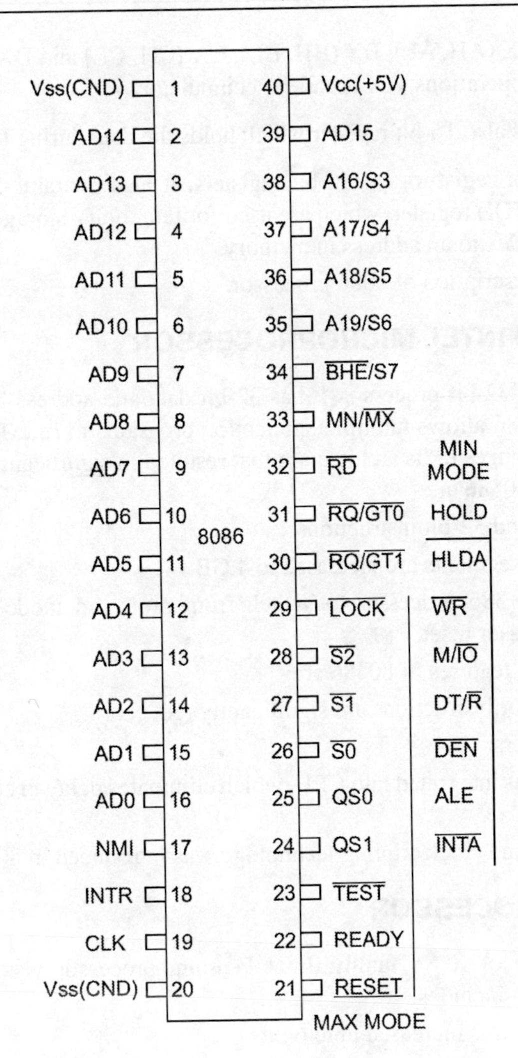

Fig. 12.4 Pin Diagram of 8086

Execution Unit is the unit in which the actual processing of data takes place. It consists of ALU, Flag Register, General Purpose Registers, Stack Pointer, Base Pointer and Index Registers.

Arithmetic Logic Unit (ALU) performs arithmetic and logic functions. Multiplication and Division functions are directly available in 8086.

Flag Register of 8086 is a 16-bit register which has additional information of overflow, direction, Interrupt & TRAP as compared to 8085.

General Purpose Registers (GPRs) are used to store the data. 8086 has 8 GPRs AH, AL, BH, BL, CH, CL, DH, DL. H represents High order and L represents LoW order byte. The acceptable register pairs to hold 16-bit data are AX (AH, AL), BX (BH, BL), CX (CH, CL) and DX (DH, DL). AL is also called accumulator. For 16-bit operations, AX is called accumulator.

Stack-Pointer of 8086 is also 16-bit register which holds the data during the execution of a program.

Base Pointer and Index registers are 16-bit registers. It also contains 16-bit source index (SI) and 16-bit destination index (DI) register, which are used for temporary storage of data. The index registers are used to point to or index to an address in memory.

This gives the brief description of 8086 processor.

12.4 80386/80486 INTEL MICROPROCESSOR

Intel 80386 was the first 32-bit processor. It has 32-bit data and address bus and can address upto 2^{32} bytes of memory. Intel 386 allows multiple application programs to run. The successor to 80386, Intel 80486 included many changes to its architecture that resulted in significant performance improvements. Some of the features of 80386 are:

- 32-bit registers and 32-bit instructions
- Size of memory segments are increased to 4 GB.
- It is possible in 386 processor to switch from protected mode back to real-mode without simulating processor reset.

Some of the improved features of 80486 are:

- Execution time for instructions are significantly reduced.
- Faster bus transfers
- Floating point was integrated into CPU, which eliminates delay in communication between CPU and FPU.
- Clock-doubling and clock-tripling technology was introduced in 80486 processors.

12.5 PENTIUM PROCESSOR

It is the fifth generation of X 86 family. Intel Pentium processor was the first superscalar CPU. Enhancements to Pentium includes:

- Data transfer rate was increased upto 64 its
- It features 8 KB code and 8KB datacaches.
- Many desktop Pentiums could work in dual-processor systems.
- Reduced CPU power consumption

These are 32-bit processors.

12.6 CORE SOLO AND DUO PROCESSORS

Intel core solo is a single core mobile processor based on Pentium. It has 32 KB instruction and data level 1 caches, 2 MB level 2 cache. It includes virtualization technology.

Core Duo processors integrate two Pentium processors on a die with some technology improvements.

Core Solo and Duo family deliver faster performance, greater energy efficiency and more multitasking.

Core Solo and Duo are 32-bit processors.

12.7 INTEL CORE i 7/i 5/i 3

The latest generation of Intel × 86, released in 2008, 2009 and 2010 respectively. These processors are based on new Nehalem microarchitecture having enhanced features like:

- Simultaneous multi-threading feature allows core i 7 to execute 8 threads at same time.
- Uses trouble boost technology
- Currently introduced (Feb. 2010) core i 3, uses 2 CPU cores.

EXCERCISES

1. What is the difference microprocessor and microcomputer?
2. Why there is a need of microprocessor?
3. How are various microprocessors classified?
4. Explain the various steps of operation of microprocessor.
5. Give the features of 8085 microprocessor.
6. Draw the pin schematic of 8086 microprocessor.
7. Explain the use of different condition flags in 8085.
8. Describe different control signals used by 8085.
9. Explain the architecture of 8085 microprocessor.
10. Give the detail of addressing modes of 8085 microprocessors.
11. List the set of instruction in 8085 with the help of suitable examples.
12. What is the role of interrupt control in case of 8085 microprocessor?
13. Explain the architecture of 8086 in detail with the help of suitable block diagram.
14. Draw the pin schematic of 8086 microprocessor.
15. Name the four segment registers used in 8086 processor.
16. What is the role of general purpose registers used in 8086 processor?
17. List some features of 80386/80486 Intel Microprocessor.
18. What is Pentium processor?
19. What is the basic difference among 8-bit, 16-bit and 32-bit microprocessors?
20. Define Core Solo, Core Duo processors and Intel Core i7/i5/i3.

Index

sinhalagnojita@gmail.com

kavita gupta ConsenSys Ventures

Mira Wilczek CEO Cogo Labs

Tanzeem Choudhury (Cornell), Prof Comp + Info Science

Health Rhythms, Inc

Oliver Smith Strategy director, Health Moonshot

@ Telefonica Innovation α

MIT Metal ion transporter

JGI (USDOE)

crispr (protospacer)

IMG/M

img.jgi.doe.gov/mer/

" /vr/